Thermal Science

About the Author

Erian A. Baskharone, Ph.D., is a Professor Emeritus of Mechanical and Aerospace Engineering at Texas A&M University, and a member of the Rotordynamics/Turbomachinery Laboratory Faculty. He is a member of the ASME Turbomachinery Executive Committee. After receiving his Ph.D. degree from the University of Cincinnati, Dr. Baskharone became a Senior Engineer with Allied-Signal Corporation (currently Honeywell Aerospace Corporation), responsible for the aerodynamic design of various turbofan and turboprop engines. His research covered a wide spectrum of turbomachinery topics, including unsteady stator/rotor flow interaction, and the fluid-induced vibration problem in the Space Shuttle Main Engine. Dr. Baskharone's perturbation approach to the problem of turbomachinery fluid-induced vibration was a significant breakthrough. He is the recipient of the General Dynamics Award of Excellence in Engineering Teaching (1991) and the Amoco Foundation Award for Distinguished Teaching (1992).

Thermal Science

Essentials of Thermodynamics, Fluid Mechanics, and Heat Transfer

Erian A. Baskharone

McGraw Hill

New York   Chicago   San Francisco
Lisbon   London   Madrid   Mexico City
Milan   New Delhi   San Juan
Seoul   Singapore   Sydney   Toronto

The McGraw·Hill Companies

Cataloging-in-Publication Data is on file with the Library of Congress.

McGraw-Hill books are available at special quantity discounts to use as premiums and sales promotions, or for use in corporate training programs. To contact a representative please e-mail us at bulksales@mcgraw-hill.com.

Thermal Science: Essentials of Thermodynamics, Fluid Mechanics, and Heat Transfer

1 2 3 4 5 6 7 8 9 0   QDB/ QDB   1 9 8 7 6 5 4 3 2

ISBN   978-0-07-177234-1
MHID   0-07-177234-0

The pages within this book were printed on acid-free paper.

Sponsoring Editor	**Proofreader**
Michael Penn	Cenveo Publisher Services
Acquisitions Coordinator	**Indexer**
Bridget Thoreson	Robert Swanson
Editorial Supervisor	**Production Supervisor**
David E. Fogarty	Pamela A. Pelton
Project Manager	**Composition**
Ridhi Mathur, Cenveo Publisher Services	Cenveo Publisher Services
Copy Editor	**Art Director, Cover**
Kevin Campbell	Jeff Weeks

To my two talented sons Daniel and Christian and to my precious wife Magda—a book like this could have never materialized without her expertise and unlimited support

Brief Contents

Preface   xiii

Part I   **Thermodynamics   1**

1   **Definitions   3**

2   **Properties of Pure Substances   11**

3   **Properties of Ideal Gases   25**

4   **Basic Laws of Thermodynamics   37**

5   **Energy Conversion by Cycles   67**

6   **Power-Absorbing Cycles: Refrigerators and Heat Pumps   73**

7   **Gas Power Cycles   83**

Part II   **Fluid Mechanics: A Control Volume Approach   105**

8   **Flow-Governing Equations   107**

9   **External and Internal Flow Structures   111**

10   **Rotating Machinery Fluid Mechanics   149**

11   **Variable-Geometry Turbomachinery Stages   207**

12   **Normal Shocks   215**

13   **Oblique Shocks   223**

14   **Prandtl-Meyer Flow   229**

15   **Internal Flows: Friction, Pressure Drop, and Heat Transfer   241**

16   **Fanno Flow Process for a Viscous Flow Field   253**

17   **Rayleigh Flow   271**

Part III   **Heat Transfer   281**

18   **Heat Conduction   283**

19   **Heat Convection   303**

20 Heat Exchangers 323

21 Heat Radiation 343

References 361

Appendix 363

Index 457

Contents

Preface   xiii

Part I   **Thermodynamics**   1

1   **Definitions**   3

1–1   System   4
1–2   Boundary   4
1–3   Property   4
1–4   State   4
1–5   Equilibrium   5
1–6   Quasi-Static Process   6
1–7   System/Surroundings Interaction Modes   7
1–8   Examples of Nonregainable Work   7
1–9   Moving Boundary Work   7
1–10   The State Postulate   8
1–11   Zeroth Law of Thermodynamics: Definition of Temperature   10

2   **Properties of Pure Substances**   11

2–1   Phases of a Pure Substance   12
2–2   The Total Energy   12
2–3   Kinetic Energy   13
2–4   Potential Energy   13
2–5   Internal Energy   13
2–6   Enthalpy   14
2–7   Equilibrium Diagrams   15
2–8   Specific Heats   22
2–9   The Ideal Rankine Cycle   23
Problems   24

3   **Properties of Ideal Gases**   25

3–1   Ideal Gas Relationships   26
3–2   Equation of State   27
3–3   Modes of System/Surroundings Interaction   27
3–4   Work   29
3–5   Details of Moving Boundary Work   29
3–6   Polytropic Processes   31
3–7   Gravitational Work   32
3–8   Shaft Work   32
3–9   Compressibility Factor: A Measure of Deviation from the Ideal Gas Behavior   32
3–10   Other Equations of State   34
Problems   36

4   **Basic Laws of Thermodynamics**   37

4–1   First Law of Thermodynamics   38
4–2   Second Law of Thermodynamics   39
4–3   Irreversibility   44
4–4   Efficiency of Work-Producing Heat Engines   47
4–5   Efficiencies of Reversible and Irreversible Heat Engines   48
4–6   Second Law in Terms of Reversible Cycles   48
4–7   Irreversible and Reversible Processes   49
4–8   Clausius Inequality: A Statement of the Second Law of Thermodynamics   49
4–9   The T-ds Equations   50
4–10   Entropy Change for Ideal Gases   51
4–11   The T-s Diagram   52
4–12   Isentropic Processes   52
4–13   Entropy Change for a Pure Substance   53
4–14   The Increase-in-Entropy Principle   61
4–15   Entropy Change for Compressed Liquids   62
4–16   Carnot Cycle   62
Problems   64

5 Energy Conversion by Cycles 67

5–1 Heat-To-Work Conversion 68
5–2 The Rankine Cycle and *T-s* Representation 68
5–3 Ideal Rankine Cycle Analysis 69
5–4 Rankine Cycle Thermal Efficiency 70
5–5 Methods of Efficiency Enhancement: Regeneration and Reheat 70

6 Power-Absorbing Cycles: Refrigerators and Heat Pumps 73

6–1 Energy Conservation for a Reversed Cycle 74
6–2 Performance Measures 74
6–3 Vapor-Compression Refrigeration Cycle 78
6–4 Choice of the Working Medium 80
Problem 81

7 Gas Power Cycles 83

7–1 Air-Standard Assumptions 84
7–2 Otto Cycle 84
7–3 Mean Effective Pressure 85
7–4 Diesel Cycle: Ideal Cycle for Compression-Ignition Engines 87
7–5 The Ideal Brayton Cycle 89
7–6 Thermal Efficiency of Brayton Cycle 90
7–7 Real-Life Brayton Cycle 90
7–8 Isentropic Efficiency of a Process 98
7–9 Brayton Cycle with Regeneration 101
Problems 103

Part II   Fluid Mechanics: A Control Volume Approach 105

8 Flow-Governing Equations 107

8–1 Continuity Equation 108
8–2 Energy Equation 109
Problems 110

9 External and Internal Flow Structures 111

9–1 Basic External Flow Structure 112
9–2 External Flows: Boundary Layer Buildup 125
9–3 Potential Flow Fields 128
9–4 Introduction of the Velocity Potential and Stream Function 131
9–5 Compressibility of a Working Medium: The Definition of Sonic Speed 135
9–6 Compressibility of the Flow Field: Definition of the Mach Number 136
9–7 Introduction of the Critical Mach Number 140
9–8 Isentropic Flow Through Varying-Area Passages 144
Problems 147

10 Rotating Machinery Fluid Mechanics 149

10–1 Classification of Turbomachinery Components 150
10–2 Velocity Diagrams 152
10–3 Sign Convention 154
10–4 Compressor- and Turbine-Rotor Directions of Rotation 154
10–5 Axial Momentum Equation 156
10–6 Radial Momentum Equation 157
10–7 Cross-Flow Area Variation 157
10–8 Total Pressure Variation Across Multistage Turbomachines 158
10–9 Variable-Geometry Stators 159
10–10 Design-Related Variables 162
10–11 Euler's Equation 166
10–12 Introduction of the Total Relative Properties 166
10–13 Incidence and Deviation Angles 170
10–14 Means of Assessing Turbomachinery Performance 170
10–15 Supersonic Stator Cascades 180
10–16 Sign Convention Governing Radial Turbomachines 185
Problems 203

11 **Variable-Geometry Turbomachinery Stages** 207

11–1 Definition of a Variable-Geometry Turbomachine 208

11–2 Examples of Variable-Geometry Turbomachines 208

Problems 208

12 **Normal Shocks** 215

12–1 Introduction 216

12–2 Shock Analysis 216

12–3 Normal Shock Tables 219

12–4 Shocks in Nozzles 220

Problems 221

13 **Oblique Shocks** 223

13–1 Oblique Shock Tables and Charts 227

13–2 Boundary Condition of Flow Direction 227

Problems 228

14 **Prandtl-Meyer Flow** 229

14–1 Argument for Isentropic Turning Flows 230

14–2 Pressure and Entropy Changes Versus Deflection Angles for Weak Oblique Shocks 231

14–3 Isentropic Turns from Infinitesimal Shocks 231

14–4 Analysis of Prandtl-Meyer Flow 235

14–5 Prandtl-Meyer Tables 238

Problems 239

15 **Internal Flows: Friction, Pressure Drop, and Heat Transfer** 241

15–1 Assumptions 242

15–2 Mass Conservation 242

15–3 Pressure Drop and Wall Shear Stress 242

15–4 Friction Factors 244

15–5 Mechanical Energy Conservation 244

15–6 Energy Conservation 245

Problems 252

16 **Fanno Flow Process for a Viscous Flow Field** 253

16–1 Introduction 254

16–2 Analysis for a General Fluid 254

16–3 Limiting Point 256

16–4 Momentum Equation 257

16–5 Working Equations for an Ideal Gas 257

16–6 Reference State and Fanno Tables 258

16–7 Friction Choking 260

16–8 Moody Diagram 262

16–9 Losses in Constant-Area Annular Ducts 265

Problems 270

17 **Rayleigh Flow** 271

17–1 Introduction 272

17–2 Analysis for a General Fluid 272

17–3 Working Equations for an Ideal Gas 277

Problems 278

Part III **Heat Transfer** 281

18 **Heat Conduction** 283

18–1 Introduction 284

18–2 Simple One-Dimensional Problems 285

18–3 Equivalent Thermal Network 286

18–4 Network Resistance for Heat-Convection Boundary Conditions 286

18–5 Multilayer Plane Walls 287

18–6 Thermal Contact Resistance 289

18–7 Transient Heat Conduction 293

Problems 301

19 **Heat Convection** 303

19–1 Introduction 304

19–2 Convection Heat Transfer Coefficient 305

19–3 Natural Convection 309

19–4 Lumped Parameter Analysis 318

19–5 Forced Convection Analysis 321

Problems 322

20 Heat Exchangers 323

20-1 Introduction 324
20-2 Log Mean Temperature Difference 325
20-3 Overall Heat Transfer Coefficient 329
20-4 Fouling Factor 331
20-5 Analysis of Heat Exchangers 334
20-6 The Similitude Principle 337
20-7 Dynamic Similarity 337
20-8 Dimensionless Parameters 338
Problems 341

21 Heat Radiation

21-1 Thermal Radiation: Blackbody 344
21-2 Planck's Law 346
21-3 Wien's Displacement Law 346

21-4 View Factor Relations 348
21-5 Radiation Functions 348
21-6 Radiation Properties 350
21-7 Total Radiation Properties 350
21-8 Radiation Shape Factor 352
21-9 Directional Radiation Properties 352
Problems 360

References 361

Appendix

Tables and Charts 363

Index 457

Preface

In writing this book, special care was given to the fact that it targets undergraduate students who are not majoring in mechanical engineering. Conciseness, therefore, was a must, as the student is introduced to the major subject of thermal science. In all three subdisciplines of thermodynamics, fluid mechanics, and heat transfer, emphasis is placed on the basics of each subject as clearly as possible. As the worldwide interest in energy generation/conversion grows daily, it becomes essential for all engineering students to gain a keen understanding of the basic principles of thermal science, and this is what is behind this book's importance.

In covering the thermodynamics topic, attention is focused on the three laws of thermodynamics and their applications in real-life problems. Special emphasis is placed on the well-known thermodynamic cycles of Carnot, Rankine, and Brayton, as well as those related to internal combustion engines. In all cases, the working medium (be that liquid or gas) will naturally undergo substantial changes in pressure and temperatures as it proceeds from one thermodynamic state to the next. The medium, in the Rankine cycle, will even change its phase from liquid to wet saturated steam to superheated vapor. In all cases, the pure substance (be that water or Freon) is carefully monitored, and its thermodynamic properties defined.

In our coverage of the open version of the Brayton cycle, the simple turbojet engine is introduced first. The topics of regeneration and reheat techniques, as they generally apply to the Rankine cycle, are discussed in detail. Next, important derivatives of the the turbojet engine, namely the turbofan and turboprop engines are introduced, with the distinction of the turboprop engine, being categorically a turboshaft engine, emphasized. In the latter category, attention is focused on the fact that the turbine-produced power is utilized not only in providing the thrust force but also in driving a propeller. Note that the cross-oceanic air carriers are exclusively turbofan engines. Turboprop engines, on the other hand, are utilized in short missions (e.g., Phoenix to Los Angeles), for they are most efficient in this distance range.

In covering the fluid mechanics subject, the student is introduced to the principle of control volume, with part of the boundary being flow-permeable. Under focus in the entire section are the flow-governing equations, including those of mass, linear momentum, and energy.

Clearly presented in this section of the book is the topic of flow behavior through turbomachinery components, such as compressors and turbines. Owing to the author's industrial experience, this topic was covered to reveal a wide range of industrial applications. Such important phenomena as subsonic-supersonic flow conversion and flow "choking" are comprehensively covered through meaningful examples. The consequences of friction and boundary layer buildup are introduced under the classical topic of Fanno flow structure, with interesting applications in the turbomachinery area. The other major topic is that of the Rayleigh flow pattern, which addresses heat exchange effects on flow behavior and is presented with applications concerning annular combustors. The examples and graphical illustrations in this case bring the student closer to the physical problem at hand.

The third part of the book presents the three modes of heat transfer: conduction, convection, and radiation. Special attention in this case is paid to the various practical applications of each mode. The purpose here is to deepen the student's comprehension of how basic principles of heat transfer apply to real-life practical problems.

On the subject of heat conduction, such real-life problems as contact resistance in a solid with multiple layers are discussed, with the subtopic of electrical analogy highlighted. On the topic of heat convection, the distinction between free (or natural) and forced convection is clearly made, with practical applications. Also covered in detail is the problem of heat exchangers, a real industry-related problem in terms of sizing and analysis. Aside from all factors governing radiation heat transfer, the topic of shape factor, which is basically how a heat-emitting surface "sees" another, is particularly emphasized.

Note to Instructors

If you are an instructor using this book as a textbook, please be aware that an Instructor's Manual with chapter problem solutions and exam problems is available at www.mhprofessional.com/thermalscience. Please visit this site for information about how you can download this material to assist your teaching.

Erian A. Baskharone

PART I Thermodynamics

Thermodynamics is the mother of the general discipline of thermal science. In addition to the zeroth law of thermodynamics, through which temperature, as a property, is defined, we discuss two major laws: the first law of thermodynamics, which stands for an energy conservation principle, and the second law of thermodynamics, which takes into account the system degradation sources during a real-life process. This leads to the definition of a new property, called *entropy*, with which the so-called irreversibilities of a process are gauged.

The working medium, in this context, is either treated as an ideal gas such as air at a "sufficiently" low pressure, or a pure substance such as water and refrigerant 12. The latter can exist in one, two, or even three phases: namely, solid (meaning ice), liquid, or water vapor, depending on the conditions to which the water substance is exposed, or dry superheated vapor. Under specific situations, such as saturated liquid and saturated vapor co-existing, both the temperature and pressure are dependent on one another. In this case, a new property termed the *steam quality* (or dryness factor) is needed to identify the state of that mixture.

At the heart of the whole thermodynamics topic is what is referred to as a system in either the closed kind (e.g., a rigid container or the piston/cylinder device in Figure 1–1) or the open one (e.g., a flowing medium through a diffuser as in Figure 1–2). Of course, open systems embrace a virtually unlimited number of other configurations, such as pipes and sudden-enlargement ducts.

1

Definitions

Chapter Outline

1–1 System **4**

1–2 Boundary **4**

1–3 Property **4**

1–4 State **4**

1–5 Equilibrium **5**

1–6 Quasi-Static Process **6**

1–7 System/Surroundings Interaction Modes **7**

1–8 Examples of Nonregainable Work **7**

1–9 Moving Boundary Work **7**

1–10 The State Postulate **8**

1–11 Zeroth Law of Thermodynamics **10**

1–1 System

A thermodynamic system is generically defined as a collection of matter contained within a boundary that is either solid or even fictional for closed and open systems, respectively.

1–2 Boundary

A boundary is what separates the system from the so-called universe (Figure 1–1).

1–3 Property

A property is any characteristic of the system. However, such characteristics as color do not constitute thermodynamic properties. The term property includes two distinct families:

- Extensive properties, which depend on the contents (meaning mass) of the system.
- Intensive properties, which are independent of the mass, mostly the property magnitude per unit mass, such as the specific volume. Obvious intensive properties are temperature and pressure.

1–4 State

This is the condition of the system as defined by all of its properties. The question here arises: How many, in the least, are required to define the state of a particular system? For a single-phase substance, two independent properties suffice. An example of two dependent properties is the volume and specific volume (defined as the volume per unit mass). As for a two-phase substance, an additional property is required and is commonly termed the quality of the mixture (defined as the mass of vapor in the mixture divided by the entire mixture mass in the coexisting vapor and liquid phase).

Figure 1–1

A rising piston within a cylinder.

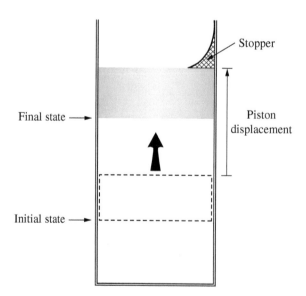

Figure 1–2

Open and closed systems.

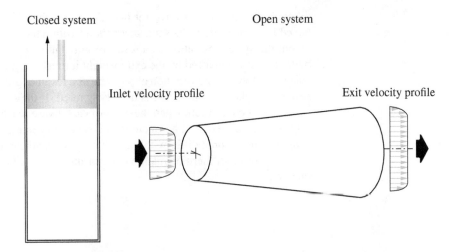

Figure 1–2

Open and closed systems.

1–5 Equilibrium

Consider a system that is not undergoing any process. At the beginning of a process, such as that of a rising piston inside a cylinder (Figure 1–3), the system possesses a set of properties that completely define the state of the system. At a given state, all of the system's properties have fixed magnitudes. If the value of any property changes, the state will change to a different one, as well.

Thermodynamics, as a topic, deals with equilibrium states. *Equilibrium* implies a state of balance. In an equilibrium state, there are no unbalanced potentials (or driving force) within the system. A system in equilibrium experiences no change when it is isolated from the surroundings.

There are many kinds of equilibrium, and a given system is not in thermodynamic equilibrium unless the conditions of all relevant types of equilibrium are satisfied. For example, a system is in a state of thermal equilibrium if the temperature has the same magnitude throughout the entire system. That is, the system involves no temperature

Figure 1–3

Expansion through a
piston/cylinder apparatus.

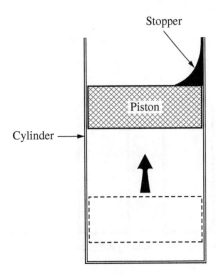

differentials (the driving force for heat flow) anywhere. Mechanical equilibrium is attained if the pressure is the same at any point within the system. The pressure may vary within the system elevation as a result of gravitational effects. But the pressure at the bottom layer is balanced by the extra weight it must carry, and therefore there is no imbalance of forces. The variation of pressure as a result of gravity in most thermodynamic systems is relatively small and is usually ignored. If the system involves two phases, it is a phase of equilibrium when the mass of each phase reaches an equilibrium level and stays there. Furthermore, a system is in a state of chemical equilibrium if its chemical composition does not differ with time, that is, no chemical reaction is occurring. A system will not be in thermodynamic equilibrium unless all of these relevant criteria are satisfied.

1–6 Quasi-Static Process

This is a process where the system progresses from one equilibrium state to another. An example of such a process is shown in Figure 1–4, where the piston-applied weights are gradually increased over a rather long (theoretically infinite) period of time.

Figure 1–4

Example of a quasi-static process.

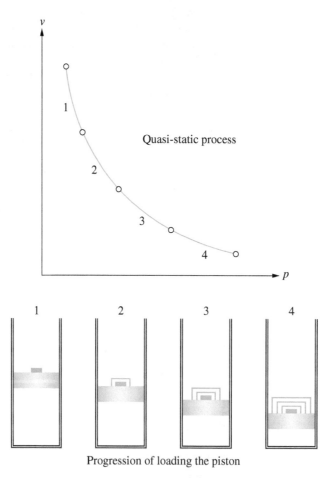

Progression of loading the piston

Figure 1–5

Two examples of nonregainable work.

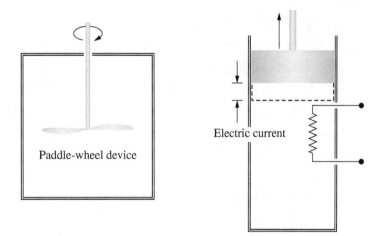

Paddle-wheel device

Electric current

1–7 System/Surroundings Interaction Modes

These are two, namely heat transfer and work, interactions that can take place. The former will occur due to a finite nonzero temperature differential between the system and its surroundings. As for work, we have two distinct categories. For a closed system, such as a piston-cylinder device, work will be exerted on or by the system due to expansion or compression of the boundary, respectively. Figure 1–5 shows such system/environment exchanges for two closed systems. For open systems, work is simply shaft work that is done due to the existence of a turbomachinery component such as a pump, a compressor, or a turbine.

1–8 Examples of Nonregainable Work

Figure 1–5 shows two types of processes where the work done is unregainable. The first involves electric current controlled from outside the system. The second is one that involves a paddle-wheel apparatus.

1–9 Moving Boundary Work

This is the most common type of work where the displacement of a piston, say, within a cylinder, will give rise to a so-called displacement type of work. The following expression quantifies this type of work (Figure 1–1):

$$W_{1-2} = \int_1^2 p \, dV$$

where p is the instantaneous magnitude of pressure and dV is the differential volume change due to the piston's movement.

EXAMPLE 1–1   An equal amount of work of 50 kJ/kg is supplied to the two systems shown in Figure 1–5. The torque exerted in the first system is 415 N.m., and the piston radius, in the second case, is 14.5 cm. Calculate the rotational speed and piston displacement in these two cases, respectively.

Solution   (a)

$$w_s = \tau\omega$$

which yields:

$$\omega = 120.5 \text{ rad}/s = 6,904 \text{ r/min}$$

(b)

$$A_{\text{piston}} = \pi r^2 = 0.066 \text{ m}^2$$

We now proceed to calculate the piston displacement as follows:

$$w = p\Delta v = pA_{\text{piston}}\Delta h$$

which yields the piston displacement:

$$\Delta h = \frac{w}{pA_{\text{piston}}} = 3.8 \text{ m}$$

1–10   The State Postulate

As noted earlier, the state of a given system is defined by all of its properties. But we know from experience that we do not need to specify all properties to define a state. Once a sufficient number of properties is specified, the rest assume specific magnitudes correspondingly. That is, specifying a certain number of properties is sufficient to define a thermodynamic state. The number of such properties is given by the so-called state postulate principle:

"The state of a simple compressible, single-phased system is totally defined by two independent intensive properties."

A system is referred to as a simple compressible system in the absence of electric, magnetic, and gravitational energy as well as surface tension. These effects are due to external force fields and are negligible for most engineering applications. Otherwise, an additional property needs to be specified for such effects that are significant. If the gravitational effects are supposed to be included, for example, the elevation (z) needs to be specified in addition to the two properties needed to define a state. The state postulate requires that the two properties be independent to define a unique state. Two properties are independent if one is allowed to vary while the other is held constant. Temperature and specific volume, for instance, are always independent of one another.

EXAMPLE 1–2   The system in Figure 1–6 is composed of a trapped amount of gas inside a piston-cylinder device. The piston, which weighs 4.0 kg, is held in place by the stiffness force of a compressed spring which provides a force of 60 Newtons. The cross-sectional area of the piston is 35 m². If the local atmospheric pressure is 0.95 bar, calculate the gas pressure.

Solution   Considering the piston balance, we have:

$$F_{\text{spring}} + p_{\text{atmospheric}}A_{\text{piston}} - p_{\text{gas}}A_{\text{piston}} + m_{\text{piston}}g = 0$$

which, upon substitution, yields:

$$p_{\text{gas}} = 1.234 \text{ bars}$$

Figure 1–6

Piston balance under the stiffness of a compressed spring.

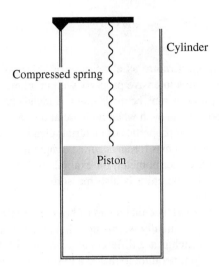

Compressed spring

Cylinder

Piston

EXAMPLE 1–3   Referring to the previous example, the following changes are made:

- The gas pressure is 1.2 bars.
- The piston is allowed to move up against a spring force which has a stiffness factor (k) of 1800 N/m (Figure 1–7).

Now calculate the piston's displacement.

Solution   Again, considering the piston balance, we have:

$$m_{\text{piston}} g + k_{\text{spring}} \Delta z + p_{\text{atmospheric}} A_{\text{piston}} - p_{\text{gas}} A_{\text{piston}} = 0$$

which yields:

$$\Delta z = 2.7 \text{ cm}$$

Figure 1–7

Different configuration of the same problem.

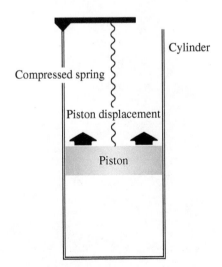

Compressed spring

Cylinder

Piston displacement

Piston

Zeroth Law of Thermodynamics: Definition of Temperature

We all recognize temperature as a measure of "hotness" at a given point within the system. This leaves this intensive property with no numerical magnitude. Furthermore, our sense of this property may be misleading. A metal chair, for instance, will feel much colder than a wooden one even when both are at the same temperature.

Fortunately, several properties of materials change with temperature in a repeatable way, and this forms the basis for an accurate temperature measurement. The commonly used mercury-in-glass thermometer, for example, is based on the expansion of mercury with temperature. Temperature is also measured by using several other temperature-dependent properties.

It is common experience that a cup of hot coffee left on the table eventually cools off and a cold drink eventually warms up. That is, when a body is brought into contact with another body which is at a different temperature, heat is transferred from the body which is at a higher temperature to that at a lower one, until both bodies attain the same temperature. At this point, the heat transfer terminates, and the two bodies are said to have reached a state of thermal equilibrium. The equality of temperature is the only requirement for thermal equilibrium.

The zeroth law of thermodynamics states that if two bodies are in a state of thermal equilibrium with a third body, then they are also in thermal equilibrium with each other. It may seem silly that such an obvious fact is referred to as one of the basic laws of thermodynamics. However, it can't be concluded from the other laws of thermodynamics, and it serves as a basis for the validity of temperature measurements. By replacing the third body with a thermometer, the zeroth law of thermodynamics can be restated as: two bodies are in thermal equilibrium if both have the same temperature reading even if they are not in contact with one another.

2

Properties of Pure Substances

Chapter Outline

2–1   Phases of a Pure Substance   **12**

2–2   The Total Energy   **12**

2–3   Kinetic Energy   **13**

2–4   Potential Energy   **13**

2–5   Internal Energy   **13**

2–6   Enthalpy   **14**

2–7   Equilibrium Diagrams   **15**

2–8   Specific Heats   **22**

2–9   The Ideal Rankine Cycle   **23**

A pure substance is one that is property-homogeneous. Examples of that include water, nitrogen, and lithium.

A pure substance does not have to be of a single chemical element. Air, for instance, would constitute a pure substance as long as the mixture of gases is homogeneous.

A mixture of two or more phases of a pure substance is still a pure substance. For example, a mixture of liquid water and superheated (gaseous) vapor is not homogeneous since part of the container would be occupied by the liquid while the other is occupied by vapor.

2-1 Phases of a Pure Substance

We all know that under standard magnitudes of temperature and pressure, copper exists in its solid phase, water in the liquid phase, and oxygen in the gaseous phase. However, under different magnitudes, temperature and pressure, a substance could be a mixture of two or even three of its phases.

The substance molecules in a solid are arranged in a three-dimensional pattern (lattice) that is repeated throughout. Because of the small distances between molecules in this phase, the attractive forces of molecules on each other are large and keep molecules at fixed positions.

The molecules' spacings in the liquid phase are not that much different from their counterparts in the solid phase. The distance between molecules generally experiences a slight increase as the solid turns into liquid, with water being a rare exception.

In the gas phase, the molecules are far apart from one another, and a molecular order is nonexistent. Gas molecules move around in random, continually colliding with one another and the wall of the container. Particularly at low magnitudes of density, the intermolecular forces are very small, and collision is the only mode of interaction between molecules. Molecules in the gaseous phase are not at a considerably higher energy level than they are in a liquid. Therefore, a gas must release a large amount of its energy before it can condense or freeze.

2-2 The Total Energy

Work and heat are defined as modes of energy transfer across the system boundary. Heat is defined as the energy transfer across the boundary due to a temperature differential between the system and its surroundings. Work is defined as the energy transfer arising from an effect that is solely equivalent to a force acting through a distance. Neither heat nor work are properties, and they cannot be represented by exact differentials since their values depend on the process path followed during a change of state.

Many forms of energy, other than heat and work, are equally important. In a classification of all different forms of energy that play a role in thermodynamic systems, distinguishing between microscopic and macroscopic forms of energy is helpful. Microscopic forms of energy are those related to the energy possessed by the individual molecules and to the interaction between molecules and to the interaction between molecules that comprise the system at hand. Macroscopic forms of energy, on the other hand, are related to the gross characteristics of a substance on a scale that is large compared to the mean free path of molecules.

The total energy (E) is a property of a system and is defined as the sum of all macroscopic forms of energy plus the total of the microscopic forms as follows:

$$E = E_{\text{microscopic}} + E_{\text{macroscopic}}$$

A thermodynamic analysis usually includes a determination of the change in the total energy of a system during a process or a series of processes. However, only rarely does a system experience significant changes in more than only a few of many different forms of energy that sum up to the total energy of the system.

2-3 Kinetic Energy

The kinetic energy of a mass m with velocity V is defined as follows:

$$E_k = \frac{1}{2}mV^2$$

The kinetic energy and the velocity of the center of the mass are physical properties and, as such, must be measured with respect to some physical external frame of reference. Often the most convenient frame of reference is one that is stationary relative to the earth. For this reference frame, a quantity of mass with no motion "relative" to the earth has a relative velocity of zero, and its kinetic energy would obviously be zero as well.

Thermodynamic analyses are most often concerned with determining the change in kinetic energy from one state to another. Since the kinetic energy is a property, the change in it is independent of the process path that the system may get into. This magnitude of change is solely dependent on the mass and velocity of the system at the end states. The kinetic energy per unit mass, which is an intensive property, is consistently defined as:

$$e_k = \frac{V^2}{2}$$

Obviously the preceding definition is well suited to evaluate the kinetic energy associated with the mass flow across the boundary of an open system.

2-4 Potential Energy

A quantity of mass (m) possesses potential energy in a gravitational field with acceleration of gravity g_c by virtue of its elevation z above some arbitrary datum. The potential energy is defined as follows:

$$E_p = mg_c z$$

From which the intensive property can be expressed as follows:

$$e_p = g_c z$$

2-5 Internal Energy

The energy associated with a substance on a molecular scale can consist of several forms. Molecules possess kinetic energy due to their individual mass and velocity as they move about a linear path. The molecules also possess vibrational and rotational angles as they rotate and vibrate as a consequence of their random motion, which is another form of energy associated with the intermolecular forces between molecules. The sum of all of these molecular or microscopic energies is called the *internal energy* of a substance.

Internal energy is a thermodynamic property. In other words, the internal energy change during a process depends solely on the end states. The symbol for internal energy

Figure 2–1

Determination of the specific heats.

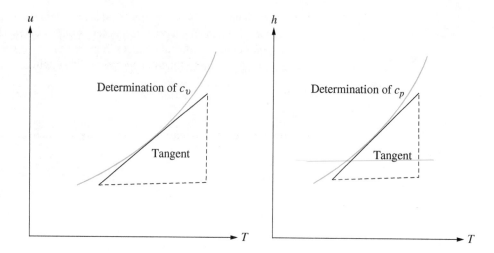

of a quantity of mass is U. This is an extensive property of which the corresponding intensive property is defined as follows:

$$u = \frac{U}{m}$$

and its variation with temperature is shown in Figure 2–1.

2–6 Enthalpy

In the analysis of open systems (known as control-volume analysis), the combination:

$$h = u + pv$$

is rather important as it combines the internal energy with that which is required to "push" the flow across any cross-flow plane in the fluid stream, and in other applications as well.

This is the same property shown versus the temperature in Figure 2–1. The tangents in this figure define two new intensive properties; namely the specific heats under constant volume and constant pressure, respectively, both of which will be discussed next. Note that the preceding combination is a new property termed enthalpy. As seen, the value of enthalpy at any state depends on other thermodynamic properties. Therefore it is, as well, a valid property.

The constant-volume specific heat c_v, by reference to Figure 2–1, is defined as follows:

$$c_v \equiv \left(\frac{\partial u}{\partial T} \right)_v$$

and the constant-pressure specific heat is defined by:

$$c_p \equiv \left(\frac{\partial h}{\partial T} \right)_p$$

These are very useful intensive properties. They are independent of any system/surroundings interaction. Figure 2–1 shows that c_v is the slope of the constant-volume line on a u versus T plot, whereas c_p is the slope of a constant-pressure line on an h versus T plot. The ratio of specific heats is particularly important variable and is only a function of temperature:

$$\gamma \equiv \frac{c_p}{c_v}$$

γ varies, for instance, across a simple turbojet engine from approximately 1.4 across the compressor to approximately 1.33 in the turbine.

2-7 Equilibrium Diagrams

All known substances can exist in several phases, based on the temperature and pressure. Referring to Figure 2–2, a mixture of more than one phase of a substance in equilibrium is also common. Because of this complex behavior, a single relationship between the pressure, temperature, and specific volume that is valid for all possible states of a substance cannot be developed. However, the qualitative aspects of these substances can be discussed in order to gain insight into their p-v-T behavior.

Consider the following simple experiment devised to monitor the changes in temperature and specific volume of a substance such as water as it is heated at constant pressure. For this purpose, imagine a transparent cylinder initially filled with (liquid) water at 20°C and maintains a pressure of 101.35 kPa (the standard ambient pressure) by means of a piston of constant weight as shown in Figure 2–2.

As heat is transferred to the water, the system temperature will begin to increase and, at the same time, water will expand. In other words, the specific volume of the cylinder's contents will increase. Since the cylinder is closed, the expansion will also cause the piston to move.

If energy is continually added to the water, a plot of the temperature as a function of specific volume could be constructed for this process. A sketch of the results is shown in Figure 2–2. Note that the water temperature continues to increase up to a state (the saturated liquid state) at which the monitor will register the first appearance of a

Figure 2–2

Constant pressure lines on the p-v diagram.

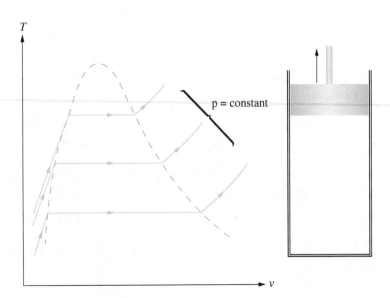

vapor bubble. During subsequent heating, the specific volume continues to increase as well, up to a state where the water temperature reaches the magnitude of 100°C. The fact that water at atmospheric pressure begins to evaporate is common knowledge, with the appearance of the first vapor bubble signaling the beginning of a process during which the pressure will remain constant.

As more energy is transferred to the now two-phase working medium, the temperature will remain constant as well. At some state, a point is reached where approximately one-half of the mixture is liquid water while the other half (by mass) is vapor. Eventually, the liquid component is all gone. Further heating will cause the temperature to increase. Upon continuing the heating process, the vapor inside the cylinder then is referred to as superheated vapor.

The p-T Diagram

Figure 2–3 shows the *p-T* diagram of a pure substance. This diagram is often called the *phase diagram* since all three phases are shown separated from one another by three lines. The sublimation line separates the solid and the vapor regions, the vaporization line separates the liquid and vapor regions, and the melting (or fusion) line separates the solid and liquid regions. These three lines meet at the triple point, where all three phases would coexist in equilibrium. The vaporization line ends at the critical point, above which no distinction can be made between the liquid and vapor phases.

The p-v-T Surface

As earlier indicated, the state of a simple compressible substance is determined and fixed by any two independent intensive properties. Once the two appropriate properties are determined, all other properties become dependent on them. Remembering that any equation with two independent variables in the form $z = z(x, y)$ represents a surface in space, we can represent the *p-v-T* behavior of a substance as a surface in space, as shown

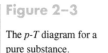

Figure 2–3

The *p-T* diagram for a pure substance.

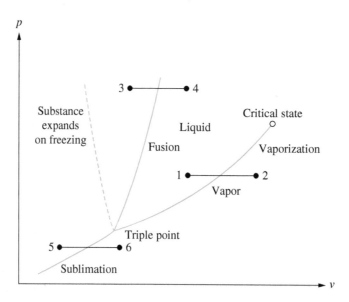

Figure 2–4

A *p-v-T* surface for a substance that expands on freezing.

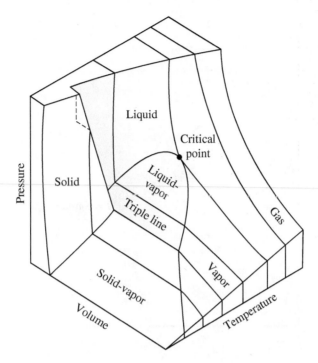

in Figure 2–4. Here T and v may be viewed as the independent variables (the base) and p as the dependent variable (height).

Going back to Figure 2–2, and focusing on the horizontal segments in this figure as well as in Figure 2–4, a new property needs to be introduced, the *steam quality* (x) or dryness factor, and is defined as follows:

$$ x = \frac{m_g}{m_{\text{mixture}}} = \frac{m_g}{m_g + m_f} $$

where m_g and m_f refer to the masses of saturated liquid and saturated vapor, respectively.

Knowing all necessary data during the experiment, one can similarly construct a pressure-specific volume chart.

Should the preceding experiment be repeated at a different (say higher) magnitude of pressure, the distance between the two saturation points would begin to shrink.

In fact, under a specific pressure and temperature combination, the convergence from liquid to vapor states will shrink down to a point (the critical state). Upon repetition of the same experiment, the locus of the saturation lines will come to form a "dome," with the L + V states all existing inside the dome as shown in Figure 2–5.

Figure 2–5 also shows two constant temperature lines on the *p-v* diagram, where the upper line is associated with the higher of the two temperatures. Note that, away from the liquid-vapor dome, on the right-hand side, the constant-temperature lines are hyperbolic in shape, as will be discussed later in conjunction with ideal gases.

Figure 2–5

Constant-temperature line on the p-v diagram.

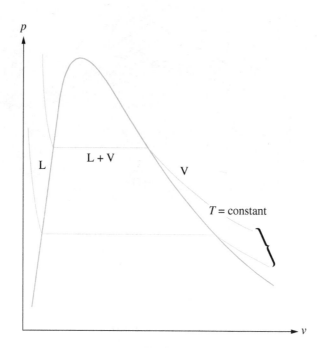

EXAMPLE 2–1

Referring to Figure 2–6, saturated water vapor is kept at 40 bars. The vapor undergoes a constant-pressure process until the volume is doubled. The vapor is then returned to the saturated vapor status through a constant-volume process. Calculate the net change in enthalpy ($h_3 - h_1$).

Solution

Referring to the water table in the Appendix at 40 bars, we get:

$$v_1 = 0.049 \text{ m}^3/\text{kg}$$

$$h_1 = 2801 \text{ kJ/kg}$$

Now we move to state 2:

$$p_2 = 40 \text{ bars (given)}$$

$$v_2 = 0.098 \text{ m}^3/\text{kg}$$

$$T_2 = 600 \text{ C}$$

$$h_2 = 2800.5 \text{ kJ/kg}$$

Now we focus on state 3:

$$v_3 = 0.098 \text{ m}^3/\text{kg}$$

Finally:

$$h_3 - h_1 = 0.5 \text{ kJ/kg}$$

Figure 2–6

Constant-pressure and
constant-volume
processes.

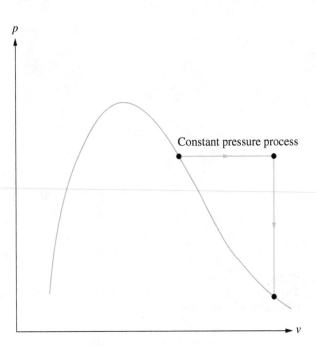

Figure 2–7

Wet steam conversion
into the superheated vapor
region.

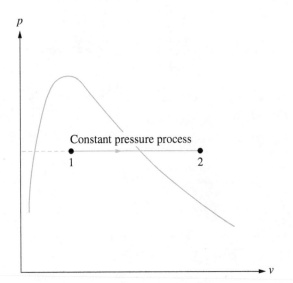

EXAMPLE 2–2 Referring to Figure 2–7, water wet steam is kept under a pressure of 20 bars and a quality of 0.4. The steam then expands until the volume is tripled. Calculate the change in internal energy.

Solution Let us consider the initial state:

$$u_1 = x_1 u_g + (1 - x_1)u_f = 1584.0 \text{ kJ/kg}$$

$$v_1 = x_1 v_g + (1 - x_1)v_f = 0.06 \text{ m}^3\text{/kg}$$

$$v_2 = 3v_1 = 0.018 \text{ m}^3\text{/kg}$$

$$p_2 = 20 \text{ bars}$$

$$u_2 = 3203.5 \text{ kJ/kg}$$

Finally:

$$u_2 - u_1 = 1619.5 \text{ kJ/kg}$$

EXAMPLE 2–3

Referring to Figure 2–8, superheated water vapor is maintained at 25 bars and 700°C. The vapor undergoes a constant-volume process until the pressure is 10 bars. Calculate the final steam quality.

Solution Using the superheated water vapor tables:

$$v_1 = 0.017 \text{ m}^3\text{/kg}$$

$$h_1 = 3777.5 \text{ kJ/kg}$$

Now we focus on the final state:

$$p_2 = 10 \text{ bars}$$

$$v_2 = 0.017 \text{ m}^3\text{/kg} = x_2 v_g + (1 - x_2)v_f$$

which yields:

$$x_2 = 0.09$$

Figure 2–8

Constant-volume conversion of superheated vapor.

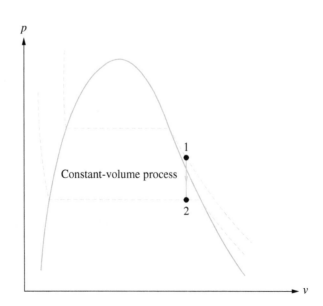

Figure 2–9

Constant-pressure
condensation process.

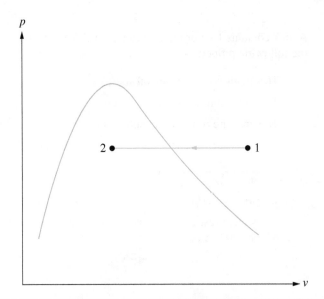

EXAMPLE 2–4

Referring to Figure 2–9, superheated water vapor is initially at a pressure of 25 bars and a temperature of 500°C. The vapor undergoes a constant-pressure condensation process until the volume is one-half of its initial magnitude. Calculate the steam final quality.

Solution

$$v_1 = 0.14 \text{ m}^3/\text{kg}$$

Now we focus on the final state:

$$v_2 = 0.07 \text{ m}^3/\text{kg}$$

$$x_2 = \frac{v_2 - v_f}{v_g - v_f} = 0.875$$

EXAMPLE 2–5

State the phase or phases of water that may exist at the given states:

(a) $T = 215°C$, $p = 2.0$ MPa

(b) $T = 240°C$, $x = 0.4$

(c) $T = 260°C$, $v = 0.40$ m³/kg

Solution

(a) At $T = 215°C$, and using the water tables in the Appendix, we find that:

$$p_{\text{sat.vapor}} = 0.021 \text{ bar.}$$

Now, since $p > p_{\text{sat.vapor}}$, the water substance exists in the superheated gas domain.

(b) Since the steam quality has a finite magnitude (0.4) between 0 and 1.0, the state is that of a wet liquid-vapor mixture.

(c) Searching the saturated-gas leg of the L + V dome, we find that at $T = 260°C$, $v_{\text{sat}} = 0.0205$ m³/kg. Therefore, that state exists in the subcooled liquid subregion (by just comparing the specific volumes).

EXAMPLE 2–6 A tank contains 1 kg of liquid water and 0.1 kg of water vapor at 200°C. Determine the following properties:

  (a) The quality of the water substance
  (b) The total volume of the container
  (c) The pressure of the water substance

Solution

  (a) $x = \dfrac{m_g}{m_g + m_f} = 0.091$

  (b) $V = mv = 1.1[xv_g + (1 - x)m_f] = 14.291 \text{ m}^3$

  (c) The water pressure is that associated with $T = 200°C$ within the L + V dome, which is 0.015 bar.

EXAMPLE 2–7 A rigid tank with a volume of 2.5 m³ contains 5 kg of a saturated liquid-vapor mixture of a water at 75°C. Now the water substance is slowly heated. Determine the temperature at which the liquid in the tank is completely vaporized.

Solution

$$v_1 = \frac{V_1}{m} = 0.5 \text{ m}^3/\text{kg}$$

$$v_1 = x_1 v_g + (1 - x_1)v_f$$

Using the tables we get $x_1 = 0.119$
  It follows that:

$$u_1 = x_1 u_g + (1 - x_1)u_f = 571.2 \text{ kJ/kg}$$

Searching the saturated vapor leg of the L + V dome, we find:

$$v_{\text{vapor}_2} = 0.3829 \text{ m}^3/\text{kg}$$

also

$$u_2 = 2545.0 \text{ kJ/kg}$$

At this state we find:

$$T_2 = 140°C$$

2–8 Specific Heats

The state postulate (presented earlier) was used to conclude that the state of a simple compressible substance is determined by the values of two independent intensive properties. As a result, the internal energy of a simple compressible fluid could be considered a function of temperature and specific volume, i.e.,

$$u = u(T, v)$$

For a "small" process where the end states are sufficiently close to one another, we get:

$$du = \left(\frac{\partial u}{\partial T}\right)_v dT + \left(\frac{\partial u}{\partial v}\right)_T dv$$

The first partial derivative in this expression is a thermodynamic property termed the specific heat at constant volume (c_v). Similarly the property enthalpy can be treated as a function of temperature and pressure, with the following similar relationship:

$$dh = \left(\frac{\partial h}{\partial T}\right)_p dT + \left(\frac{\partial h}{\partial p}\right)_T dv$$

The first derivative on the right-hand side is referred to by the specific heat under constant pressure (c_p). Both specific heats are graphically shown in Figure 2–1 as tangents to u versus T and h versus T curves, respectively.

2–9 The Ideal Rankine Cycle

Figures 2–10 and 2–11 show the different components and their functions in an ideal Rankine cycle. First the water substance (usually saturated liquid) is compressed up to

Figure 2–10

Components comprising the Rankine cycle.

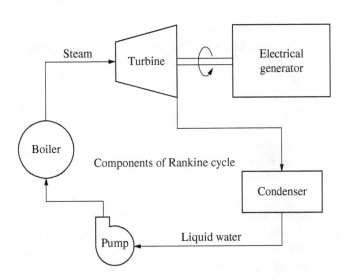

Components of Rankine cycle

Figure 2–11

Components' functions.

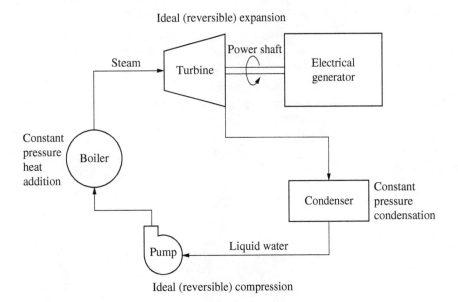

Figure 2–12

Constant-pressure process under the piston's weight.

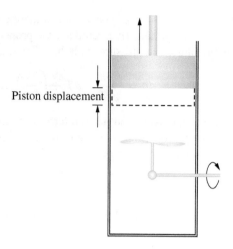

Piston displacement

a high pressure magnitude through the pump component. In the boiler, heat is added at constant pressure. The substance (usually superheated vapor) then expands across the steam turbine to a significantly low pressure. Finally, and across the condenser, the substance (usually wet steam) undergoes a condensation process back to the initial state.

PROBLEMS

2–1 Water steam is kept under a pressure of 5.0 bars and a quality of 0.9. Shaft work is imparted (Figure 2–12) at constant volume through a paddle-wheel device until the pressure is 10 bars. Calculate the final temperature.

2–2 Two kilograms of water substance at 200°C and 300 kPa is contained in a weighted piston-cylinder device. As a result of heating under constant pressure, the temperature rises to 400°C. Determine the change in volume (ΔV), internal energy, and enthalpy.

2–3 Determine the pressure and specific volume of a water substance at 20°C that has a specific internal energy of 1200 kJ/kg.

2–4 A rigid tank contains 50 kg of saturated liquid water at 200°C. Determine the pressure in the tank as well as the specific volume.

2–5 Water vapor at 5.0 bars and 500°C follows a constant-pressure (Figure 2–10) process, until its temperature increases to 800°C. Determine the enthalpy and internal energy changes that occur during the process.

2–6 Dry saturated steam at 200°C is heated in a constant-pressure process. Determine the amount of work per kilogram of the steam performed on the surroundings if the final temperature of the steam is 500°C.

2–7 Find the quality of a liquid-vapor mixture where $v_f = 0.00101$ m³/kg, $v_g = 0.00526$ m³/kg, and the total mass is 2.0 kg, which occupies a volume of 0.01 m³.

3

Properties of Ideal Gases

Chapter Outline

3–1   Ideal Gas Relationships   **26**

3–2   Equation of State   **27**

3–3   Modes of System/Surroundings Interaction   **27**

3–4   Work   **29**

3–5   Details of Moving Boundary Work   **29**

3–6   Polytropic Processes   **31**

3–7   Gravitational Work   **32**

3–8   Shaft Work   **32**

3–9   Compressibility Factor: A Measure of Deviation from the Ideal Gas Behavior   **32**

3–10   Other Equations of State   **34**

Figure 3–1

Validity of the ideal gas
status to superheated
water vapor.

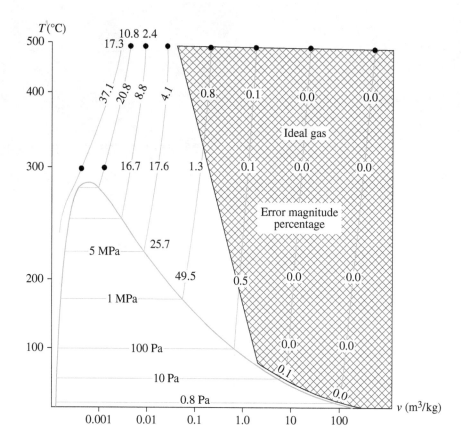

An ideal gas is any gas under theoretically zero pressure. This would naturally constitute a vacuum, a case which is rather silly to talk of. Engineers, on the other hand, define an ideal gas to be one under a "sufficiently" low pressure. Using the loose term "sufficiently," engineers proceed to consider 14 or even more pressure magnitudes to define an ideal gas.

Figure 3–1 shows the subregion on the p-v diagram for water where superheated vapor may very well be treated as an ideal gas, for the error in most of this subregion is less than or equal to 1 percent.

Going back to the engineers' definition of a sufficiently low pressure, a fluid particle traversing the compressor and the same particle leaving the turbine of the same turbojet engine is considered to comprise an ideal gas.

3–1 Ideal Gas Relationships

The internal energy and enthalpy of an ideal gas are both functions of only temperature, that is,

$$u = u(T)$$

$$h = h(T)$$

In this case, and no matter how far apart the end states are, one can express the internal energy and enthalpy changes, during a process, as follows:

$$\Delta u = c_v \Delta T$$

$$\Delta h = c_p \Delta T$$

3-2 Equation of State

For an ideal gas, a rather important relationship applies for the pressure, temperature, and specific volume. This so-called equation of state is:

$$pv = RT$$

or:

$$\frac{p}{\rho} = RT$$

where ρ is the gas density. The symbol R in the preceding relationships refers to the so-called gas constant, where:

$$R = \frac{R^*}{M}$$

with R^* being the universal gas constant and is equal to 8315 kJ/mole, and M being the gas molecular weight. In terms of the specific heats, we have:

$$c_v = \frac{1}{\gamma - 1} R$$

$$c_p = \frac{\gamma}{\gamma - 1} R$$

Again it is emphasized that an ideal gas doesn't have to exist under zero pressure or even close to that.

3-3 Modes of System/Surroundings Interaction

Heat energy can be exchanged between the system and its surroundings. Noteworthy here is the fact that the definition of surroundings cannot be the whole universe minus the system. We cannot, and this is a silly proposition, expect any thermal change on the land of the moon just because our system is emitting heat, at a finite magnitude, to the surroundings. In other words, each system will have a finite space that is adjacent to it and within which the emitted energy will practically present itself, not the rest of the entire universe. Focusing on heat transfer, energy can be exchanged in one of three modes.

Heat Conduction

This is the heat transfer mode from the more energetic particles of a substance to the adjacent less energetic ones. Conduction can take place in solids, liquids, or gases. In gases and liquids, conduction is due to the collision of molecules during their random motion. In solids, it is due to the combination of vibrations of molecules in a lattice and the energy transport by free electrons.

It is observed that the rate of heat conduction $\dot{Q}$ through a layer of constant thickness Δx is proportional to the temperature difference ΔT across the layer, normal to the direction of heat transfer, and is inversely proportional to the layer thickness, that is,

$$\dot{Q}_{\text{cond}} = k_t A \frac{\Delta T}{\Delta x}$$

where the constant of proportionality k_t is referred to as the thermal conductivity coefficient. Per unit area, the conduction heat transfer $\dot{q}_{\text{cond}}$ can be expressed as follows:

$$\dot{q}_{\text{cond}} = k_t \frac{\Delta T}{\Delta x}$$

Heat Convection

Convection is the mode of heat energy transfer between a solid surface and the adjacent gas or liquid in the vicinity of the surface. Convection is labeled forced if the fluid is forced to flow over a surface by external means, such as the case of a pump or a blower. This, and the free-convection mode, will be discussed at length in Part III of this book.

The rate of heat transfer by convection $\dot{Q}_{\text{conv}}$ is determined from Newton's law (of cooling), which can be expressed as follows:

$$\dot{Q}_{\text{conv}} = h A (T_s - T_f)$$

where h is the convection heat transfer coefficient, A is the surface area through which heat transfer takes place, T_s is the surface temperature, and T_f is the system's bulk mean temperature away from the surface. Per unit cross-sectional area, the amount of convection heat transfer can be expressed as follows:

$$\dot{q}_{\text{conv}} = h (T_s - T_f)$$

Heat Radiation

Radiation is the heat energy emitted by matter in the form of electromagnetic waves as a result of changes in the electric configurations of the atom or a molecule. Unlike conduction and convection, radiation doesn't require the presence of a medium. Energy transfer by radiation is the fastest, for it occurs at the speed of light. An example is the heat radiated from the sun.

The maximum rate of radiation that can be emitted from a surface at an absolute temperature T_s is given by the Stefan-Boltzmann blackbody law as follows:

$$\dot{Q}_{\text{emit,max}} = \sigma T_s^4$$

where the Stefan-Boltzmann constant σ is equal to $5.67 \times 10^{-8} \text{ W/}m^3 K^4$. The idealized surface that emits radiation at the maximum rate is termed a blackbody. The radiation emitted by real-life surfaces is less by comparison at the same temperature, and can be expressed as follows:

$$\dot{Q}_{\text{emit}} = \epsilon \sigma A T_s^4$$

where ϵ is the so-called surface emissivity, and has a magnitude between zero and unity.

3–4   Work

This is an energy transfer between the system and its surroundings that is not a function of any temperature differential. An example of work is an electric wire penetrating the system boundary to switch on a light bulb. When it is energized, the bulb begins heating the substance inside the system. The key point here is that the lighting of this bulb did not occur as a result of interaction between the system and its surroundings; therefore, the energy transfer here is that of work.

Heat and work energies are directional quantities, and thus the complete description of heat and work interactions have to abide by a strict sign convention:

- Heat transfer to the system is positive.
- Work done on the system is also positive.

3–5   Details of Moving Boundary Work

One form of mechanical work is that encountered by expansion or compression of a gas in a piston-cylinder device or any deformable-boundary system. This is commonly referred to as boundary work. Consider the gas enclosed in the piston-cylinder device in Figure 3–2. Let us refer to the piston cross-sectional area by A and to the volume contained in the cylinder by V. The boundary work during an incrementally small change of state can be expressed as follows:

$$W_b = Fds = pAds = pdV$$

Throughout the entire process, the total amount of boundary work can be written as follows:

$$W_b = \int_1^2 pdV$$

Figure 3–2

Moving boundary work.

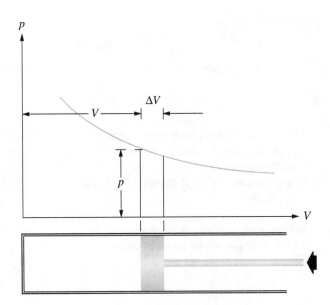

Figure 3–3

Two examples of
nonregainable work.

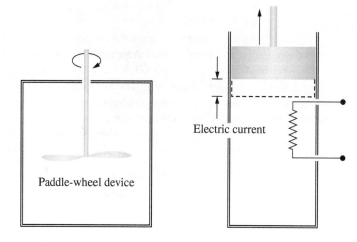

Paddle-wheel device

Electric current

Example 3–1

Figure 3–3 shows two different systems where work is being input in two different
ways for a total of two minutes:

System A

A spinning paddle-wheel device where the shaft speed is 1000 rpm and under a
torque of 68.4 N·m.

System B

An electric wire operating under a voltage of 180 V, and a current of 25 A.
   Calculate the input amount of work in each case.

Solution

System A

$$W_{in} = \tau \omega \Delta t = 859.5 \text{ kJ}$$

System B

$$W_{in} = IV\Delta t = 540.0 \text{ kJ}$$

Example 3–2

The volume of the passenger compartment of an aircraft is 2100 m³. Automatic
equipment maintains the air inside the plane at 98 kPa and a temperature of 23°C.
Calculate the mass of the air inside the plane. Determine the percentage of increase
in the mass of the air if the pressure is increased to 101 kPa and the temperature
drops to 20°C.

Solution

For the given pressure and temperature, the air is assumed to behave as an ideal gas,
which abides by the equation of state:

$$pv = RT$$

A more convenient form of the equation of state for this problem:

$$pV = mRT$$

Solving for the mass of air, we get:

$$m = \frac{pV}{RT} = 2423 \text{ kg}$$

If the pressure of the air is increased and the temperature decreased, the air mass must increase because the volume of air remains constant. Denoting the initial condition with the subscript 1 and the final condition by 2, we find that the ratio of the initial mass of air in the compartment to the final mass is:

$$\frac{m_2}{m_1} = \frac{p_2}{p_1}\frac{T_1}{T_2} = 1.041$$

The percent increase in the mass of the air is:

$$\text{Percent Increase} = \frac{m_2 - m_1}{m_1} = \frac{m_2}{m_1} - 1 = 4.1\%.$$

3-6 Polytropic Processes

Referring to Figure 3–4, during the expansion and compression processes of gases, the pressure and specific volume are often related through the following general expression:

$$p = Cv^{-n}$$

Substitution in the expression for boundary work yields:

$$w_b = \frac{p_2 v_2 - p_1 v_1}{1 - n}$$

Figure 3–4

Family of polytropic processes.

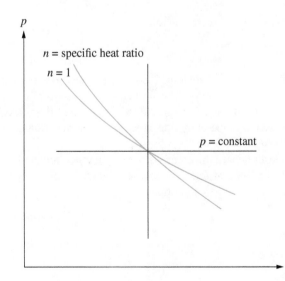

3-7 ## Gravitational Work

This is defined as the work done by or against a gravitational field. In a gravitational field, the force acting on the body is

$$F = mg_c$$

where m is the mass and g_c is the local gravitational acceleration. Then, the work required to raise this body from level z_1 to z_2 is:

$$W_g = \int_1^2 F \, dx = mg_c(z_2 - z_1)$$

3-8 ## Shaft Work

Energy transmitted with a rotating shaft is very common in engineering applications, particularly in turbomachinery bladed components. The torque τ applied to the shaft is assumed constant here. In this case, the work done can be expressed as follows:

$$\dot{W}_s = 2\pi\omega\tau$$

where ω is the rotational speed (in radians/s) and τ is the applied torque in N·m

3-9 ## Compressibility Factor: A Measure of Deviation from the Ideal Gas Behavior

The ideal gas equation of state is rather simple and is thus very convenient to use. Real gases, however, deviate from the ideal gas behavior significantly at states near the saturated-vapor line and the critical point. This deviation at a given temperature and pressure can accurately be accounted for by the introduction of a correction factor known as the compressibility factor Z, which is defined as follows:

$$Z = \frac{pv}{RT}$$

or:

$$pv = ZRT$$

Z can also be expressed as follows:

$$Z = \frac{v_{\text{actual}}}{v_{\text{ideal}}}$$

where $v_{\text{ideal}} = \frac{RT}{p}$. Obviously $Z = 1$ for ideal gases. For real gases, Z can be greater than or less than unity. The farther away Z is from unity, the more the gas deviates from the ideal gas behavior. Added to this, and by reference to Figure 3–5, note that the constant-temperature line experiences an inflection point right at the critical point as shown in Figure 3–6.

Gases behave differently at a given temperature and pressure, but they behave very much the same at temperatures and pressures once these are normalized using their critical temperatures and pressures:

$$p_R = \frac{p}{p_{cr}}$$

$$T_R = \frac{T}{T_{cr}}$$

Figure 3–5

Compressibility factor as
a function of reduced
pressure.

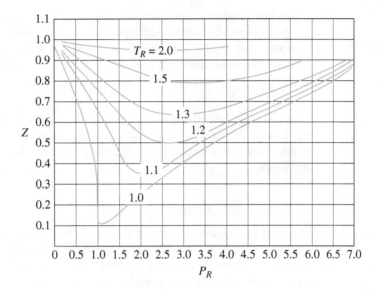

Figure 3–6

Critical isotherm of a pure
substance has an
inflection point.

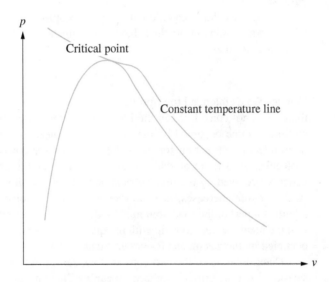

In this case, p_R and T_R are referred to as the reduced pressure and temperature, respectively. The Z factor for all gases is approximately the same at the same reduced pressure and temperature. This is called the *principle of corresponding states*.

The use of a compressibility chart requires knowledge of critical point data, and the results obtained are accurate to within a few percentage points.

The following observations can be made from the generalized compressibility chart:

1   At very low reduced pressures, the gases behave ideally regardless of temperature.

2   At high temperature (meaning a reduced temperature of more than 2), the ideal gas behavior can be assumed with good accuracy regardless of the pressure (except for magnitudes of reduced pressure that are much greater than unity).

3   The deviation of a gas from the ideal-gas behavior is greatest in the vicinity of the critical point (Figure 3–1).

When p and v or T and v are given instead of p and T, the generalized compressibility chart can still be used to determine the third property, but it would involve a tedious trial-and-error approach. Therefore, it is very convenient to define one more reduced property, the *reduced specific volume* v_R, which is defined as follows:

$$v_R = \frac{v_{actual}}{RT_{cr}/p_{cr}}$$

Note that v_R is defined differently from p_R and T_R. It is related to T_{cr} and p_{cr} instead of v_{cr}. Lines of constant v_R are also added to the compressibility chart, and this enables one to determine T as a function of p without having to resort to a time-consuming iterative procedure.

3–10   Other Equations of State

The ideal-gas equation of state is very simple, but its range of applicability is limited (if you want to be really accurate). It is desirable to have equations of state that represent the p-v-T behavior of substances accurately over a larger range with no limitations. Such equations are naturally more complicated. Several equations have been proposed for this purpose.

The van der Waals equation of state was proposed in 1873, and it has two constants, which are determined from the behavior of a substance at the critical point. This equation is given as follows:

$$\left(p + \frac{a}{v^2}\right)(v - b) = RT$$

Van der Waals intended to improve the ideal-gas equation of state by including two of the effects not considered in the ideal-gas model: the intermolecular attraction forces and the volume occupied by the molecules themselves. The term $\frac{a}{v^2}$ accounts for the intermolecular attraction forces, and b accounts for the volume occupied by the gas molecules. In a room at standard atmospheric pressure and temperature, the volume actually occupied by the molecules is only about one-thousandth of the room volume. As the pressure increases, the volume occupied by the molecules becomes an increasingly significant part of the total volume. Van der Waals proposed to correct this by replacing v in the ideal-gas relationship with the quantity $(v - b)$, where b represents the volume occupied by the gas molecules per unit mass.

Determination of the two constants appearing in this equation is based on the observation that the critical isotherm (meaning the constant-temperature line) passing by the critical-state point on a p-v diagram has an inflection point at the critical point (Figure 3–6). Thus, the first and second derivatives of p with respect to v at the critical point must be zero. That is,

$$\left(\frac{\partial p}{\partial v}\right)_{T=T_{cr}} = 0$$

and

$$\left(\frac{\partial^2 p}{\partial v^2}\right)_{T=T_{cr}} = 0$$

By performing the differentiation and eliminating v_{cr}, the constants a and b are determined to be:

$$a = \frac{27 R^2 T_{cr}^2}{64 p_c r}$$

and

$$b = \frac{RT_{cr}}{8p_{cr}}$$

The accuracy of the van der Waals equation of state is often inadequate, but it can be improved by using the values of a and b that are based on the actual behavior of the gas over a wider range instead of a single point. Despite its limitations, this equation has a historical value in that it was one of the first attempts to model the behavior of real gases.

Example 3–3   Predict the pressure of nitrogen gas at $T = 175$ K and $v = 0.00375$ m^3/kg on the basis of the ideal-gas equation of state, and then the van der Waals equation

Solution   Using the ideal-gas equation of state:

$$p = \frac{RT}{v} = 13,860 \text{ Pa}$$

Now using van der Waals's equation, we first determine the constants for nitrogen:

$$a = 0.175 \ m^6 \cdot \text{k·Pa}/kg^2$$

$$b = 0.00138 \ m^3/\text{kg}$$

Finally, we get the pressure by substituting in van der Waals's equation as follows:

$$p = \frac{RT}{v - b} - \frac{a}{v^2} = 9465.0 \text{ kPa}$$

The error here is 31.7%

Example 3–4   Determine the specific volume of water vapor at $p = 40$ MPa and $T = 500°$C. Use the ideal-gas equation of state and compare the predicted values of the specific volume using the tables.

Solution   Using the ideal-gas equation of state [$R_{water} = 461.5$ kJ/(kg K)]

$$v = \frac{RT}{p} = 0.0089 \text{ m}^3/\text{kg}$$

The tabulated value for the specific volume is:

$$v = 0.0056 \text{ m}^3/\text{kg}$$

Comparing the above two magnitudes, we get:

$$\text{Error} = 58.6\%$$

Example 3–5   Calculate the specific volume of nitrogen for a pressure of 3.0 MPa and a temperature of 165 K:

(a) Assume ideal gas behavior.

(b) Use the compressibility factor.

Solution

(a) Assuming ideal gas behavior of nitrogen [$R = 296.8$ J/(kg K)]

$$v = \frac{RT}{p} = 0.0163 \text{ m}^3/\text{kg}$$

(b) The nitrogen pressure in atmospheres is:

$$p = 29.6 \text{ atm}$$

From the compressibility factor (Figure 3–5) we get:

$$Z \approx 0.87$$

and the specific volume is:

$$v = \frac{ZRT}{p} = 0.0142 \text{ m}^3/\text{kg}$$

$$\text{Error} = 14.8\%$$

PROBLEMS

3–1   A rigid tank contains 10 kg of air at 150 kPa and 20°C. More air is added to the tank until the pressure and temperature rise to 250 kPa and 30°C, respectively. Determine the amount of air added to the tank.

3–2   A 1 m³ tank containing air at 25°C and 5 bars is connected through a valve to another tank containing 5 kg of air at 35°C and 2 bars. Now the valve is opened, and the entire system is allowed to reach thermal equilibrium with the surroundings which are at 20°C. Determine the volume of the second tank and the final equilibrium pressure of air.

3–3   Air is compressed from 1.36 bars and 21°C to 10.2 bars in a gas turbine. This compressor is so operated that the air temperature remains constant. Calculate the change in specific volume of the air as it passes through the compressor.

3–4   Determine the specific volume of nitrogen gas at 10 MPa and 150 K based on:
(*a*) The ideal gas equation of state
(*b*) The compressibility-factor chart

Compare the two magnitudes.

3–5   Determine the specific volume of superheated water vapor at 1.6 MPa and 225°C using the same two tools as in Problem 1.

3–6   A 3.27 m³ tank contains 100 kg of nitrogen at 225 K. Determine the pressure in the tank using:

- The ideal-gas equation of state
- The van der Waals equation

4

Basic Laws of Thermodynamics

Chapter Outline

4–1   First Law of Thermodynamics **38**

4–2   Second Law of Thermodynamics **39**

4–3   Irreversibility **44**

4–4   Efficiency of Work-Producing Heat Engines **47**

4–5   Efficiencies of Reversible and Irreversible Heat Engines **48**

4–6   Second Law in Terms of Reversible Cycles **48**

4–7   Irreversible and Reversible Processes **49**

4–8   Clausius Inequality: A Statement of the Second Law of Thermodynamics **50**

4–9   The T-ds Equations **51**

4–10   Entropy Change for Ideal Gases **52**

4–11   The T-s Diagram **52**

4–12   Isentropic Processes **53**

4–13   Entropy Change for a Pure Substance **53**

4–14   The Increase-in-Entropy Principle **62**

4–15   Entropy Change for Compressed Liquids **62**

4–16   Carnot Cycle **62**

4–1 **First Law of Thermodynamics**

The first law of thermodynamics is a statement of energy conservation. This conservation principle states that the algebraic sum of all energy transfers across the system boundary must be equal to the change of energy inside the system. Since heat and work are the only forms of energy that may penetrate the boundary, we can write the first law in the following differential form:

$$\delta Q - \delta W = dE$$

The minus sign appears with the work term because of the sign convention adopted for work (presented earlier). This expression defines the change in the energy content. In the absence of electric and magnetic energies as well as surface tension, this energy quantity consists of three terms:

- The internal energy U represents the energy possessed by the molecules of the substance by virtue of their microscopic kinetic and potential energies.
- The macroscopic kinetic energy which is the kinetic energy of the system due to its motion.
- The macroscopic potential energy represents the potential energy of the system due to its existence in a gravitational field.

Thus the following expression can be written:

$$E = me = U + KE + PE$$

where:

$$\text{Kinetic energy} \quad KE = m\frac{V^2}{2}$$
$$\text{Potential energy} \quad PE = mgz$$

where z is the system elevation above a specific datum. In the absence of appreciable changes in kinetic and potential energies, the first law of thermodynamics can be stated, on a unit mass basis, as follows:

$$q - w = u_2 - u_1$$

This is the first law of thermodynamics as it applies to a closed system.

Example 4–1

A piston-cylinder device initially contains 0.4 m³ of air at 100 kPa and 80°C. The air is now compressed to 0.1 m³ in such a way that the temperature remains constant. Determine the work done during this process.

Solution

At the specified conditions, air can be treated as an ideal gas since it is at a high temperature and low pressure relative to critical-point values ($T_{cr} = -147$ C, $p_{cr} = 3390$ kPa for nitrogen, the main constituent of air). For an ideal gas at constant temperature, T_O:

$$pV = mRT_O = C$$

or:

$$p = \frac{C}{V}$$

where C is a constant. By simple substitution, we get:

$$W = \int_1^2 p\,dV = C\int_1^2 \ln\frac{V_2}{V_1} - p_1 V_1 \ln\frac{V_2}{V_1} = -55.45 \text{ kJ}$$

4-2 Second Law of Thermodynamics

Reversible Process

A process of an isolated system undergoing a finite change of state is termed *reversible* if the system can be restored to its initial state, except for changes Dp of an order smaller than the maximum change Δp that occurs during the process in question. The phrase "smaller than" means less than any finite quantity (in practical terms zero).

The extension of the preceding definition to a nonisolated system follows. A process of any system is reversible if the process could be performed in at least one way such that the system and surroundings can be restored to their respective initial states.

An example of a reversible process is the slow adiabatic expansion of a simple system. Consider a simple system consisting of a quantity of air that is confined by a piston in an insulated cylinder. Let E and V be the energy and volume of the system at the initial state. According to the state postulate, these two properties are sufficient to determine the state of the system. Let the air expand adiabatically with no friction by letting the piston move slowly to a position where the volume of the system is $(V + \Delta V)$. In general, such a process will involve a decrease in pressure from p to $(p - \Delta p)$. The work involved in the process cannot be greater than $p\Delta V$ (because the force on the piston is never greater than p times the piston cross-sectional area) and cannot be less than $(p - \Delta p)\Delta V$. Thus we may write:

$$(p - \Delta p)\Delta V < W < p\Delta V$$

Now, letting ΔV approach zero, the work approaches $p\Delta V$ within a term of smaller order as compared to the work itself. According to the first law of thermodynamics, the change of energy of the system for any adiabatic process is given by:

$$dU = -\delta W$$

We may summarize the effects of the preceding process as follows:

1 Effects internal to the system:
   - A change in volume from V to $(V + dV)$
   - A change of energy from E to $(E + dE)$
   - A change of pressure from p to $(p - dp)$
2 Effects external to the system:

   The rise of a weight by an amount that produces δW.

Let us now start from the final state of the process and cause the piston to move slowly back to a position where the volume is V. The change of volume in the second process will be:

$$dV' = -dV$$

The corresponding work $\delta W'$ will again be equal to pdV' within second-order quantities. The change in internal energy will, within second-order quantities, be given by:

$$dU' = -\delta W' = -p\,dV' = p\,dV = -dU$$

Since the energy and volume of a simple system determine its state, the initial state and thus the initial pressure of the first process must have been restored within terms of second order as compared to the changes which took place. Therefore, the effects of the second process are:

1. Effects internal to the system:
   - A change of volume from $V + dV$ to V
   - A change of internal energy from $U + dU$ to U
   - A change of pressure from $p - dp$ to p
2. Effects external to the system:

The lowering of a weight by an amount that gives rise to $-\delta W$ to within second-order terms.

Thus, we have shown that both internal and external effects of a slow adiabatic expansion of a simple system may be undone, except for differences of second order. In other words, a slow adiabatic frictionless expansion is reversible.

Heat Reservoir

A heat reservoir is a system in a stable equilibrium state such that, when subjected to finite heat interactions, its temperature remains unchanged. In practice, a reservoir can be achieved either by having a system that is very large as compared to other systems with which it may interact, or by providing compensating interactions in order to maintain the system at a fixed state, as in a thermostatically-controlled constant-temperature bath. Another possibility is a system consisting of two phases such as H_2O in a state consisting of two coexisting but finitely different phases, such as liquid water and ice.

Carnot Cycle for a Pure Substance

Shown in Figure 4–1 is a schematic of a typical heat engine. By combining several reversible processes for a system, we may construct a reversible heat engine. An example of such an engine is the Carnot engine (Figure 4–1). A cylinder containing a quantity of gas, such as air, confined by a piston is brought into good communication with a large heat reservoir S at temperature T_1. Once mutual equilibrium is established, the following processes are executed:

Process 1–2
With the fluid in state 1, the heat source is removed from the cylinder so that the fluid is adiabatically isolated, and the piston is allowed to move out slowly to position 2. During this process, both the pressure and temperature will fall. When the temperature reaches that of a second large heat reservoir at T_2, the motion of the piston is stopped.

Process 2–3
The reservoir Z is now brought into good communication with the cylinder, and the piston is made to move back toward its original position so slowly that the air

Figure 4–1

Carnot cycle for a pure substance.

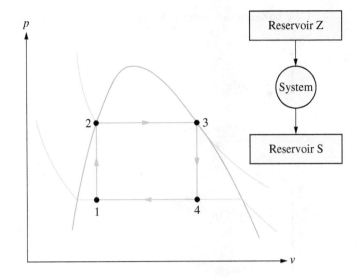

remains at the temperature of the reservoir at all times. At any pressure during this compression, the temperature and volume of the air will be less than those at the same pressure during the expansion process 1–2, because of the flow of heat out to the sink.

Process 3–4
When the piston reaches some point 3, the reservoir Z is removed so that the air is thermally isolated once more, and the compression process is continued. Without the cooling action of the reservoir, the temperature will rise. When it reaches T_1 again, the piston is stopped.

Process 4–1
The cylinder is brought into good communication with the reservoir S, and the piston is allowed to slowly move outward. This is where the cycle is complete.

Now the adiabatic expansion process 1–2 may be executed again, and the cycle repeated as often as desired. This is an example of a Carnot cycle as it applies to a pure substance, which may be described as consisting of two isothermal processes connected by two adiabatic and reversible processes. A characteristic of the Carnot cycle is its reversibility. All of its processes can be performed both in the forward direction and in the reverse order.

Figure 4–2 represents the same Carnot cycle, but for an ideal gas as the working medium. The representation here is on both the p-v and T-v diagrams.

Classical Statements of the Second Law of Thermodynamics

Clausius Statement

It is impossible to construct a device that operates in a cycle and produces no effect other than the production of work exchanging heat with just one heat reservoir.

Kelvin-Planck Statement

It is impossible to construct a device that operates in a cycle and produces no effect other than the production of work exchanging heat with a single reservoir.

Figure 4–2

Representation of an
ideal-gas Carnot cycle.

Cycle on the pressure volume diagram

Cycle on the temperature volume diagram

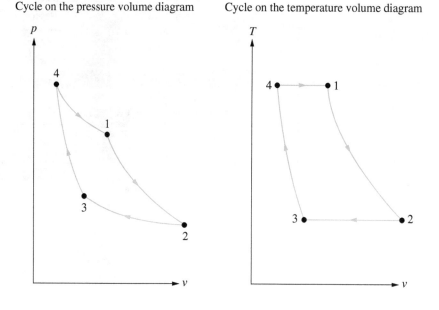

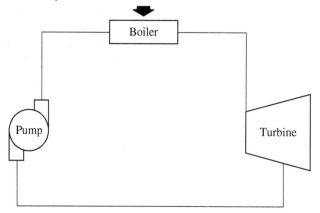

Perpetual Motion Machines of the Second Kind

Real-life engines should abide by the first and second laws of thermodynamics. An engine that violates the first law of thermodynamics is termed a perpetual motion machine of the first kind (PMM1). Engines that violate the second law of thermodynamics are collectively called perpetual motion machines of the second kind (PMM2). Figure 4–3

Figure 4–3

Externally reversible heat
engines between two
reservoirs.

Perpetual motion machine of the second kind

shows an example of a PMM2, where the engine is operating continuously on a closed cycle that is composed of a pump, a combustor, and a turbine. Such a machine is impossible to build, for it is exchanging heat with one heat source while producing a continuous amount of work. A real-life engine will have to interact with two reservoirs, a heat source and a heat sink.

Efficiency of Reversible Heat Engines

All externally reversible heat engines operating between the same two reservoirs will have the same thermal efficiency. The validity of this statement can be shown by assuming that it is not true and showing that such an assumption would be a direct violation of the second law of thermodynamics. Figure 4–4 shows two externally reversible heat engines, A and B, operating between the same two reservoirs. Let us first assume that engine A is the more efficient and then set $Q_{HA} = Q_{HB}$, thus supplying the same amount of heat to each engine. The work output of engine A, in view of our assumption, will be greater than the output of B, since A has the higher efficiency. Let us now reverse engine B, since it is externally reversible. Engine A produces more work than is required to drive a device such as a heat pump, thus, we get a net work output from the combination. The high-temperature reservoir can be eliminated, since Q_{HA} and Q_{HB} are equal in magnitude and opposite in direction relative to this reservoir. The net result is a device operating on a cycle and producing work while exchanging heat with a single reservoir. This is a direct violation of the Kelvin-Planck statement of the second law of thermodynamics (note that such a machine would be a PMM2 which, according to the second law of thermodynamics, is an impossible machine) and is, therefore, unattainable. Everything that we did in arriving at this configuration was legitimate except for the initial assumption that one of the heat engines was more efficient than the other. This assumption was clearly incorrect, so we can conclude that all externally reversible heat engines operating between the same two reservoirs have the same thermal efficiency.

The thermal efficiency of an externally reversible heat engine must therefore be a function of only the temperatures of these two reservoirs, since the only relevant property of a reservoir is its temperature.

The thermal efficiency of an externally reversible heat engine must therefore be a function of the temperatures of these two reservoirs. Thus:

$$\eta_{th} \equiv \frac{W}{Q_H} = \frac{Q_H - Q_L}{Q_H} = 1 - \frac{Q_L}{Q_H} = f_1(T_H, T_L)$$

Figure 4–4

Schematic of a heat engine.

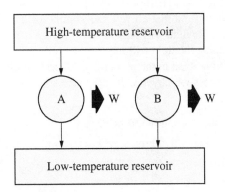

or:

$$\frac{Q_L}{Q_H} = f_2(T_H, T_L)$$

For an externally reversible cycle, it can be shown that the ratio $\frac{Q_L}{Q_H}$ is simply equal to the ratio $\frac{T_L}{T_H}$. That is,

$$\frac{Q_L}{Q_H} = \frac{T_L}{T_H}$$

Then:

$$\eta_{th} \equiv \frac{W}{Q_H} = 1 - \frac{Q_L}{Q_H} = 1 - \frac{T_L}{T_H}$$

This is the Carnot engine efficiency. No heat engine can have a thermal efficiency higher than the Carnot efficiency when operating between the same two reservoirs. The proof of this statement is obtained by the same general technique used to prove that all externally reversible heat engines have the same efficiency when operating between the same two reservoirs. First an assumption is made that a heat engine can have a thermal efficiency higher than that of the Carnot engine. This higher-efficiency engine is used to drive a reversed Carnot engine, with the result that a violation of the Kelvin-Planck statement of the second law of thermodynamics occurs. Since this is an impossibility, then the primary assumption is in error.

4-3 Irreversibility

A process is irreversible if: (1) a system in a stable state could be made to change to another allowed state with the sole external effect being the rise of a weight, or (2) a so-called perpetual motion machine of the second kind (PMM2, a cycle that is in violation with the second law of thermodynamics) could be devised.

In view of the equivalence of the second law of thermodynamics and the impossibility of a PMM2, the above-mentioned two criteria are entirely equivalent to one another. Either one can be applied directly to many simple processes. However, both can become cumbersome, and it is often better to employ other criteria developed next in this section.

Some Irreversible Processes

Let us consider a closed adiabatic cylinder separated into two halves by a piston of zero mass which is held fixed by means of a locking mechanism. Let a fluid, such as air, be contained in the left-hand half of the cylinder, while the other half is evacuated. Let U_1 and V_1 be the initial internal energy and volume of the fluid at a stable state 1. If we unlock the piston, the fluid will expand freely until it reaches the far right-side end of the cylinder. Eventually, the fluid will reach a new stable state 2 of volume V_2. Since neither work nor heat is involved in the process, the energy of the system does not change. The effects of the process which we will call process A, are as follows:

1 Internal Effects:
   - An increase of volume from V_1 to V_2
   - A zero change in internal energy ($U_2 = U_1$)
2 External Effects:
   None

If the preceding process were reversible, a process B could be found such that its effects would be as follows:

1   Internal Effects:
    - A decrease of volume from V_2 to V_1
    - A zero change of energy
2   External Effects:
    None

We will call process B the inverse of process A. We emphasize, however, that whereas process B is assumed to undo all effects of the original process A, it need not follow the reverse path with equal and opposite interactions in every detail.

We will now show that A is irreversible. Consider a process C having an initial state that is identical to the final state of process B, which is also the initial state of process A. Process C is to be a real process which we know can be carried out, as distinguished from B which is purely hypothetical. In process C we resist the piston motion outward by means of a piston rod. The gas expands slowly from volume V_1 to volume V_2 while a weight is raised outside. The combined process B–C has resulted in a change from a stable state 2 at V_2 to another state 3 at V_3 with the sole external effect being the rise of a weight. The internal energy in state 3 is less than that in state 2 by the work done in lifting the weight.

An increase or decrease of energy at constant volume is a change to an allowed state. It follows, therefore, that the combined process B–C was a change from the stable state 2 to an allowed state for which the sole external effect was the rise of a weight. Since this is a violation of the second law of thermodynamics, the process B is impossible and process A, as a result, is irreversible. The alternative proof in terms of the PMM2 concept should also be evident.

Internal and External Irreversibilities

The term "internal irreversibility" refers to the degradation effects within the system, such as friction and lack of equilibrium during a thermodynamic process. External irreversibility, on the other hand, refers to the finite nonzero temperature differential between the system and a reservoir during a heat-exchange process.

For example, we may interpose between the hot reservoir at T_1 and the cold reservoir at T_2 of Figure 4–5, a reversible-cycle engine R like the Carnot engine. Then heat may be delivered to T_2 by engine R while heat is extracted from T_1 and a weight is raised by R in the environment. Since R is reversible, the effects of this process can be completely reversed; that is, heat could be taken from T_2, the weight could be lowered, and heat

Figure 4–5

Heat engine B reversed and driven by engine A.

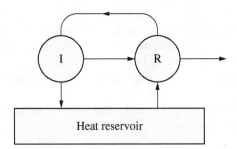

delivered to T_1, each effect being of the same magnitude but opposite in sign to that of the original process.

Thus, the same process in T_2 may be considered irreversible if it is caused by a direct heat interaction with T_1, or reversible if we choose the interacting environment of T_2. We see, therefore, that the irreversibility of the original process was external to T_2 in that it had to do with the kind of environment with which T_2 was interacting. By contrast, if temperature gradients were established in T_2 during the heating process, the irreversibility would be internal to T_2. Similarly, we can see that the irreversibility of the unrestrained expansion was internal to the fluid system and could not be eliminated by altering the environment.

The distinction between reversibility in general and what we will call internal reversibility corresponds to the distinction between the two definitions given earlier. A process is reversible if anything involved can be restored to its initial state. A process in a particular system is internally reversible if it can be performed in at least one way with an arbitrarily selected environment such that the system and the environment can be restored to their respective initial states.

The concept of external reversibility is merely the application of the definition of internal reversibility to the environment as a system. Thus, if a process involves internal reversibility alone or external reversibility alone, it will be an irreversible process with regard to an isolated system. If, on the other hand, it involves both kinds of reversibilities, the process is then termed as reversible (or totally reversible) process with regard to an isolated system.

Reversible Process as a Limiting Case

Any process which involves the motion of a boundary of a system (such as the piston displacement in a piston-cylinder apparatus) must be executed infinitely slowly if it is to be reversible. For, if part of the fluid should be accelerated by a finite amount as the boundary moves, the force applied by it for inward motion would be greater than that associated with the reverse motion, and more work would be taken from the environment during the inward motion than could be restored to it during the reverse motion.

Any process which involves a transfer of heat must also be executed infinitely slowly to be reversible. For if the system should be heated at a finite rate, it must be at a lower temperature than the heat source in order to overcome resistance to the transfer of heat. It was shown earlier in this section that no process which involves heat transfer across a finite interval of temperature can be reversible.

Reversibility, then, demands that some kinds of processes be executed with infinite slowness. It is obvious that this condition will not be encountered in a real-life situation. On the other hand, the reversible processes discussed earlier, which included unrestrained expansion, friction, transfer of heat across a finite temperature differential, and finite velocity of a piston, are common occurrences. For example, it is impossible to move a piston in a cylinder or even to move one part of a gaseous system relative to another part without setting up shearing forces, namely friction forces. It is always possible by lubrication to reduce friction, and it is always possible by reducing relative velocities to reduce the friction. Generally, we may reduce friction as near to zero as we please, but for no finite motion executed in a finite period of time will friction vanish. The frictionless process is thus a limiting case, one which actual processes may be made to approximate.

So, reversible processes are limiting cases of actual processes. We may, with difficulty, bring an actual process as near to reversibility as we please, but we cannot make it entirely reversible. The reversible process is the limiting process of a series of actual processes.

A reversible process is a slow one which passes only through equilibrium states, if it occurs, or through quasistatic states. For if the system should assume a nonequilibrium state in a heat interaction, it would proceed toward an equilibrium state. The reverse of this part of the process could not occur without external effects which had no counterparts in the original process.

Since the states passed through in a reversible process are equilibrium states, they are states for which properties can be determined accurately. For this reason, the work and heat interactions in a reversible process may be known with greater precision for any actual process.

4–4 Efficiency of Work-Producing Heat Engines

According to the second law of thermodynamics, a heat engine exchanging heat with just one reservoir while producing work (termed perpetual motion machine of the second kind, or a PMM2) can never exist. Defining the efficiency of a heat engine, we find that an engine with an efficiency of unity is a PMM2 and, therefore, impossible. We will next prove that a heat engine with an efficiency greater than unity is, in effect, a PMM2 and is, therefore, impossible. We will prove the following theorem: The efficiency of a work-producing heat engine operating between two systems at stable states is always less than unity.

Proof

Let us assume that a work-producing heat engine (Figure 4–6) could be devised for which:

$$\eta_{th} > 1$$

Then since

$$\eta = \frac{Q_1 + Q_2}{Q_1}$$

We have:

$$\frac{Q_1 + Q_2}{Q_1} > 1$$

Figure 4–6

A thermodynamic cycle.

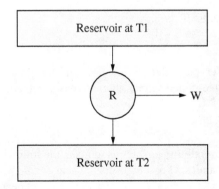

and

$$\frac{Q_2}{Q_1} > 0$$

In addition, since the engine is producing work, we have:

$$Q_1 + Q_2 > 0$$

The first inequality (above) requires that Q_1 and Q_2 should be both positive or both negative, but the latter inequality rules that out. We, therefore, get:

$$Q_1 > 0$$

and

$$Q_2 > 0$$

That is, heat must flow to the engine from the cold reservoir as well as from the hot one. We may now allow the heat Q_2 to flow directly from the hot reservoir to the cold one so that the cold reservoir suffers no change in state. Since the result is a PMM2 (an impossible engine), just as it is for η_{th} equal to unity, the theorem is proved.

4-5   Efficiencies of Reversible and Irreversible Heat Engines

We will prove the following theorem: The efficiency of an irreversible engine operating between a given pair of reservoirs at stable states is less than that of any reversible engine operating between the same two reservoirs.

Proof

Let us assume that an irreversible engine I has an efficiency that is greater than the efficiency of a reversible engine R working between the same two reservoirs. Let W_I denote the work done by engine I when it receives heat Q_I and W_R the work done by engine R when it receives the same amount of heat. Our assumption reduces to:

$$W_I > W_R$$

We now reverse the operation of engine R so that the hot reservoir receives zero net heat, and the combined system I-plus-R exchanges heat with a single reservoir. If

$$W_I - W_R > 0$$

then a PMM2 results. If

$$W_I - W_R = 0$$

then the cold reservoir suffers no change, and the process is reversible. Engine I is, therefore, reversible. Since neither alternative is acceptable, the theorem is proved.

4-6   Second Law in Terms of Reversible Cycles

The concept of reversibility enables us to make the following restatement of the second law of thermodynamics: "For any reversible heat engine which may exchange heat with a single reservoir in a stable state, the net work and net heat transfer in a cycle are zero."

In terms of symbols, we have:

$$\oint_{\text{rev}} \delta Q = \oint_{\text{rev}} \delta w = 0$$

Proof

We will take as proposition A the second law in the following form: A PMM2 is impossible. We will take as proposition B the preceding equation. The proof will consist of showing that, if not A then not B and, if not B then not A. If not A, then a PMM2 is possible. Since, however, it is always possible to cause a system in a stable state to absorb work adiabatically (by frictional rubbing, for example), it follows that a PMM2 is reversible. We therefore have:

$$\oint_{\text{rev}} \delta W > 0$$

and not B.

If not B, then we have at least one reversible cyclic process for which either

$$\oint_{\text{rev}} \delta Q = \oint_{\text{rev}} \delta W > 0$$

or:

$$\oint_{\text{rev}} \delta Q = \oint_{\text{rev}} \delta W < 0$$

Since these two processes are reversible, the same heat engine which can do one can also do the other. The first of these is a PMM2, and, therefore, proposition A would not hold. Thus we have if not B then not A, and the proof that A and B are equivalent is complete.

Based on simple observations, thermodynamic processes can only happen in one direction. For example:

- Gas contained in a piston-cylinder device can naturally expand under the transfer of heat energy, but it will not give back the same amount of heat if the piston goes back to its initial position.

- A container with a hot medium can cool down under ambient temperature, but it will not be hotter under the same ambient-temperature environment.

Thus thermodynamic processes seem to have directions in which they naturally occur, but there are no reciprocal effects in the opposite directions.

4-7 Irreversible and Reversible Processes

Any exchange of heat with the surroundings is through a finite nonzero temperature differential, a case that constitutes external irreversibility.

The system itself is imperfect internally. Examples of such degrading effects include, for instance, friction and lack of equilibrium in progressing from one state to the next. These features represent internal irreversibilities.

A process which has no such factors is referred to as a totally reversible process. Reversible processes are frictionless by definition, for such factors as friction are inherent in any real-life process. Also, any exchange of heat must arise as a result of a finite nonzero temperature difference between the system and its surroundings.

Clausius Inequality: A Statement of the Second Law of Thermodynamics

This alternate statement of the second law of thermodynamics is the most popular, for it leads to the definition of a new property called entropy.

For a real-life thermodynamic cycle (Figure 4–7), the following inequality applies:

$$\oint \frac{\delta Q}{T} \leq 0$$

The equality sign in this expression will apply only to a reversible process, that is,

$$\oint \left(\frac{\delta Q}{T}\right)_{rev} = 0$$

The way the preceding relationship appears suggests that the integral (above) represents the change in a property. This property is referred to as the entropy S, that is,

$$dS = \left(\frac{\delta Q}{T}\right)_{rev}$$

Now we might elect to represent a process on a frame of reference where one of the axes represents the entropy. On a unit mass basis, the foregoing relationship can be rewritten as follows:

$$ds = \left(\frac{\delta q}{T}\right)_{rev}$$

Let us now consider a cycle that is composed of an irreversible and then a reversible path, with the latter bringing the system back to its initial state. Let us now apply Clausius

Figure 4–7

Constant-volume lines are steeper.

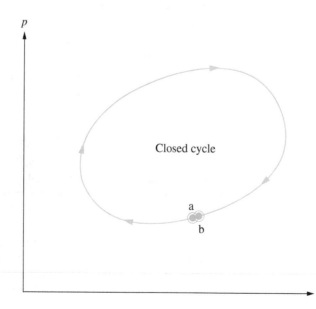

inequality to this cycle:

$$\oint \frac{\delta q}{T} = \int_1^2 \left(\frac{\delta q}{T}\right)_A + \int_2^1 \left(\frac{\delta q}{T}\right)_B < 0$$

Since the process B is reversible, this process can be reversed, i.e.,

$$\int_{1 \ A}^2 \frac{\delta q}{T} + \int_{1 \ B}^2 \frac{\delta q}{T} < 0$$

or:

$$ds > \left(\frac{\delta q}{T}\right)_{\text{irrev}}$$

4-9 The *T-ds* Equations

In the previous discussion of entropy, the change in this property across a thermodynamic process was evaluated by conceiving a reversible process connecting the two end states. This is possible because entropy is a legitimate thermodynamic property with a change that depends only on the end states.

Let us now consider a closed system containing a simple compressible substance that undergoes a reversible process. On a unit mass basis, the first law of thermodynamics gives rise to the following expression:

$$\delta q = \delta w + du$$

Comment

Note that a mathematical phrase in the form of, say, dw, by definition, is invalid for the specific work and is by no means a property. Now:

$$\delta w = pdv$$

with δq, in a reversible process, is related to the entropy change as follows:

$$\delta q = Tds$$

Now the statement of the first law of thermodynamics can be written as follows:

$$Tds = du + pdv$$

Now, recalling the enthalpy definition, we can write the following expression:

$$du = d(h - pv) = dh - pdv - vdp$$

under which case the first law statement assumes the following form:

$$Tds = dh - vdp$$

In summary, we can rewrite the preceding *T-ds* expression as follows:

$$ds = \frac{du}{T} + \frac{pdv}{T}$$

and:

$$ds = \frac{dh}{T} + \frac{vdp}{T}$$

4-10 Entropy Change for Ideal Gases

For an ideal gas, the following relationships apply:

$$pv = RT$$

$$du = c_v dT$$

With the two preceding expressions, the differential change in entropy can be rewritten as follows:

$$ds = c_v \frac{dT}{T} + R \frac{dv}{v}$$

Integrating both sides, and assuming c_v to be constant (i.e., the process takes place over a reasonably small temperature change), we get:

$$s_2 - s_1 = c_v \ln \frac{T_2}{T_1} + R \ln \frac{v_2}{v_1}$$

Again using the equation of state to change the independent variables in the preceding equation, we get the following two expressions:

$$s_2 - s_1 = c_p \ln \frac{v_2}{v_1} + c_v \ln \frac{p_2}{p_1}$$

$$s_2 - s_1 = c_p \ln \frac{T_2}{T_1} - R \ln \frac{p_2}{p_1}$$

4-11 The *T-s* Diagram

Being a legitimate property, the specific entropy (s) can now be used as an axis in a chart, which is shown on the temperature-entropy diagram in Figure 4–8 for an ideal gas. On this diagram, two families of constant-pressure and constant-volume lines are plotted.

Figure 4–8

Horizontal intercepts of two constant-pressure lines.

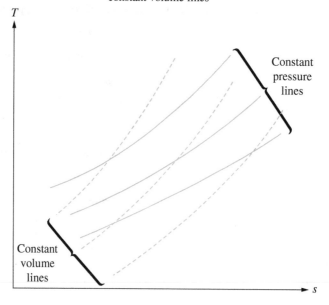

Constant pressure lines are less steep than constant volume lines

Figure 4–9

The double-leg process in the example.

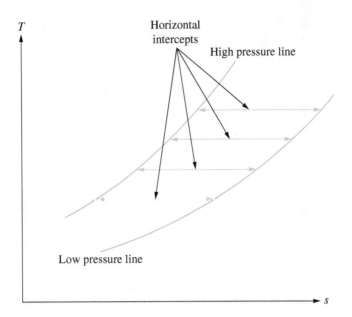

4-12 **Isentropic Processes**

These are adiabatic and reversible processes within which the entropy change is zero. There are several interproperty relationships which are associated with such processes,

$$\frac{p_2}{p_1} = \left(\frac{T_2}{T_1}\right)^{\frac{\gamma}{\gamma-1}}$$

$$pv^\gamma = \text{constant}$$

$$\frac{p}{\rho^\gamma} = \text{constant}$$

where 1 and 2 are the two end states and γ is the specific heat ratio.

4-13 **Entropy Change for a Pure Substance**

Within the liquid-vapor dome, entropy can be calculated by simple interpolation knowing the steam quality x as follows:

$$s = xs_g + (1 - x)s_f$$

Example 4-2

Figure 4–10 shows a double-leg process on the T-s diagram. First, saturated refrigerant-134 liquid at a pressure of 1.4 bars is heated at constant pressure up to a temperature of 10°C. The superheated vapor is then cooled down, through a constant-volume process, to a pressure of 0.6 bar. Calculate the net magnitude of heat exchanged with the surroundings per unit mass.

Figure 4–10

A process beginning at
the water critical point.

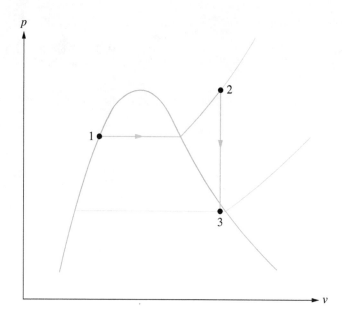

Solution Using the tables, we have:

$$u_1 = 25.7 \text{ kJ/kg}$$

$$u_2 = 234.6 \text{ kJ/kg}$$

$$s_2 = 0.94 \text{ kJ/(kg·K)}$$

$$s_3 = s_2 = u_f + x_3(u_g - u_f) = 0.94 \text{ kJ/(kg°C)}$$

which yields:

$$x_3 = 0.98$$

With this, we can proceed as follows:

$$u_3 = 7.53 \text{ kJ/kg}$$

$$q_{1-2} = u_2 - u_1 = 208.9 \text{ kJ/kg}$$

$$q_{2-3} = u_3 - u_2 = -227.1 \text{ kJ/kg}$$

$$q_{net} = -18.2 \text{ kJ/kg}$$

Example 4–3 Water substance at the critical point (Figure 4–11) undergoes an isentropic process
down to a pressure of 14 bars. Calculate the final quality.

Figure 4-11

A constant-quality
compression process.

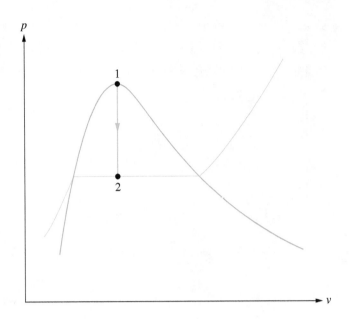

Solution

$$p_1 = 220.9 \text{ bars}$$

$$T_1 = 374.1 \text{ °C}$$

$$s_1 = 4.43 \text{ kJ/(kg·K)}$$

$$s_2 = s_1 = s_f + x_2(s_g - s_f) = 4.43 \text{ kJ/(kg·K)}$$

$$x_2 = 0.10$$

Example 4-4

Two kilograms of water liquid plus vapor mixture at a pressure of 1.0 bar and a quality
of 0.8 undergoes a constant-quality process (Figure 4–12) up to a final pressure of
10 bars. Calculate the amount of heat exchanged with the surroundings.

Solution

Using the water tables, we get:

$$u_1 = u_f + (1 - x_1)u_g = 1754.4 \text{ kJ/kg}$$

$$u_2 = u_f + (1 - x_2)u_g = 1609.9 \text{ kJ/kg}$$

$$q_{1-2} = m(u_2 - u_1) = -288.9 \text{ kJ}$$

Figure 4–12

A constant-quality process.

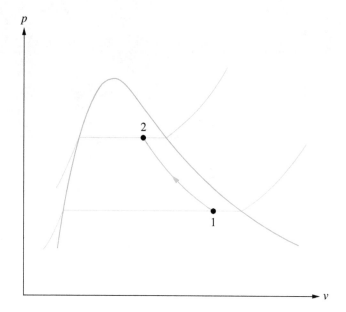

Example 4–5

One-half kilogram of air is maintained in an expandable container at 1.8 bars and 310 K. The air is allowed to expand under constant pressure until the volume is tripled. If the surroundings' temperature is 280 K, calculate the total (system + surroundings) entropy change.

Solution

$$v_1 = \frac{RT_1}{p_1} = 0.49 \text{ m}^3/\text{kg}$$

Now:

$$W_{1-2} = mRT_1 \ln \frac{v_2}{v_1} = 48{,}872 \text{ kJ}$$

$$T_2 = \frac{p_2 v_2}{R} = 922 \text{ K}$$

Now we can calculate the system entropy change as follows:

$$\Delta s_{\text{system}} = c_p \ln \frac{T_2}{T_1} - R \ln \frac{p_2}{p_1} = 1.13 \text{ kJ/(kg·K)}$$

On the other hand:

$$\Delta S_{\text{surroundings}} = \frac{Q_{\text{surroundings}}}{T_{\text{surroundings}}} = \frac{W_{1-2}}{T_{\text{surroundings}}} = 174.5 \text{ kJ/kg}$$

Finally:

$$\Delta s_{\text{total}} = m\Delta s_{\text{system}} + \Delta s_{\text{surroundings}} = 175.15 \text{ kJ/K}$$

Example 4-6

Superheated water vapor at 2 bars and 350°C expands isentropically down to a pressure of 1 bar. Calculate the end-state specific enthalpy.

Solution

$$s_2 = s_1 = 7.301 = x_2 s_g + (1 - x_2) s_f$$

which yields:

$$x_2 = 0.956$$

now:

$$h_2 = x_2 h_g + (1 - x_2) h_f = 2576.2 \text{ kJ/kg}$$

Example 4-7

One-half kilogram of saturated water liquid is maintained at 5 bars. Heat is then transferred until the temperature is 600°C at constant pressure. Heat energy is finally extracted isentropically until the pressure is 0.1 bar. Calculate the total (system + surroundings) increase in entropy if the surroundings' temperature is 288 K.

Solution

State 1

$$T_1 = 151.9 \,°\text{C}$$

$$v_1 = 0.0011 \text{ m}^3/\text{kg}$$

$$u_1 = 639.7 \text{ kJ/kg}$$

$$h_1 = 640.2 \text{ kJ/kg}$$

$$s_1 = 1.861 \text{ kJ/(kg·K)}$$

State 2

$$v_2 = 0.084 \text{ m}^3/\text{kg}$$

$$u_2 = 3299.6 \text{ kJ/kg}$$

$$h_2 = 3701.7 \text{ kJ/kg}$$

$$s_2 = 7.35 \text{ kJ/(kg·K)}$$

$$Q_{1-2} = m(h_2 - h_1) = 1530.8 \text{ kJ}$$

State 3

$$s_f = 0.649 \text{ kJ/(kg·K)}$$

$$s_g = 8.15 \text{ kJ/(kg·K)}$$

$$s_3 = s_2 = s_g + (1 - x_3) s_f$$

$$x_3 = 8.20 \text{ kJ/(kg·K)}$$

$$Q_{2-3} = 0 \ldots \text{(isentropic process)}$$

Now:

$$(\Delta s)_{\text{sys}} = s_2 - s_1 = 5.49 \text{ kJ/(kg·K)}$$

$$(\delta s)_{\text{surr}} = \frac{Q_{1-2}}{T_{\text{surr}}} = 5.32 \text{ kJ/(kg·K)}$$

Finally:

$$(\Delta s)_{\text{total}} = 10.81 \text{ kJ/(kg·K)}$$

Example 4–8   Superheated water vapor at 6 bars and 0.514 m^3/kg expands under constant pressure down to a steam quality of 0.6. Calculate the final magnitude of internal energy.

Solution   At a pressure of 6 bars and a quality of 0.6, we have:

$$u_2 = x_2 u_g + (1 - x_2)u_f = 1566.9 \text{ kJ/kg}$$

Example 4–9   Wet steam of water at a pressure of 5 bars and a quality of 0.3 undergoes a constant-volume process, up to a pressure of 7 bars. The steam is then heated at constant pressure until the water substance is in a saturated vapor state. The water substance enters an isentropic expansion process. Finally, the steam is cooled back down to the initial state. Calculate the net amount of work during the cycle.

Solution

$$v_1 = x_1 v_g + (1 - x_1)v_f = 0.11324 \text{ m}^3/\text{kg}$$

$$v_2 = v_1 = 0.11324 = x_2 + (1 - x_2)v_f$$

giving rise to the properties at state 4 as follows:

$$x_2 = 0.413$$

$$v_3 = v_g = 0.2729 \text{ m}^3/\text{kg}$$

$$v_4 = v_3 = 0.2729 = x_4 + (1 - x_4)v_f$$

which yields:

$$x_4 = 0.727$$

Now we focus on computing the specific internal energies:

$$u_1 = x_1 u_g + (1 - x_1)u_f = 1216.2 \text{ kJ/kg}$$

$$u_2 = 1470.8 \text{ kJ/kg}$$

$$u_3 = u_g = 2572.5 \text{ kJ/kg}$$

$$u_4 = 2037.0 \text{ kJ/kg}$$

$$w_{\text{net}} = h_3 - h_4 - u_2 + u_1 = 280.5 \text{ kJ/kg}$$

Example 4-10   Wet steam at a temperature of 130°C and a quality of 0.85 is heated at constant volume to a temperature of 210°C. Calculate the amount of added heat per unit mass.

Solution

$$v_1 = x_1 v_g + (1 - x_1) v_f = 0.5684 \text{ m}^3/\text{kg}$$

$$u_1 = x_1 u_g + (1 - x_1) u_f = 2240.8 \text{ kJ/kg}$$

Now we know that:

$$v_2 = v_1 = 0.5684 \text{ m}^3/\text{kg}$$

With this, we proceed to calculate the following:

$$T_2 = 210 \,°\text{C}$$

$$u_2 \approx 2646.8 \text{ kJ/kg}$$

Finally:

$$q_{1-2} = u_2 - u_1 = 406.0 \text{ kJ/kg}$$

Example 4-11   Superheated water vapor exists at a temperature and pressure of 1100°C and 0.1 bar, respectively. The vapor is heated up to a temperature of 1300°C at the same pressure. Calculate the change in enthalpy:

- By using the superheated vapor tables
- By considering the vapor to be an ideal gas

You may consider a specific heat ratio of 1.33. Also calculate the error in considering the vapor as an ideal gas.

Solution   Part 1:

$$h_1 = 4891.2 \text{ kJ/kg}$$

$$h_2 = 5409.7 \text{ kJ/kg}$$

$$\Delta h = 518.5 \text{ kJ/kg}$$

Part 2:

$$R_{\text{water}} = \frac{R^*}{M_{\text{water}}} = \frac{R^*}{18} = 0.4615 \text{ kJ/kg}$$

$$c_{p\,\text{water}} = \frac{\gamma}{\gamma - 1} R_{\text{water}} = 1.86 \text{ kJ/kg}$$

$$\Delta h = c_p \Delta T = 372.0 \text{ kJ/kg}$$

$$\text{Error} = 28.2\%$$

Example 4–12   Repeat the solution to Example 4–11, but with the pressure being 5.0 bars, compared to 0.1 bar, with the change in specific entropy being the variable that is required.

Solution   Using the tables, we have:

$$s_1 = 11.23 \text{ kJ/(kg°C)}$$

$$s_2 = 11.58 \text{ kJ/(kg°C)}$$

$$\Delta s = 0.35 \text{ kJ/(kg°C)}$$

$\underline{p = 0.1 \text{ bar}}$

$$\Delta s = c_p \ln T_2 T_1 = 0.31 \text{ kJ/kg°C)}$$

$$\text{Error} = 11.2\%$$

$\underline{p = 5 \text{ bar}}$

$$\Delta s = 0.65 \text{ kJ/(kg°C)}$$

$$\text{Error} = 52.3\%$$

Example 4–13   Air treated as an ideal gas with a specific heat ratio γ of 1.4 is kept in a piston-cylinder device under a pressure of 6.5 bars and a temperature of 610 K. First, the air is allowed to lose an amount of heat energy that is equal to 85 kJ/kg under constant pressure down to a temperature of 472 K. The air then expands isentropically to a temperature of 315 K before it is heated at constant volume back to the initial state. Calculate the amounts of work and heat energy exchanged across the boundary for each individual process.

Solution

$$v_1 = \frac{RT_1}{p_1} = 0.269 \text{ m}^3/\text{kg}$$

$$v_2 = \frac{RT_2}{p_2} = 0.208 \text{ m}^3/\text{kg}$$

Now we examine each process separately:

$$w_{1-2} = p_1(v_2 - v_1) = -39.7 \text{ kJ/kg}$$

$$q_{2-3} = 0 \ldots \text{ (isentropic process)}$$

Now we continue with the rest of the processes:

$$w_{2-3} = c_v(T_3 - T_2) = 112.6 \text{ kJ/kg}$$

$$w_{3-1} = 0 \ldots \text{ (Constant-volume process)}$$

$$q_{3-1} = c_v(T_1 - T_3) = 211.7 \text{ kJ/kg}$$

Example 4–14   Air is kept under a pressure and temperature of 3 bars and 500 K, respectively. The air is then compressed under a constant pressure until the volume reaches one-fourth of its original magnitude. The air then expands isothermally until the specific volume becomes equal to the initial magnitude. A constant-volume process then completes the cycle. Calculate the net amount of exerted specific work.

Solution

$$v_1 = \frac{RT_1}{p_1} = 4.78 \ \text{m}^3/\text{kg}$$

$$v_2 = 0.25v_1 = 1.2 \ \text{m}^3/\text{kg}$$

$$T_2 = \frac{p_2v_2}{R} = 125.4 \ \text{K}$$

$$w_{1-2} = p_1(v_2 - v_1) = -107.4 \ \text{kJ/kg}$$

Now we focus on the isothermal process:

$$w_{2-3} = \int_1^2 RT\frac{dv}{v} = 49.7 \ \text{kJ/kg}$$

Finally:

$$w_{\text{net}} = -577.0 \ \text{kJ/kg}$$

Example 4–15   Three kilograms of water steam are contained in a rigid container with a volume of 1.1 m^3 at a pressure of 3 bars. Electric power is then conveyed to the steam with a current (I) of 5.0 A and a voltage of 105 V. The final state is where the mixture is entirely saturated vapor. Calculate:

• The initial quality
• The final pressure
• The amount of time elapsed during the process

Solution

$$v_1 = \frac{V}{m}$$

Now corresponding to the initial pressure of 3 bars, we have:

$$v_1 = x_1 v_g + (1 - x_1)v_f$$

which yields a steam quality (x) of 0.61. Now we proceed to calculate the initial internal energy:

$$u_1 = x_1 u_g + (1 - x_1)u_f = 1881.0 \ \text{kJ/kg}$$

Now we look up the tables again to find the pressure and internal energy at the saturated vapor line:

$$p_2 = 5.0 \text{ bars}$$

$$u_2 = 2561.2 \text{ kJ/kg}$$

Finally:

$$IV\Delta t = m(u_2 - u_1)$$

which yields:

$$\Delta t = 3.9 \text{ seconds}$$

4-14 The Increase-in-Entropy Principle

Combining the system and its surroundings in what is theoretically referred to as the universe, one could write the total combined entropy change as follows:

$$ds_{universe} = ds_{system} + ds_{surroundings}$$

The second law of thermodynamics places a restriction on such an isolated system as follows:

$$ds_{tot} > 0$$

In other words, each and every practical process will have its own "signature" in the universe.

4-15 Entropy Change for Compressed Liquids

Let us borrow the entropy-change relationship for an ideal gas, namely:

$$\Delta s = c_v \ln\left(\frac{T_2}{T_1}\right) + R \ln\left(\frac{v_2}{v_1}\right)$$

For a compressed liquid, and no matter how high the applicable pressure might be, the change in specific volume remains virtually ignorable. This leads, in a practical sense, to the following relationship:

$$\Delta s = c_v \ln\left(\frac{T_2}{T_1}\right) = c \ln\left(\frac{T_2}{T_1}\right)$$

where the specific heat c is tabulated for virtually any pure substance in the compressed (or subcooled) liquid phase.

4-16 Carnot Cycle

As stated earlier, this cycle operates between two constant-temperature reservoirs, and is composed of the following four reversible processes:

- An isothermal (constant-temperature) expansion process during which heat is reversibly transferred to the working medium

- A reversible adiabatic (i.e., isentropic) expansion process, with the end state being that having the same temperature as the low-temperature reservoir
- A reversible isothermal compression process during which heat is reversibly extracted from the working medium to the low-temperature reservoir
- An isentropic compression process which progresses until the working medium reaches the temperature of the high-temperature reservoir

Thermal Efficiency of Carnot Cycle

Referring to the temperatures of the high and low-temperature reservoirs by T_H and T_L, and the corresponding amounts of heat transfer by q_H and q_L, respectively, then:

$$q_H = T_H(s_2 - s_1)$$
$$q_L = T_L(s_4 - s_3)$$

and since:

$$(s_4 - s_3) = -(s_2 - s_1)$$

The ratio of heat transfer quantities can be written as follows:

$$\left(\frac{q_H}{q_L}\right)_{rev} = \frac{T_H}{T_L}$$

The thermal efficiency of the Carnot heat engine is therefore given by:

$$\eta_{th,Carnot} = 1 - \frac{T_L}{T_H}$$

Example 4–16

In a piston-cylinder apparatus, 0.2 kg of air is maintained at 2 bars and 400 K. The air first undergoes a constant-volume process until the temperature reaches a magnitude of 900 K. An isothermal process then takes place until the initial magnitude of pressure is reached. An isentropic process finally closes the cycle. If the surrounding temperature is 288 K, calculate the total (system plus surroundings) increase in entropy.

Solution

For process 1–2, we have:

$$W_{1-2} = 0$$

$$v_2 = v_1 = \frac{RT_1}{p_1} = 0.57 \ m^3/kg$$

Moving on to process 2–3, we have:

$$v_3 = \frac{RT_3}{p_3} = \frac{RT_2}{p_1} = 1.29 \ m^3/kg$$

$$W_{2-3} = m \int_2^3 p \, dv = mRT_2 \int_2^3 \frac{dv}{v} = 84{,}388 \ kJ$$

Figure 4–13

A two-leg process inside
the L + V dome.

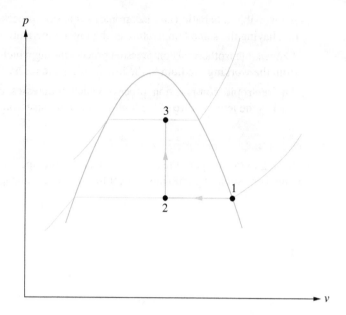

Finally we consider process 3–1:

$$W_{3-1} = mp_3(v_3 - v_1) = -28,800 \text{ kJ}$$

Now:

$$W_{net} = 55,588.0 \text{ kJ}$$
$$Q_{net} = 55,588.0 \text{ kJ}$$

Finally:

$$\Delta s_{total} = \Delta s_{system} + \Delta s_{surroundings} = 0 + \frac{Q_{net}}{T_{surroundings}} = 193 \text{ kJ/(kg K)}$$

The fact that the system undergoes a complete cycle stands behind the result that the entropy change is zero.

Figure 4–13 schematically shows a perpetual motion machine of the second kind (PMM2). Note that the presence of the condensing component, which is missing in this figure, would give rise to the Rankine cycle, which is discussed next.

PROBLEMS

4–1   An amount of air is trapped within a piston-cylinder device. The air is compressed by exerting an amount of work of 250 kJ at constant pressure of 2.0 bars between specific volumes of 4.6 and 1.2 m^3/kg. Calculate the mass of air that is trapped in the piston-cylinder device.

4–2   One-half kilogram of superheated water vapor exists at a pressure of 0.1 bars in an expandable container, and is treated as an ideal gas. If the density changes from 1.2 kg/m^3 and 1.8 kg/m^3. Calculate the total amount of work exerted.

4–3 Air exists under a pressure of 5.0 bars and undergoes a constant pressure process between the volumes of 1.2 m^3 and 3.2 m^3, respectively. Calculate the amount of specific work exerted on the air.

4–4 By reference to Figure 4–8, prove that the horizontal intercepts (i.e., Δs) between any two constant-pressure lines remain equal at whatever pressure level the system might be at.

4–5 Air is compressed isentropically from a pressure and temperature of 0.1 MPa and 20°C to a pressure of 1.0 MPa.
(a) Calculate the final temperature.
(b) Calculate the final/initial density ratio.
(c) Calculate the specific work exerted during the process.

4–6 Air is compressed polytropically from 100 kPa, 200°C, and 10.0 m^3 to a final volume of 1.5 m^3. Determine the final temperature and pressure if the polytropic exponent is:
(a) $n = 0$
(b) $n = 1.0$
(c) $n = 1.33$

Determine the amount of work exerted in the process of part c.

4–7 A quantity of 2.5 kg of air is contained in a piston-cylinder device and has an original internal volume of 0.09 m^3. The initial pressure and temperature are 2 MPa and 250°C, respectively. The air expands through a polytropic process where $n = 1.45$, until the pressure reaches 250 KPa. Calculate the final temperature and work exerted.

4–8 Saturated water vapor exists under a pressure of 1.0 bar. First the saturated vapor is condensed under constant pressure until the quality is 0.7 (Figure 4–12). Heat is then added at constant volume up to the state where the pressure is 5.0 bars. Calculate the final quality, assuming that the water substance is within the L + V dome; otherwise calculate the final temperature.

4–9 Referring to Figure 4–9, prove that the horizontal intercepts between two constant-pressure lines are constant.

4–10 Referring to Figure 4–12, a process takes place within the L + V dome with a constant steam quality of 0.8, between the pressures of 0.1 and 10.0 bars. Calculate the change in enthalpy per unit mass.

5

Energy Conversion by Cycles

Chapter Outline

5–1   Heat-to-Work Conversion   **68**

5–2   The Rankine Cycle and *T-s* Representation   **68**

5–3   Ideal Rankine Cycle Analysis   **69**

5–4   Rankine Cycle Thermal Efficiency   **70**

5–5   Methods of Efficiency Enhancement: Regeneration and Reheat   **70**

5–1 Heat-to-Work Conversion

Most practical cycles involve a fluid that experiences a series of processes that constitute a cycle. The working medium here could be a gas, a liquid, or a multiphase substance. This fluid is usually circulated through devices or components where heat is added or rejected and through components where work is being exerted on or by the fluid. The first case represents the function of a pump or a compressor, while the other represents expansion through a turbine. Should the net work be added to the cycle (meaning exerted on the working medium), it would be a heat pump or a refrigerator cycle.

5–2 The Rankine Cycle and *T-s* Representation

This is one of the most common cycles. The working medium here is a pure substance. The components of the Rankine cycle are shown in Figure 5–1, together with the *T-s* representation. Let us start with the flow station just out of the condenser (state 1) entering the pump component. Shaft power, across the pump, is supplied so that the exiting compressed liquid is at a high pressure. Next is state 2, where the flow enters the boiler, across which heat energy is added to the flowing medium. Next is state 3, where the working medium (now in the form of vapor or, at least, wet steam) enters the steam turbine, where shaft power is produced, with the exit state being 4. The last process of this cycle occurs as the two-phase (liquid plus vapor) medium enters the condenser, where the outcome is saturated liquid.

The processes comprising the Rankine cycle are idealized as follows:

- The pump process is isentropic.
- The boiler flow process is idealized as being one of constant pressure.

Figure 5–1

Real-life Rankine cycle.

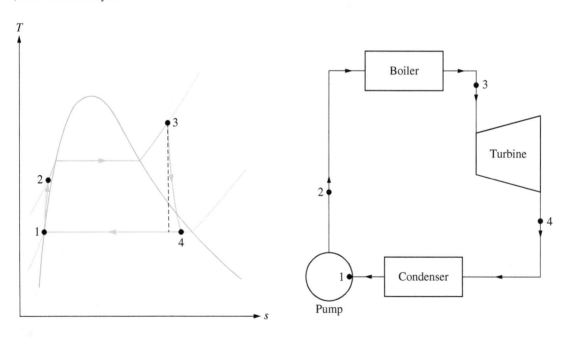

- The steam turbine power extraction is assumed to be isentropic.
- The condenser flow process is idealized as a constant-pressure process.

5-3 Ideal Rankine Cycle Analysis

In this section, the conservation-of-energy principle will be applied separately to each component, assuming steady-state, steady-flow conditions, and ignoring the effect of gravitational energy.

Referring to each component by a control volume (CV), we have:

$$(\dot{Q})_{CV} - (\dot{W})_{CV} = \left[\dot{m}\left(h + \frac{v^2}{2}\right)\right]_{ex} - \left[\dot{m}\left(h + \frac{V^2}{2}\right)\right]_{in}$$

In fact we can pursue the identification process by also ignoring the kinetic energy changes between the inlet and exit stations. This is, in most cases, a design objective in building efficient turbines.

The Pump

Applying the previously cited simplifications to the pump component, we get:

$$\dot{W}_{pump} = \dot{m}(h_2 - h_1)$$

Now recalling the *T-ds* equation:

$$T\,ds = dh - v\,dp = 0$$

Now we can derive the supplied pump power on a unit mass basis, by integration, as follows:

$$w_P = v(p_2 - p_1) = \frac{p_2 - p_1}{\rho_1}$$

The preceding expression is true in the sense that the density change across the pump is virtually ignorable.

The Boiler

Again, on a unit mass basis, the energy conservation principle simply reduces to:

$$q_B = h_3 - h_2$$

The Steam Turbine

$$w_T = h_3 - h_4$$

The Condenser

$$q_C = h_4 - h_1$$

Figure 5–1 shows a real-life Rankine cycle, with the imperfection being a result of the nonisentropic processes across both the pump and steam-turbine components.

5–4 Rankine Cycle Thermal Efficiency

This, by definition, is the net (turbine minus pump) shaft work divided by the heat energy supplied across the boiler, that is,

$$\eta_{th} = \frac{w_T - w_P}{q_B}$$

Example 5–1

An ideal Rankine cycle utilizes water as a working medium. The boiler pressure is 60 bars, and the condenser pressure is 0.01 bar. Superheated vapor enters the turbine at a temperature T_3 of 600°C and leaves the condenser as saturated liquid. Calculate the cycle's thermal efficiency.

Solution

$$v_1 = 1.01 x 10^{-3} m^3/kg$$

$$w_{pump} = v_1(p_2 - p_1) = 6.05 \text{ kJ/kg}$$

$$h_2 = h_1 + v_1(p_2 - p_1) = 197.9 \text{ kJ/kg}$$

$$q_{2-3} = h_3 - h_2 = 3460.5 \text{ kJ/kg}$$

Now we determine the turbine-exit state 4 as follows:

$$s_4 = s_3 = x_3 v_g + (1 - x_3)s_f$$

$$x_4 = 0.869$$

$$h_4 = x_4 h_g + (1 - x_4)h_f = 2271.0 \text{ kJ/kg}$$

Finally:

$$\text{Thermal efficiency } \eta_{th} = \frac{w_{net}}{q_{in}} = 39.9 \text{ percent}$$

5–5 Methods of Efficiency Enhancement: Regeneration and Reheat

The simple Rankine cycle typically has a low thermal efficiency (refer to the preceding example). A logical question then is, can any modification be made to the simple Rankine cycle to improve its efficiency? The answer lies with two terms: regeneration and reheat.

Regeneration

In practice, the concept of regeneration for the Rankine cycle consists of extracting steam from the turbine (Figure 5–2) and passing this steam through a heat exchanger (called a feed-water heater) to heat the water before it enters the boiler. The extracted steam is condensed in this heat exchanger, and the liquid is returned to the cycle (in the condenser). This extraction may occur at only one point or at many points along the expansion process. In large power-generating facilities, there may be up to nine extraction points. The exact number is a trade-off between improved thermal efficiency and increased capital cost of the hardware.

Figure 5-2

Rankine cycle with regeneration.

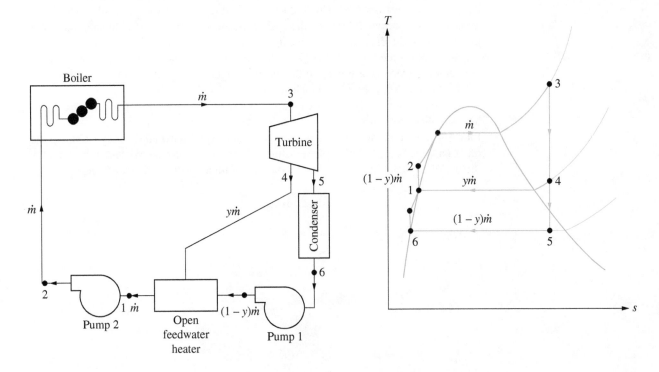

Since the extracted steam can no longer produce work in the turbine, the power output of the turbine is reduced. However, the amount of work that must be supplied is reduced by a larger amount, so that a net increase in efficiency occurs. With regeneration, the heat being supplied from the external source is being supplied at a higher average temperature.

Reheat

In a typical simple Rankine cycle, the steam leaving the turbine would have a quality of close to 80 percent. A liquid content of 20 percent is so high that the small liquid

Figure 5-3

A Rankine cycle utilizing the reheat concept.

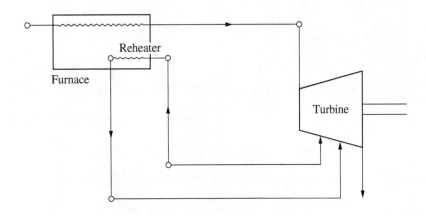

droplets would impinge on the turbine blades, causing serious blade corrosion as well as a reduction in turbine efficiency. This situation can be avoided if the turbine inlet pressure is reduced. This would move the turbine-exit state to the right, thus reducing the liquid content in the wet mixture. This reduction in turbine inlet pressure, while solving the moisture problem, would reduce the cycle efficiency. A more desirable solution to the problem is to add reheat, that is, expand the steam to a pressure well above the condenser pressure, and route the steam back to the furnace where heat is added to increase its temperature, then bring it back to the turbine (Figure 5–3) and continue the expansion until the condenser pressure is reached.

Reheat significantly increases the turbine work, but it also increases the required heat supplied from the external source. The cycle efficiency is not altered significantly for a given inlet pressure, while the turbine moisture problem can be suppressed. On the other hand, for a given tolerable moisture content in the turbine exhaust, reheat allows the use of a much higher turbine inlet pressure with a resulting higher efficiency.

6

Power-Absorbing Cycles: Refrigerators and Heat Pumps

Chapter Outline

6-1 Energy Conservation for a Reversed Cycle **74**

6-2 Performance Measures **74**

6-3 Vapor-Compression Refrigeration Cycle **78**

6-4 Choice of the Working Medium **80**

Figure 6-1

A reversed cycle with
heat transfer from a
low-temperature
reservoir.

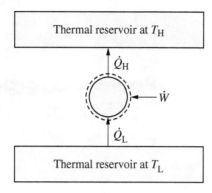

In this chapter, we explore the so-called *reversed cycles* in which work or power input moves energy from a low-temperature thermal reservoir to a high-temperature one
(Figure 6–1). In a refrigerator, one seeks to maintain the cold space at a desired low
temperature, whereas for a heat pump, the high-temperature space is the focus. Refrigerators are common in food processing and storage. Heat pumps are becoming increasingly
popular for residential and other space-heating applications. In this section, we discuss
refrigerators and heat pumps that operate on a vapor-compression cycle.

6-1   Energy Conservation for a Reversed Cycle

Consider a steady-flow device that operates on an arbitrary reversed cycle as shown in
Figure 6–2. As indicated by the dashed line in this figure, a control surface surrounds
the device. The only energy flows crossing the control surface are a power input and two
heat interactions, one with the low-temperature reservoir, and the other with the high-
temperature one. We can thus express the principle of energy conservation as follows:

$$\dot{W}_{in} + \dot{Q}_L = \dot{Q}_H$$

Although the cyclic device contained within our control surface may be quite complex,
the overall energy conservation expression describing its operation is quite simple.

6-2   Performance Measures

The coefficient of performance (COP) is used to quantify how well a reversed-cycle
device performs its job in the same way that the thermal efficiency is used to characterize
the performance of a power-producing cycle; that is,

$$\text{COP} \equiv \beta = (\text{Desired energy})/(\text{Energy that costs})$$

For a refrigerator, the energy removed from the low-temperature space is the desired
energy; thus,

$$\beta_{refrig} = \frac{\dot{Q}_L}{\dot{W}_{in}}$$

where the power input comprises the energy that costs. Applying the conservation of
energy principle again we obtain the following:

$$\beta_{refrig} = \frac{\dot{Q}_L}{\dot{Q}_H - \dot{Q}_L} = \frac{1}{\dot{Q}_H/\dot{Q}_L - 1}$$

Figure 6–2

Reversed Carnot cycle.

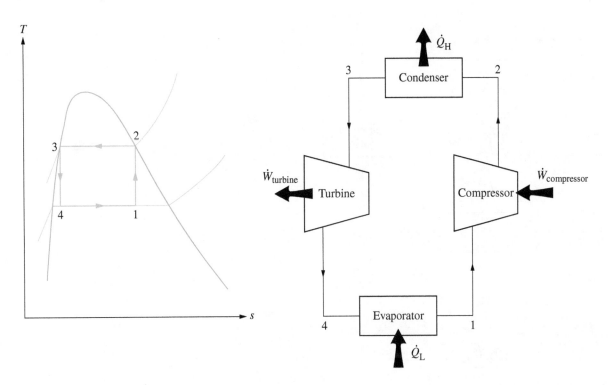

In refrigeration applications, the unit *ton* is frequently used to quantify the energy removal rate from the cold space. One ton of refrigeration equals 3.517 kW.

Similarly, the coefficient of performance for a heat pump is that delivered to the high-temperature space, meaning that:

$$\beta_{HP} = \frac{\dot{Q}_H}{\dot{W}_{in}}$$

The preceding expressions apply to both ideal and real devices.

To get a handle on the upper limits of the coefficient of performance for a device operating between two fixed temperatures, we resurrect the idea of the Carnot cycle, but now operating in reverse (Figure 6–2). The Carnot cycle, as is well-known by now, is an ideal cycle (with the highest efficiency) in which all the processes are performed reversibly (i.e., there is no friction, lack of equilibrium, or any other thermodynamically degrading factor). Figure 6–3*a* illustrates the steady-flow, reversed Carnot cycle. First the adiabatic and reversible (i.e., isentropic) compression process 1–2. Next, we have a reversible heat rejection at constant temperature (process 2–3). The third step is that of an isentropic expansion process (process 3–4). Finally comes the process 4–1 in which a reversible heat addition at constant temperature takes place.

Recognizing that the net power supplied is $\dot{W}_{in} = \dot{Q}_{compressor} - \dot{Q}_{turbine}$, we can express the coefficients of performance for the reversed Carnot cycle refrigerators and heat pumps as previously indicated. We can also relate the heat transfer rates to the

Figure 6–3

(a) Schematic diagram of a vapor compression cycle and (b) the *T-s* diagram.

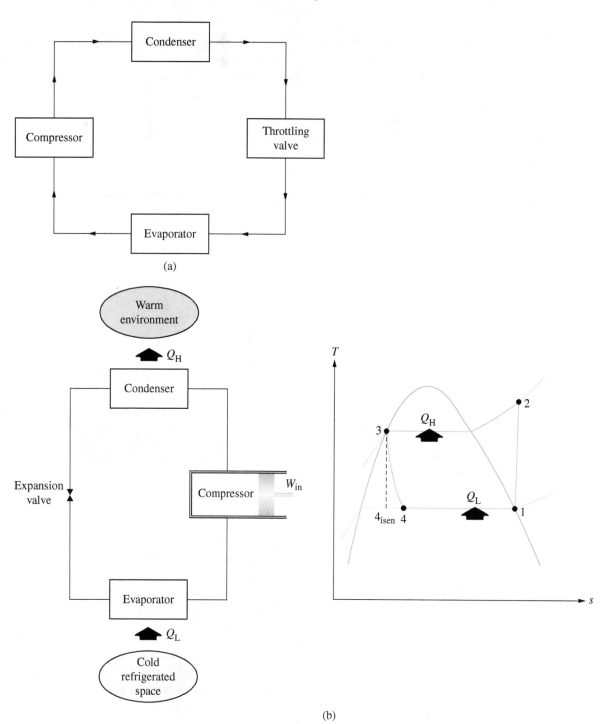

(a)

(b)

**Figure 6–3
(*continued*)**

(c) Comparison of a
refrigerator versus a heat
pump.

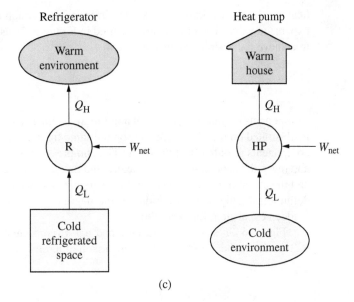

(c)

temperatures in the condenser and evaporator by applying the definition of entropy for a reversible, constant-temperature process. Expressing the heat transfer rates on a per-unit-mass basis, we have:

$$q_L = \frac{\dot{Q}_L}{\dot{m}}$$

$$q_L = T_L(s_1 - s_2)$$

and,

$$q_H = \frac{\dot{Q}_H}{\dot{m}}$$

$$q_H = T_H(s_2 - s_3)$$

Utilizing the final two expressions for the evaporator and condenser, we get:

$$\beta_{\text{refrig, Carnot}} = \frac{1}{T_H/T_L - 1}$$

and,

$$\beta_{\text{HP, Carnot}} = \frac{T_H/T_L}{T_H/T_L - 1}$$

We need to point out that, although theoretically possible, the reversed Carnot cycle (Figure 6–3*b*) is impractical for several reasons. First, compressors do not function well with wet, liquid-vapor mixtures. In practice, the refrigerant is frequently slightly superheated upon entering the compressor. Second, the power produced by a turbine in a reversed Carnot cycle would be quite small, and the complexity and expense required to recover this small amount of energy preclude the use of a turbine. In real refrigerators and heat pumps, the working fluid expands irreversibly through a simple throttling device or expansion valve. The refrigerator pictured in Figure 6–3c uses a length of capillary tube to achieve the desired expansion.

6–3 Vapor-Compression Refrigeration Cycle

Components of this cycle are shown in Figure 6–3b. The ideal vapor-compression cycle is illustrated on the T-s coordinates on the same figure, together with a schematic of the components used to effect the cycle. The compression process takes place adiabatically and reversibly, and hence isentropically, and the heat transfer processes in the condenser and the evaporator are assumed to be reversible. The throttling process, however, is, by definition, highly irreversible due to the degrading effects in the blade passage there (sudden expansion, for example).

Figure 6–3b shows a schematic and a T-s representation of a heat pump. Figure 6–3c, on the other hand, offers a comparison of a refrigerator versus a heat pump.

Cycle Analysis

We now analyze the vapor-compression cycle (Figure 6–4), and develop relationships to evaluate the refrigerator's and heat pump's coefficient of performance.

The Compressor Process (1–2)

$$\dot{W}_{\text{compressor}} = \dot{m}(h_2 - h_1)$$

Figure 6–4

T-s representation of the example problem.

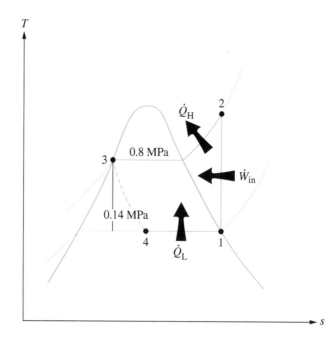

The Condenser Process (2–3)

$$\dot{Q}_H = \dot{m}(h_2 - h_3)$$

The Expansion Valve Process (3–4)

Assuming adiabatic operation, and neglecting potential and kinetic energy changes, the conservation of energy principle assumes the following form:

$$h_3 = h_4$$

Note that a throttling process is characterized as a constant enthalpy process, as indicated above, and that the downstream state for the expansion valve is in the liquid-vapor mixture region (normally called the *wet-mixture dome*). The mixture quality x_4 is readily determined from h_4 since p_4 is usually known.

The Evaporator Process (4–1)

$$\dot{Q}_L = \dot{m}(h_1 - h_4)$$

where $\dot{Q}_H$ and $\dot{Q}_L$ are the heat transfer rates extracted from and supplied to the refrigerant across the condenser and evaporator, respectively.

Coefficient of Performance

Using the relationships from the preceding analysis, we can finally attain the following two expressions:

$$\beta_{refrig} = \frac{\dot{Q}_H}{\dot{W}_{in}} = \frac{h_1 - h_4}{h_2 - h_1}$$

and

$$\beta_{HP} = \frac{\dot{Q}_H}{\dot{W}_{in}} = \frac{h_2 - h_3}{h_2 - h_1}$$

Other than the restrictions that the compression and throttling processes are adiabatic, these relationships apply to both ideal (reversible) and real (irreversible) systems. Furthermore, let us note that, although the ideal cycle presented in Figure 6–3a shows the working fluid entering the compressor as saturated vapor, the entering fluid (state 1) in a real compressor is likely to be slightly superheated to avoid any possibility of moisture. Similarly, state 3 may lie in the subcooled-liquid region in a real device.

Example 6–1   A refrigerator uses Refrigerant-134a as the working fluid and operates on an ideal vapor-compression refrigeration cycle between 0.14 and 0.80 MPa (Figure 6–4). If the mass flow rate of the refrigerant is 0.05 kg/s, determine:

(a) The rate of heat removal from the refrigerated space and the power input to the compressor

(b) The heat rejection rate to the environment

(c) The refrigerator's coefficient of performance (COP or β)

Solution   In an ideal vapor-compression refrigeration cycle, the compression process is isentropic, and the refrigerant enters the compressor as saturated vapor at the evaporator pressure. Also, the refrigerant leaves the condenser as saturated liquid at the condenser pressure. From the Refrigerant-134a tables, the enthalpies of the refrigerant at all four states are determined as follows:

$$h_1 = 236.04 \text{ kJ/kg}$$

$$s_1 = 0.9322 \text{ kJ/kg·K}$$

$$s_2 = s_1 = 0.9322 \text{ kJ/kg·K}$$

$$h_2 = 272.05 \text{ kJ/kg}$$

$$h_3 = 93.42 \text{ kJ/kg}$$

$$h_4 = h_3(\text{throttling device}) = 93.42 \text{ kJ/kg}$$

(a) The rate of heat removal from the refrigerated space and the power input to the compressor are determined as follows:

$$\dot{Q}_L = \dot{m}(h_1 - h_4) = 7.13 \text{ kW}$$

$$\dot{Q}_c = \dot{m}(h_2 - h_1) = 1.80 \text{ kW}$$

(b) The rate of heat rejection from the refrigerant to the environment is determined as follows:

$$\dot{Q}_H = \dot{m}(h_2 - h_3) = 8.93 \text{ kW}$$

(c) The coefficient of performance of the refrigerator is determined, by reference to its definition, as follows:

$$\beta_{\text{refrig}} = \frac{\dot{Q}_L}{\dot{W}_c} = 4.0$$

which means that the refrigerator removes 4.0 units of energy from the refrigerated space for each unit of the electric energy it receives.

6–4   **Choice of the Working Medium**

Figure 6–5 shows, on the p-v diagram, the critical points of both water and Refrigerant-12. Examination of this chart reveals that the liquid + vapor dome associated with the latter has a much lower range of pressure and temperature. This makes it more sensitive, at normal temperatures, to temperature variations as far as the evaporation and condensation processes are concerned.

Figure 6–5

Comparison between the critical points for water and R-12.

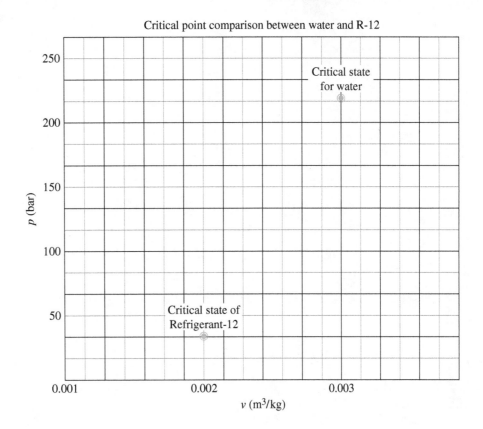

Critical point comparison between water and R-12

PROBLEM

6–1   Consider an ideal vapor-compression refrigerant cycle (Figure 6–4) operating between pressures of 0.14 and 0.80 MPa. The mass flow rate of the refrigerant (R-134a) is 0.04 kg/s (different from the magnitude in Example 6–1). Determine the rate at which energy is removed from the cold space, the rate at which energy is rejected to the surroundings, the power input to the compressor, and the cycle's coefficient of performance. Also, compare the cycle's COP with that of a reversed Carnot cycle operating between the same two pressures.

7

Gas Power Cycles

Chapter Outline

7–1   Air-Standard Assumptions   **84**

7–2   Otto Cycle   **84**

7–3   Mean Effective Pressure   **85**

7–4   Diesel Cycle: Ideal Cycle for Compression-Ignition Engines   **87**

7–5   The Ideal Brayton Cycle   **89**

7–6   Thermal Efficiency of Brayton Cycle   **90**

7–7   Real-Life Brayton Cycle   **90**

7–8   Isentropic Efficiency of a Process   **98**

7–9   Brayton Cycle with Regeneration   **101**

7-1 Air-Standard Assumptions

In gas power cycles, the working medium, namely air, is in its gaseous form throughout the cycle. Even in gas turbines, where the working medium is that of combustion products, the mixture is predominently air and is treated as such. The heat energy is provided by a burning fuel. To summarize, the air-standard assumptions are as follows:

- The working medium is air throughout all components of the cycle. Note that the working medium in the "hot" segment of the cycle is, in real life, composed of combustion products.
- All components that comprise the cycle are internally reversible.
- The combustion process is replaced by a heat-addition process from an external source.
- The exhaust process is replaced by a heat-rejection process that restores the working medium to its initial state.

7-2 Otto Cycle

Referring to Figure 7–1, the Otto cycle is the ideal cycle for spark-ignition reciprocating engines. Referring to Figure 7–2, initially both the intake and exhaust valves are closed, and the piston is in its lowest position (BDC). During the compression stroke, the piston moves upward, compressing the air-fuel mixture. Shortly before the piston reaches its highest position, the spark plug fires and the mixture ignites, increasing the pressure and temperature of the system. The high-pressure gases force the piston down which, in turn, forces the crankshaft to rotate, producing a useful work output during the expansion (or power) stroke. At the end of the stroke, the piston is at its lowest position, and the cylinder is filled with combustion products. Now the piston moves upward one more time, purging the exhaust gases through the exhaust valve (the exhaust stroke), and down a second time (the intake stroke). Notice that the pressure in the cylinder is slightly above the ambient magnitude during the exhaust stroke, and slightly below during the intake stroke.

Figure 7–1

Representation of the idealized Otto cycle.

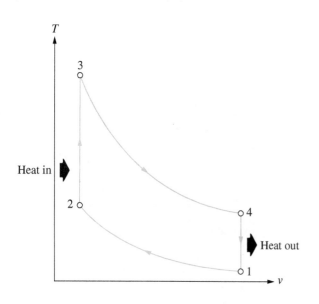
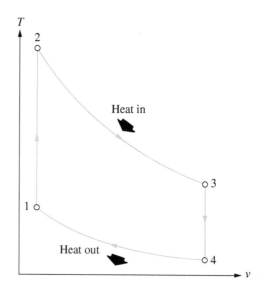

Figure 7-2

Real-life and idealized Otto cycle.

Piston and valves positions
at the end of each stroke

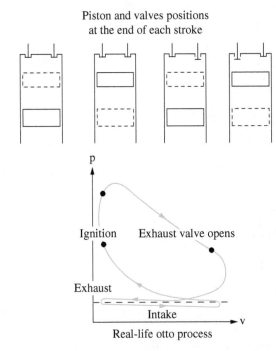

Real-life otto process

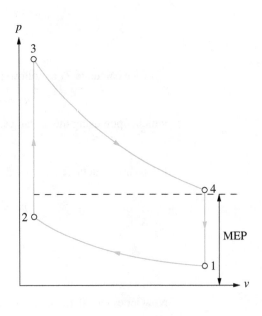

7-3   Mean Effective Pressure

This is a fictional pressure MEP which can be expressed, in light of Figure 7–2, as follows:

$$\text{MEP} = \frac{w_{\text{net}}}{v_{\max} - v_{\min}}$$

Example 7-1

An ideal Otto cycle has a compression ratio of 8 (Figure 7–3). At the beginning of the compression process, the air is at 100 kPa and 17°C, and 800 kJ/kg of heat is transferred to the air during the constant-volume heat addition process. Determine:

- The maximum temperature and pressure that occur during the cycle
- The net work output
- The thermal efficiency

Solution

$$T_1 = 17 + 273 = 290 \text{ K}$$

Using the tables:

$$u_1 = 206.9 \text{ kJ/kg}$$

Let us now move to process 1–2, which is an isentropic compression process:

$$T_2 = T_1 \left(\frac{v_1}{v_2} \right)^{\gamma - 1}$$

$$u_2 = 444.0 \text{ kJ/kg}$$

$$p_2 = p_1 \left(\frac{v_1}{v_2} \right)^{\gamma} = 16.9 \text{ bars}$$

Now we calculate T_3 as follows:

$$u_3 = u_2 + q_{in} = 1244.0 \text{ kJ/kg}$$

which, upon using the air tables, gives rise to:

$$T_3 = 1540.0 \text{ K}$$

Let us now calculate p_3:

$$p_3 = p_2 \left(\frac{T_3}{T_2} \right) = 42.5 \text{ bars}$$

$$T_4 = T_3 \left(\frac{v_3}{v_4} \right)^{\gamma - 1} = 728.5 \text{ K}$$

$$u_4 = 536.1 \text{ kJ/kg}$$

Now let us calculate the net work:

$$w_{net} = (u_3 - u_4) - (u_2 - u_1) = 470.9 \text{ kJ/kg}$$

Finally we calculate the thermal efficiency as follows:

$$\eta_{th} = \frac{w_{net}}{q_{in}} = \frac{w_{net}}{u_3 - u_2} = 58.9 \text{ percent}$$

Figure 7–3

Otto cycle for
Example 7–1.

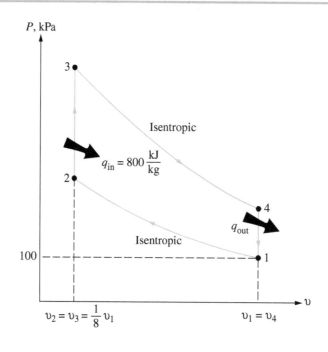

Figure 7–4

The ideal Diesel cycle.

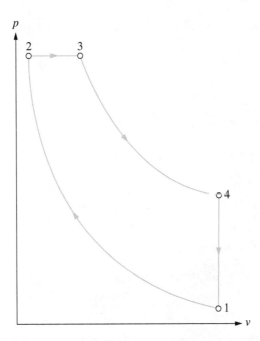

Figure 7–4

The ideal Diesel cycle.

7–4 Diesel Cycle: Ideal Cycle for Compression-Ignition Engines

The compression-ignition (CI) engine is one in which the air is compressed to a temperature that is above the auto ignition magnitude of the fuel, and combustion starts on contact as the fuel is injected into the hot air. Therefore, the spark plug and carburetor are replaced by a fuel injector in diesel engines. An additional volumetric ratio here is the so-called *cutoff ratio* which, by reference to Figure 7–4, is the ratio between the volumes at the end and beginning of the heat addition process, as shown in the figure.

Example 7–2

An ideal Diesel cycle with a compression ratio of 17.0 and a cutoff ratio of 2.0 has a temperature of 313 K and a pressure of 1 bar at the beginning of the isentropic compression process. Assuming a constant specific heat ratio γ of 1.4, calculate the cycle's thermal efficiency.

Solution

First we consider the isentropic compression process as follows:

$$T_2 = T_1\left(\frac{v_1}{v_2}\right)^{\gamma-1} = 972.0 \text{ K}$$

$$p_2 = p_1\left(\frac{v_1}{v_2}\right)^{\gamma} = 52.8 \text{ bars}$$

Since the heat addition process is one of constant pressure, and using the equation of state, we get:

$$T_3 = T_2\left(\frac{v_3}{v_2}\right) = 1944.0 \text{ K}$$

Let us now calculate the magnitudes of internal energy throughout the cycle:

$$u_1 = 224.0 \text{ kJ/kg}$$

$$u_2 = 735.0 \text{ kJ/kg}$$

$$u_3 = 1628.0 \text{ kJ/kg}$$

Before we calculate p_4, let us consider the constant-volume compression process 4–1:

$$p_4 = p_1 \left(\frac{v_1}{v_4}\right)^{\gamma} = 0.019 \text{ bar}$$

Now, let us consider the isentropic expansion process 3–4:

$$T_4 = T_3 \left(\frac{p_4}{p_3}\right)^{\frac{\gamma-1}{\gamma}} = 201.7 \text{ K}$$

$$u_4 = 142.6 \text{ kJ/kg}$$

Finally we calculate the thermal efficiency as follows:

$$\eta_{th} = \frac{w_{net}}{q_{in}} = 59.4 \%$$

Example 7–3

In an air-standard internal combustion engine, the inlet air pressure is 1.2 bars and the inlet temperature is 310 K. The compression ratio is 12.5. Heat is added at the magnitude of 782 kJ/kg at constant temperature. If the combustor pressure ratio is 1.7, calculate the cycle's thermal efficiency assuming an average specific heat ratio γ magnitude of 1.36.

Solution

We have:

$$p_2 = p_1 \left(\frac{v_1}{v_2}\right) = 14.4 \text{ bars}$$

$$T_2 = T_1 \left(\frac{p_2}{p_1}\right)^{\frac{\gamma-1}{\gamma}} = 599.0 \text{ K}$$

Now:

$$q_{in} = c_v (T_3 - T_2)$$

which yields:

$$T_3 = 1580.0 \text{ K}$$

Now we consider the power stroke 3–4:

$$T_4 = T_3 \left(\frac{p_3}{p_4}\right)^{\frac{\gamma-1}{\gamma}} = 749.7 \text{ K}$$

Now we finally calculate the cycle's thermal efficiency as follows:

$$\eta_{th} = \frac{w_{net}}{q_{in}} = 55.2 \text{ percent}$$

Figure 7–5

Representation of the dual cycle.

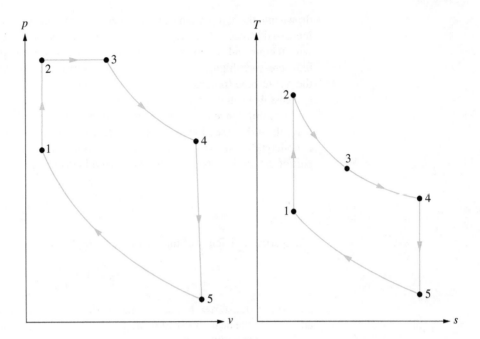

7–5   The Ideal Brayton Cycle

The ideal gas cycle comes in two forms, namely the closed (e.g., power plant) and the open (e.g., jet engine) versions. Here, as in the Rankine cycle, all the processes are reversible.

A schematic of a gas turbine power plant is shown in Figure 7–8, side-by-side with a simple turbojet engine. The power plant cycle consists of four processes. First air is

Figure 7–6

Ideal Brayton cycle.

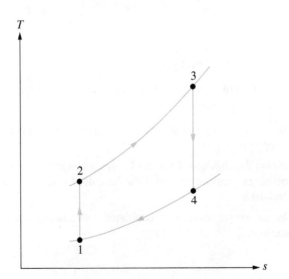

drawn into the compressor section, where the pressure and temperature of the air are both increased (process 1–2). Upon departing the compressor, the fuel is added to and mixed with the air and the mixture is ignited in the combustor (process 2–3). The resulting high-pressure, high-temperature combustion products (mostly air) now expand through the gas turbine (process 3–4), after which the flow is passed through a heat exchanger (process 4–1) all the way back to the compressor.

The difference between this and the idealized open Brayton cycle is that the gases leave the turbine section to an exhaust device (nozzle or exhaust diffuser). Nevertheless, in both types of Brayton cycles (although it is less than a cycle in the case of a jet engine), part of the turbine-specific work is utilized in driving the compressor.

7–6   Thermal Efficiency of Brayton Cycle

The thermal efficiency of the idealized cycle is:

$$\eta_{th} = 1 + \frac{q_L}{q_H}$$

No work is transferred to the working medium during the heat-exchange processes 2–3 and 4–1, so the heat absorbed from the high-temperature reservoir and the heat rejected to the low-temperature reservoir are given by:

$$q_L = q_{4-1} = h_1 - h_4$$

Now by simple substitution, we can obtain the following expression for the thermal efficiency:

$$\eta_{th} = 1 + \left(\frac{h_1 - h_4}{h_3 - h_2}\right)$$

Now applying the so-called *cold air-standard assumption* (simply stating that c_p remains constant at all times), we get the final thermal efficiency expression:

$$\eta_{th} = 1 + \left(\frac{T_1 - T_4}{T_3 - T_2}\right)$$

7–7   Real-Life Brayton Cycle

Figure 7–7 again shows the Brayton cycle with the exception of introducing the degrading real-life effects, namely:

- The compressor process is no longer isentropic, but is rather adiabatic and with a net entropy production.
- The combustion process is no longer a constant-pressure process, for there is a net decline in pressure (actually the so-called *total* or *stagnation pressure*) across the combustor.
- The turbine process is also associated with a net increase in entropy, for it is no longer isentropic.

Figure 7-7

Real-life Brayton cycle.

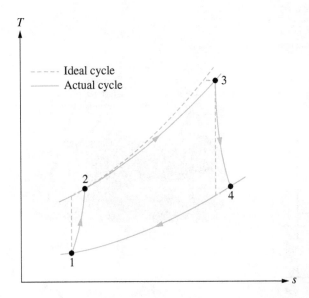

Example 7-4

Figure 7–8 shows a simple turbojet engine where the compressor inlet conditions are 1 bar and 340 K. The compressor pressure ratio is 12.0 and the combustor pressure loss is 1.4 bars. If the compressor and turbine isentropic efficiencies are 0.78 and 0.84, respectively, with specific heat ratios being 1.4 and 1.33, respectively, calculate the engine thermal efficiency, if the turbine inlet temperature is 1380 K.

Solution

$$\eta_C = 0.78 = \frac{(\Delta h)_{\text{ideal}}}{(\Delta h)_{\text{actual}}}$$

and from the air tables,

$$h_2 = 617.3 \text{ kJ/kg}$$

which yields:

$$T_2 = 610.0 \text{ K}$$

Also corresponding to a T_3 of 1380 K, we have:

$$h_3 = 1491.4 \text{ kJ/kg}$$

Now:

$$\eta_{\text{th}} = 0.84 = \frac{(\Delta h)_{\text{actual}}}{(\Delta h)_{\text{ideal}}}$$

which yields:

$$T_4 = 889.2 \text{ K}$$

Figure 7–8

Figure 7–8

Representation of two Brayton cycle examples.

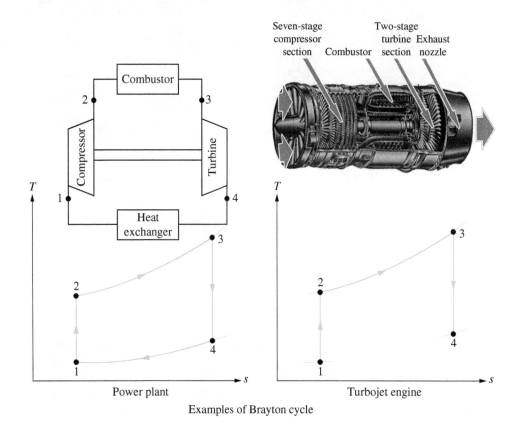

Examples of Brayton cycle

From the air tables, we get:

$$h_4 = 921.0 \text{ kJ/kg}$$

Finally:

$$\eta_{th} = \frac{w_{net}}{q_{in}} = \frac{(h_3 - h_4) - (h_2 - h_1)}{q_{in}} = 76 \text{ percent}$$

Example 7–5   The simple turbojet engine in Figure 7–8 operates across a pressure ratio of 12. The compressor inlet conditions are 0.4 bar and 260 K, and the pressure across the combustor drops by 6 percent. The compressor and turbine efficiencies are 76 and 88 percent, respectively, and the corresponding specific heat ratios are 1.4 and 1.33.

If the turbine inlet temperature is 1200 K, calculate the discharge velocity. Assume an isentropic exhaust nozzle process.

Solution   Using the expression for the compressor efficiency, we get:

$$T_2 = 464.3 \text{ K}$$

which, upon checking the air tables, yields:

$$h_2 = 467.0 \text{ kJ/kg}$$

Referring to the combustor:

$$p_3 = 0.94 p_2 = 4.78 \text{ bars}$$

For a simple turbojet engine:

$$w_{\text{turbine}} = w_{\text{compressor}}$$

or:

$$\eta_T c_{p_T}(T_3 - T_4) = \frac{1}{\eta_C} c_{p_C}(T_2 - T_1)$$

Upon substitution, we get:

$$w_T = W_C = 270.0 \text{ kJ/kg}$$

$$T_4 = 835.0 \text{ K}$$

$$h_4 = 860.0 \text{ kJ/kg}$$

Now recognizing that the nozzle flow process is isentropic, and taking into account the fact that the turbine-specific work is equal to that of the compressor, we get:

$$p_4 = 1.58 \text{ bars}$$

Now:

$$T_5 = 745.7 \text{ K}$$

$$h_5 = 761.0 \text{ kJ/kg}$$

Finally we can calculate the exit velocity as follows:

$$h_4 - h_5 = \frac{V^2}{2}$$

which yields:

$$V_5 = 445.0 \text{ m/s}$$

Shown in Figure 7–9 is the real-life Brayton cycle for a simple turbojet engine. This time, the combustor pressure (actually total pressure) loss is illustrated.

Figure 7–9

Progression of processes
across a turbojet engine.

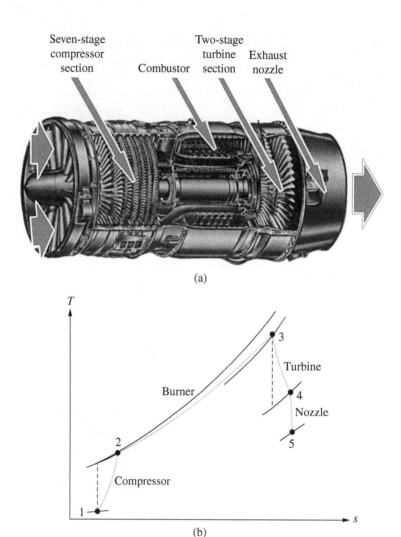

Seven-stage
compressor
section

Two-stage
turbine
section

Combustor

Exhaust
nozzle

(a)

(b)

Example 7–6

Air enters a 76 percent efficient compressor of a simple turboprop engine (Figures 7–10 and 7–11) at 1.3 bars and 310 K, with an exit pressure of 14 bars. On the other side of the engine, namely the turbine, the products of combustion are assumed to be predominantly air at 13.2 bars and 1280 K, with the turbine isentropic efficiency being 88 percent. Calculate the turbine-exit pressure in two ways:

- By assuming γ magnitudes of 1.4 and 1.33 across the compressor and turbine, respectively
- By assuming an average magnitude of 1.36 across both components

Solution

<u>Part 1</u>

$$w_C = 477.4 \text{ kJ/kg}$$

On the other hand:

$$w_T = W_C = 477.4 \text{ kJ/kg}$$

Figure 7–10

Example of a turboshaft
system: A turboprop
engine.

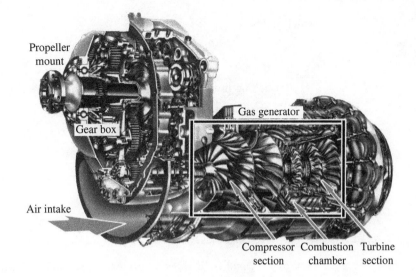

Figure 7–11

Thermodynamic cycle in a typical turboprop engine.

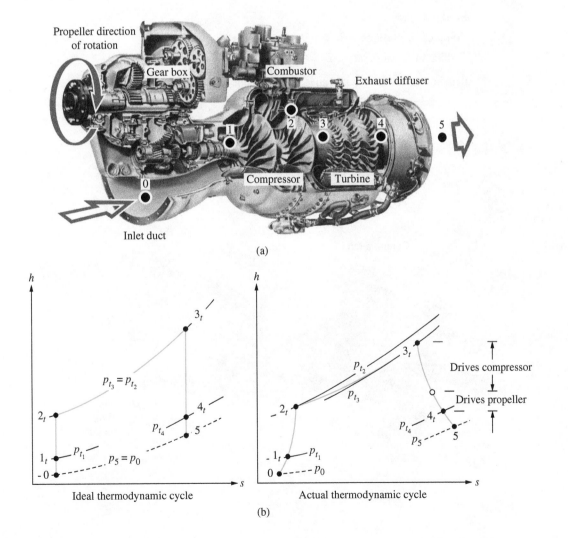

(a)

(b)

which yields:

$$p_4 = 2.92 \text{ bars}$$

Part 2

Proceeding in the same manner with a unified specific heat ratio of 1.36 this time, we get:

$$p_4 = 3.08 \text{ bars}$$

The error involved here is calculated to be 0.5 percent.

Example 7–7

Figures 7–12 and 7–13 show a turbofan engine and its ideal representation on the *T-s* diagram. At an altitude of 25,000 ft, the following operating conditions apply:

- $T_1 = 185 \text{ K}$
- $p_1 = 0.4 \text{ bar}$
- Compressor pressure ratio = 12
- Turbine inlet temperature $T_3 = 1100 \text{ K}$

Calculate the fan's rotational speed if the torque exerted on it is 1.5 N·m.

Figure 7–12

A turbofan engine with separate exhaust nozzles.

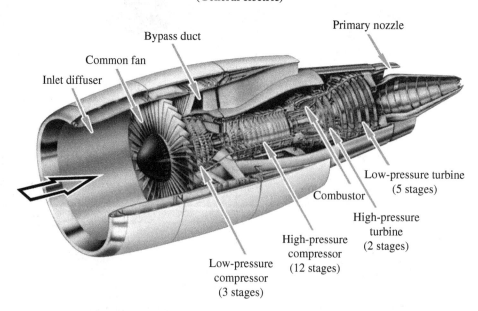

CF6-80C2 Turbofan engine for the boeing 767
(General electric)

Figure 7–13

Brayton cycle representation for a turbofan engine.

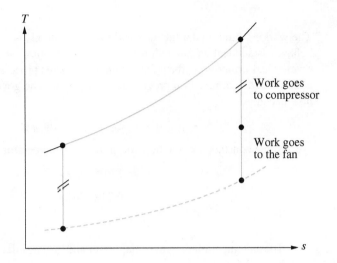

Solution

$$p_2 = 4.8 \text{ bars}$$

$$T_2 = T_1 \left(\frac{p_2}{p_1} \right)^{\frac{\gamma-1}{\gamma}} = 376.3 \text{ K}$$

$$w_C = c_p(T_2 - T_1) = 191.3 \text{ kJ/kg}$$

$$w_T = c_p T_3 \left(\frac{p_3}{p_1} \right)^{\frac{\gamma-1}{\gamma}} - 1 = 1137.3 \text{ kJ/kg}$$

$$w_F = w_T - w_C = 946 \text{ kJ/kg}$$

$$\omega_F = \frac{w_F}{\tau_F} = 630.6 \text{ rad/s} = 18,067 \text{ r/min}$$

Note, by reference to Figure 7–13, that the turbine output shaft work is sufficient to run both the compressor and the fan.

Example 7–8

A stationary power plant operating on the ideal Brayton cycle has a pressure ratio of 8.0. The gas temperature is 300 K at the compressor inlet station and 1300 K at the turbine inlet. Utilizing the cold air-standard assumptions, determine the following:

(a) The gas temperature at the compressor and turbine exit stations

(b) The back (compressor-to-turbine) work ratio

(c) The cycle's thermal efficiency

Solution The working fluid, under the air-standard assumptions, is assumed to be air, which behaves as an ideal gas, and all four processes that make up the cycle are internally reversible. Furthermore, the combustion and exhaust processes are replaced by heat addition and heat rejection processes, respectively. The governing energy equation now becomes:

$$q - w_{shaft} = h_{exit} - h_{inlet}$$

We assume constant specific heat for air at room temperature, and thus take:

$$c_p = 1.005 \text{ kJ/(kg K)}$$

$$c_v = 0.718 \text{ kJ/(kg K)}$$

$$\gamma = 1.4$$

(a) The air temperatures at the compressor and turbine inlet stations are determined by applying the energy equation to processes 1–2 and 3–4, as follows:

$$T_2 = T_1 \left(\frac{p_2}{p_1} \right)^{\frac{\gamma-1}{\gamma}} = 543.4 \text{ K}$$

$$T_4 = T_3 \left(\frac{p_4}{p_3} \right)^{\frac{\gamma-1}{\gamma}} = 717.7 \text{ K}$$

(b) To find the back work ratio, we need to find the work input to the compressor and that out of the turbine:

$$w_{compressor} = h_2 - h_1 = c_p(T_2 - T_1) = 244.6 \text{ kJ/kg}$$

$$w_{turbine} = h_3 - h_4 = c_p(T_3 - T_4) = 585.2 \text{ kJ/kg}$$

Thus, the back work ratio (BW) can be calculated as follows:

$$BW = \frac{w_{compressor}}{w_{turbine}} = 0.418$$

$$q_{in} = c_p(T_3 - T_2) = 760.4$$

$$w_{net} = w_{turbine} - w_{compressor} = 362.7 \text{ kJ/kg}$$

Thus, the thermal efficiency can now be calculated as follows:

$$\eta_{th} = \frac{w_{net}}{q_{in}} = 44.8 \text{ percent}$$

7–8 Isentropic Efficiency of a Process

The *T-s* diagram is useful for comparing actual to ideal processes (Figures 7–14 and 7–15). In many processes, such as the compression of a gas in a piston-cylinder apparatus, the actual process involves very little heat transfer, so the process is essentially adiabatic. Thus, the idealization of this type of process is a reversible adiabatic (or isentropic)

Figure 7–14

Isentropic versus actual compression processes.

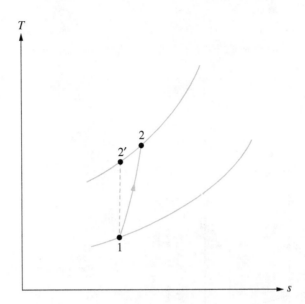

process. The actual process is then compared to the ideal process by a parameter called the *isentropic efficiency* (adiabatic in some textbooks).

$$\eta_C = \frac{w_{id}}{w_{actual}} = \frac{\Delta h_{id}}{\Delta h_{actual}}$$

Both the actual and ideal (compression and expansion) processes are illustrated in Figures 7–15 and 7–16. The actual process is represented by a continuous line while the ideal one is represented by the broken line.

Figure 7–15

Isentropic versus actual expansion process.

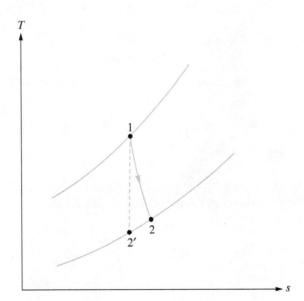

Figure 7–16

T-s diagram for an ideal Brayton cycle with regeneration.

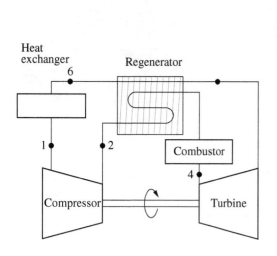

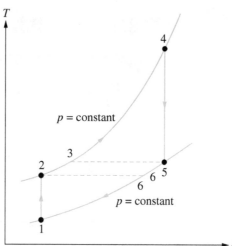

Example 7-9

A system of 2.0 kg of air expands adiabatically from a pressure of 8.0 bars and a temperature of 200°C to a pressure of 1.0 bar. If the isentropic expansion efficiency is 85 percent, calculate the air temperature after the expansion and the actual work done.

Solution

Applying the first law of thermodynamics:

$$Q_{1-2} - W_{1-2} = U_2 - U_1 \text{ (adiabatic process)}$$

$$0 - W_{1-2} = U_2 - U_1$$

$$W_{1-2} = U_1 - U_2$$

$$W_{1-2} = mc_v(T_1 - T_2)$$

There must be a net increase in entropy across the actual process (which is shown in Figure 7–17).

$$\eta = \frac{W_{\text{actual}}}{W_{\text{ideal}}} = 0.85$$

but:

$$(W_{1-2})_{\text{actual}} = mc_v(T_1 - T_{2s})$$

where the subscript *s* signifies the term *isentropic*, and since the process at hand is isentropic, then:

$$\left(\frac{p_2}{p_1}\right)^{\frac{\gamma-1}{\gamma}} = \frac{T_{2s}}{T_1}$$

$$T_{2\text{ideal}} = 261.2K$$

Checking the air tables, we get $c_v = 0.7165$ kJ/kg·K. Thus:

$$W_{1-2_s} = 303.8 \text{ kJ}$$

It follows that:

$$W_{1-2} = 258.2 \text{ kJ}$$

We find T_2 from:

$$W_{1-2} = mc_v(T_1 - T_2)$$

which yields a value for T_2 of 293.0 K.

7–9   **Brayton Cycle with Regeneration**

In gas turbine engines, the temperature of the exhaust gases leaving the turbine is often considerably higher than the temperature of the air leaving the compressor. Therefore, the high-pressure air leaving the compressor can be heated by transferring heat to it from the hot exhaust gases in a counter-flow heat exchanger, which is also known as a *regenerator*. A sketch of the gas turbine engine utilizing a regenerator and the corresponding *T-s* diagram are shown in Figure 7–16.

The thermal efficiency of the Brayton cycle increases as a result of regeneration (Figure 7–17) since the portion of energy of the exhaust gases that is normally rejected to the surroundings is now used to preheat the air entering the combustor. This, in turn, decreases the heat input (thus fuel) requirements for the same net work output. Note, however, that the use of the regenerator is recommended only when the turbine exhaust

Figure 7–17

Effect of regeneration.

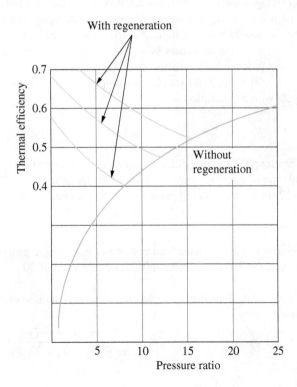

temperature is substantially higher than the compressor exit temperature. Otherwise, heat will flow in the reverse direction (to the exhaust gases), decreasing the efficiency. This situation is encountered in gas turbine engines operating at very high pressure ratios.

The highest temperature occurring within the regenerator is T_4, the temperature of the exhaust gases leaving the turbine and entering the regenerator. Under no conditions can the air be preheated in the regenerator to a temperature above this value. Air normally leaves the regenerator at a lower temperature, T_5. In the limiting (ideal) case, the air will exit the regenerator at the inlet temperature of the exhaust gases T_4. Assuming the generator is well insulated and any changes in the kinetic and potential energies are negligible, the actual and maximum heat transfers from the exhaust gases to the air can be expressed as follows:

$$q_{\text{regen}_{\text{actual}}} = h_5 - h_2$$

$$q_{\text{regen}_{\text{maximum}}} = h_4 - h_2$$

The extent to which a regenerator approaches an ideal regenerator is called the *effectiveness*, E, and is defined as follows:

$$E = \frac{h_5 - h_2}{h_4 - h_2}$$

When the cold air-standard assumptions are utilized, the preceding expression reduces to:

$$E = \frac{T_5 - T_2}{T_4 - T_2}$$

It is obvious that a regenerator with a higher effectiveness will save a greater amount of fuel since it will preheat the air to a higher temperature prior to combustion. However, achieving a higher effectiveness requires the use of a larger regenerator, which carries a higher price tag and involves a larger (total) pressure loss. Therefore, the use of a regenerator with a very high effectiveness cannot be justified economically unless the savings from the fuel costs exceed the additional expenses involved. Most regenerators used in practice have effectiveness below 85 percent.

Under the cold air-standard assumptions, the thermal efficiency of an ideal Brayton cycle with regeneration is:

$$\eta_{\text{th, regen}} = 1 - \left(\frac{T_1}{T_3}\right)(r_p)^{\gamma-1}\gamma$$

Therefore, the thermal efficiency of an ideal Brayton cycle with regeneration depends on the ratio of the minimum to maximum temperatures as well as the pressure ratio. Figure 7–17 shows the effect of regeneration on the cycle's efficiency.

Example 7–10

Determine the thermal efficiency of the gas turbine power plant (Figure 7–17) in Example 7–10 if a regenerator with an efficiency of 80 percent is installed.

Solution

We first calculate the air enthalpy at the regenerator exit using the definition of effectiveness:

$$\epsilon = \frac{h_5 - h_2}{h_4 - h_2} = \frac{T_5 - T_2}{T_4 - T_2}$$

Which yields:

$$T_5 = 682.8 \text{ K}$$

Thus:

$$q_{in} = h_3 - h_5 = c_p(T_3 - T_5) = 620.3 \text{ kJ/kg}$$

This represents a saving of 140.1 kJ/kg from the heat input requirements. The addition of a regenerator does not affect the net work output of the plant. Thus:

$$\eta_{th} = \frac{w_{net}}{q_{in}} = 54.9 \text{ percent}$$

PROBLEMS

7–1 An ideal Otto cycle has a compression ratio of 7.5. At the beginning of the compression process, the air is at 100 kPa and 17°C, and 800 kJ/kg of heat is transferred to the air during the constant-volume heat addition process. Determine:
(a) The maximum temperature and pressure attained during the cycle
(b) The net work output
(c) The cycle's thermal efficiency
(d) The cycle's mean effective pressure

7–2 An ideal Diesel cycle with a compression ratio of 17.0 and a cutoff ratio of 2.0 has a temperature of 40°C and a pressure of 100 kPa at the beginning of the compression process. During the constant-pressure heat addition, heat is transferred to the air from a reservoir with a temperature of 1800°C. During the constant-volume process, heat is rejected to the environment, which is at 25°C and 100 kPa. Use the cold air-standard assumptions, and assume that γ is 1.4, to calculate the following:
(a) The temperature and pressure of the gas at the end of the isentropic-compression process and at the end of the combustion process.
(b) The cycle's thermal efficiency.
(c) The maximum thermal efficiency of any cycle operating between the two heat reservoirs.

7–3 Figure 7–5 shows the so-called *dual cycle*, which is a close approximation of the actual performance of a compression-ignition engine, and it consists of a two-segment heat addition process. The first is a constant-volume process and the second a constant-pressure process. The compression, expansion, and heat-rejection processes are identical to the air-standard Otto and Diesel cycles. The cycle has a compression ratio of 8. At the beginning of the compression process, the air is at 100 kPa and 17°C. During the constant-volume burning process, 150 kJ/kg of heat is transferred to the air. Next, 200 kJ/kg of heat is transferred during the constant-pressure process. Calculate the cycle's thermal efficiency.

7–4 An ideal Diesel cycle has a compression ratio of 19, with the heat energy added being 1200 kJ/kg. The air at the beginning of the cycle has a volume of 2.5 L, and the engine runs at 2000 r/min. Determine the power output of the engine and its thermal efficiency.

7–5 At sea-level takeoff, the turboprop engine in Figures 7–10 and 7–11 is operating ideally. The design point is defined as follows:

- $\dot{m} = 3.5$ kg/s
- $T_1 = 288$ K

- $p_1 = 1$ bar
- Compressor pressure ratio $= 14$
- $T_3 = 1250$ K

If the propeller shaft speed is 22,000 r/min, calculate the torque exerted on the propeller.

7-6 A stationary power plant that is operating on an ideal Brayton cycle has a pressure ratio of 8. The gas temperatures at the compressor and turbine inlet stations are 330 K and 1300 K, respectively, and the plant inlet pressure is 1.0 bar. Determine the cycle's thermal efficiency.

7-7 Compute the rate of temperature change of 0.1 kg of air as it is being compressed adiabatically with a power input of 1.0 kW.

7-8 Saturated water vapor at $p = 4.0$ bars is expanded reversibly and adiabatically in a piston-cylinder device to half of its original volume.
(a) Sketch the process on a p-v diagram.
(b) Calculate the quality of the water substance at the final state.
(c) Calculate the amount of work done.

7-9 A constant-quality process takes place between the two pressures 0.01 bar and 5.0 bars. If the constant quality is 0.8, calculate the change in specific enthalpy during the process.

7-10 A stationary power plant operating on an ideal Brayton cycle with regeneration has a pressure ratio of 8.0. The gas temperature is 300 K and 1300 K at both the compressor and turbine inlet stations, respectively. If the regenerator is 80 percent effective, determine the cycle's thermal efficiency.

7-11 An ideal Brayton cycle with regeneration utilizes air as the working medium. The air enters the compressor at 101 kPa and 37°C. The compressor pressure ratio is 12.0. The regenerator has an efficiency of 70 percent. Calculate:
(a) The specific work input to the compressor
(b) The specific work output of the turbine
(c) The amount of heat energy input during the combustion process, as well as that rejected in the exhaust process

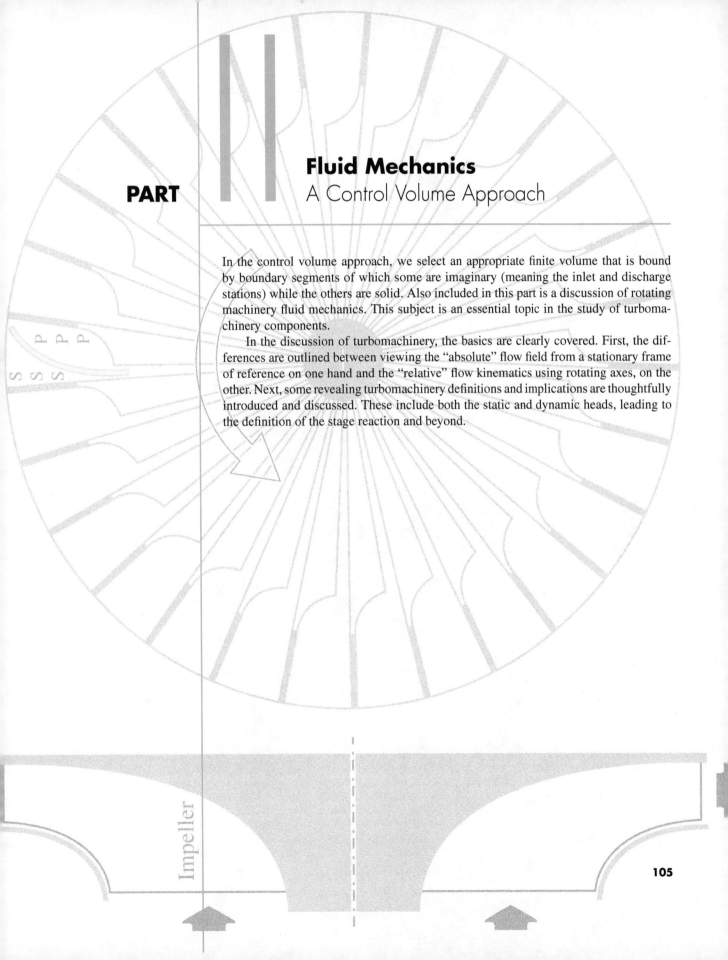

PART II

Fluid Mechanics
A Control Volume Approach

In the control volume approach, we select an appropriate finite volume that is bound by boundary segments of which some are imaginary (meaning the inlet and discharge stations) while the others are solid. Also included in this part is a discussion of rotating machinery fluid mechanics. This subject is an essential topic in the study of turbomachinery components.

In the discussion of turbomachinery, the basics are clearly covered. First, the differences are outlined between viewing the "absolute" flow field from a stationary frame of reference on one hand and the "relative" flow kinematics using rotating axes, on the other. Next, some revealing turbomachinery definitions and implications are thoughtfully introduced and discussed. These include both the static and dynamic heads, leading to the definition of the stage reaction and beyond.

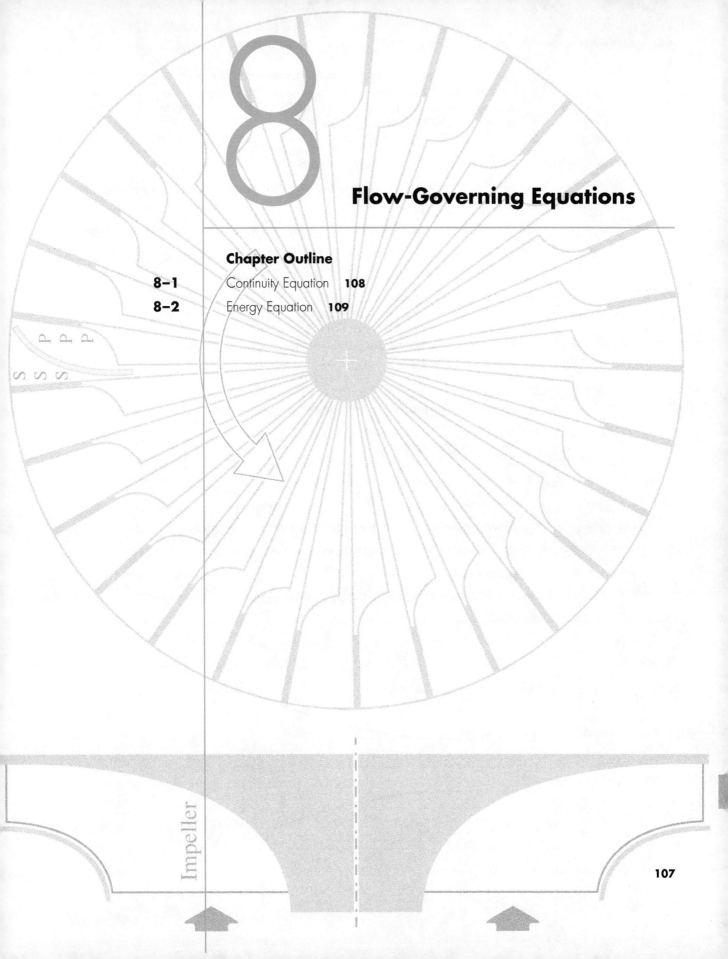

8

Flow-Governing Equations

Chapter Outline

8–1   Continuity Equation   **108**

8–2   Energy Equation   **109**

Impeller

These are the regular conservation laws, including the conservation of mass (continuity equation), as well as those of momentum and energy. Application of these principles at the flow-permeable boundary segments for a simple flow problem will result in the magnitudes of the unknown variables.

8–1 Continuity Equation

Referring to the (subsonic) diffuser in Figure 8–1, the continuity equation can be written as follows:

$$\int_{A_{ex}} \rho V \, dA = \int_{A_{in}} \rho V \, dA$$

Example 8–1

Consider the (subsonic) diffuser in Figure 8–1 and the profiles of velocity at its inlet and exit stations, where the inlet and exit radii are 0.04 m and 0.07 m, respectively. The velocity at inlet and exit stations are parabolic, with the maximum velocity at the inlet being 210 m/s. Now calculate the mass flow rate and the maximum velocity V_{maxexit} at the exit station. Assume a constant density across the diffuser of 1.2 kg/m^3.

Solution

Applying the continuity equation at the inlet station, we get:

$$\dot{m} = \text{constant} = \rho \int_{A_{in}} V \, dA = \rho \int_0^{0.96R} V_{\max} \left[1 - \left(\frac{r}{R} \right)^2 \right] 2\pi r \, dr = V_{\max} \rho \pi R^2$$

or:

$$\dot{m} = 0.212 \text{ kg/s}$$

Now applying the continuity equation at the exit station we get:

$$\dot{m} = 0.212 = 2 \times 0.667\pi R^2 \rho V_{\max}$$

which yields:

$$V_{\text{maxexit}} = 39.3 \text{ m/s}$$

Figure 8–1

Inlet and exit velocity profiles for a subsonic diffuser.

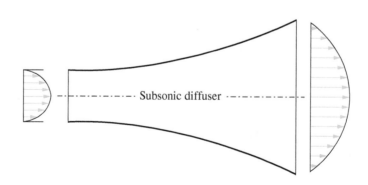

Subsonic diffuser

Figure 8–2

A nozzle with uniform
inlet and parabolic exit
velocity profiles.

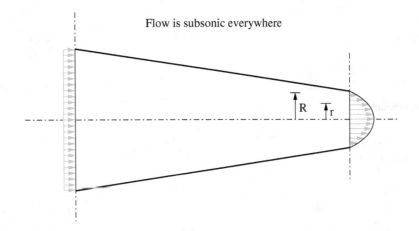

Flow is subsonic everywhere

8-2   Energy Equation

This is another conservation principle. Now following the guidelines of the first law of thermodynamics, and referring to any flow passage (i.e., those in Figures 8–1 and 8–2), the energy equation can be expressed for a flow process as follows:

$$q - w_s = \Delta\left(\frac{p}{\rho}\right) + \Delta\,KE + \Delta\,PE$$

or:

$$q - w_s = \Delta h + \Delta\,KE + \Delta\,PE$$

Example 8–2

Referring to a subsonic nozzle such as that in Figure 8–2, where the average inlet and exit velocity profiles are both averaged, consider the following conditions:

- The density is constant at 0.8 kg/m³.
- The inlet and exit maximum velocities are 105 m/s and 185 m/s, respectively.
- The inlet and exit radii are 0.08 m and 0.06 m, respectively.
- The amount of heat added is 164.0 kJ/kg.
- The change in elevation can be ignored.

Calculate the pressure drop across the nozzle.

Solution   Applying the energy-conservation equation yields the following:

$$q_{\text{in}} = \frac{\Delta p}{\rho} + \frac{1}{2}\left(\int_0^{V_{\text{ex}}} \rho\right) dA - \left(\int_0^{V_{\text{in}}} \rho\right) dA$$

Substitution in this equation yields:

$$\Delta p = 1.31 \text{ bars}$$

Figure 8–3

Geometry of a stream filament.

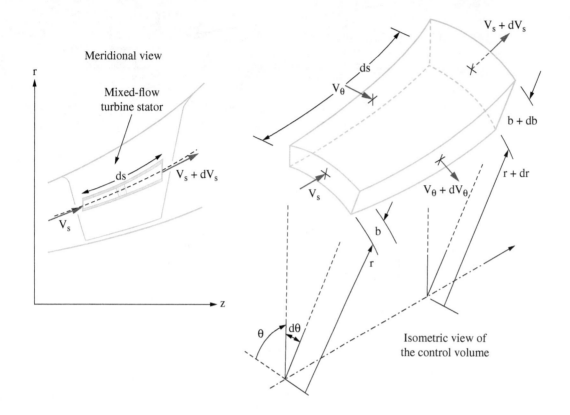

PROBLEMS

8–1   The circular-cross-section nozzle shown in Figure 8–2 handles liquid water (assumed incompressible). In a totally reversible process, the water flow stream proceeds from a radius of 0.6 m to one that is 0.2 m at exit. The velocity distribution at the inlet is uniform, while that at the exit is parabolic and is given by:

$$V_2 = V_{2\max}\left[1 - \left(\frac{r}{R}\right)^2\right]$$

where $V_{2\max} = 8$ m/s. Applying the continuity equation (in terms of volumetric discharge), calculate the inlet velocity.

8–2   A simplified method of analyzing cascade flows, away from the solid surfaces, is to cast the governing equations in the s-θ frame of reference as shown in Figure 8–3, where s is the distance along the stream tube in the meridional view. In this case, the stream tube thickness $[b = b(s)]$ and radius $[r = r(s)]$ are introduced in the governing equations for the purpose of generality. The upper and lower sides of the stream tube are assumed to be surfaces of revolution. Assuming a constant static density (ρ), prove that the differential form of the continuity equation in this case comes out to be:

$$\frac{1}{b}\frac{\partial}{\partial s}(rbV_s) + \frac{\partial V_\theta}{\partial \theta} = 0$$

External and Internal Flow Structures

Chapter Outline

9–1   Basic External Flow Structure   **112**

9–2   External Flows: Boundary Layer Buildup   **125**

9–3   Potential Flow Fields   **128**

9–4   Introduction of the Velocity Potential and Stream Function   **131**

9–5   Compressibility of a Working Medium: The Definition of Sonic Speed   **135**

9–6   Compressibility of the Flow Field: Definition of the Mach Number   **136**

9–7   Introduction of the Critical Mach Number   **140**

9–8   Isentropic Flow Through Varying-Area Passages   **144**

Impeller

Basic External Flow Structure

An external flow is a flow over the outside surface of an object. Common examples are the flow around a vehicle, a truck, and an aircraft as they speed along. Depending on the geometry, external flows can be very simple or quite complicated. The simpler flows, such as those over thin flat plates and streamlined bodies (Figure 9–1), can be treated as boundary layer flows. Flows that are required to make sharp turns, however, exhibit a variety of behaviors.

Boundary Layer Concept

In this section we explore boundary layer flows in detail. Our desire to understand the physics of such flows is grounded in a need to be able to calculate drag and heat transfer for engineering purposes.

We begin our discussion with a definition:

> A boundary layer flow is a flow in which the effects of viscosity and/or thermal effects are concentrated in a thin region near a surface of interest.

The key word in this definition is *thin*.

If the surface over which the velocity boundary layer grows is hotter or colder than the free stream fluid, a thermal boundary layer will develop. Note that the hydrodynamic (velocity) and thermal (temperature) boundary layer thicknesses are functions of the distance x from the plate leading edge, i.e.,

$$\delta_H = \delta_H(x)$$

$$\delta_T = \delta_T(x)$$

Furthermore, both of these layers are thin. This thickness, a key characteristic of a boundary layer, enables us to simplify the analysis of real flows by treating the situation as two distinct regimes: one close to the surface, where frictional and thermal effects are important, and a second regime outside the boundary layer, where the effects of friction and heat transfer are unimportant. Since Prandtl's discovery of the boundary

Figure 9–1

Structure of an external flow.

Undisturbed streamline

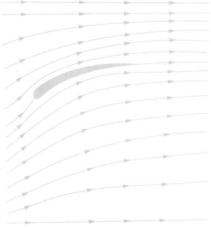

Undisturbed streamline

layer, tremendous strides have been made in understanding real flows and using this understanding in the discipline of engineering design. We will continue our discussion of boundary layers after we look at other types of external flows.

Bluff Bodies, Separation, and Wakes

Objects that are not thin in the flow direction are termed bluff bodies often greatly disturb the free-stream flow. Note that a flat plate at zero incidence angle (or angle of attack) is a thin object. Examples of such bluff bodies are cars, trucks, and buildings. The flow passing around a bluff body cannot follow the sharp contours and, therefore, it separates from the surface. Figure 9–2 shows separation of the boundary layer from the upper surface of an airfoil that is situated at a finite angle of attack. Separation causes a loss of lift on an aircraft wing and thus it is to be avoided in this application. Modern aircraft have complex, variable-geometry wing structures to prevent flow separation while simultaneously creating high lift forces. Depending on the specific geometry, a separated flow may, in turn, create a large disturbance downstream of the bluff body called the body's wake (Figures 9–3, 9–4, and 9–5). You are probably familiar with the wakes produced by ships, and have probably been aware of the wake of fast-moving trucks and cars as they pass by. When a flat plate is inclined at some

Figure 9–2

Velocity and pressure coefficients over a symmetrical airfoil.

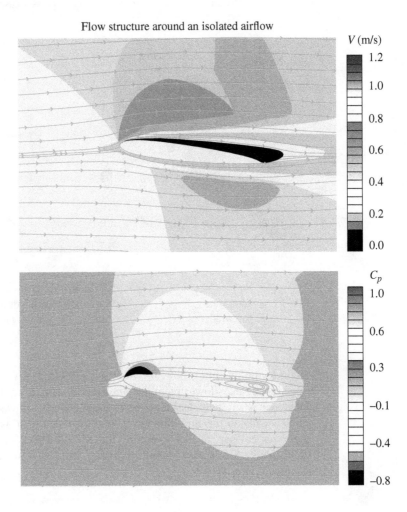

Flow structure around an isolated airflow

Figure 9–3

Sources of losses over an airfoil cascade.

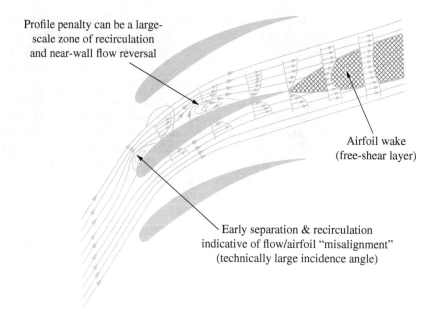

Profile penalty can be a large-scale zone of recirculation and near-wall flow reversal

Airfoil wake (free-shear layer)

Early separation & recirculation indicative of flow/airfoil "misalignment" (technically large incidence angle)

Boundary-layer mixing region in a compressor cascade.

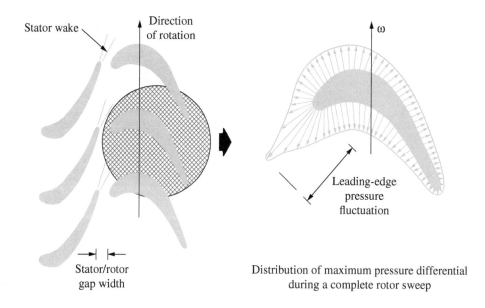

Stator wake

Direction of rotation

ω

Stator/rotor gap width

Leading-edge pressure fluctuation

Distribution of maximum pressure differential during a complete rotor sweep

angle to the incoming flow, the flow separates and a wake is formed. Note that the huge disturbance downstream is completely dissimilar to the flow over a plate at zero incidence where boundary layer flows dominate. The flow separation phenomenon applies equally to the flow through a Venturi meter such as that in Figure 9–6, where the divergent part of it is operating under an adverse, or unfavorable, pressure gradient, where the (static) pressure is rising, causing flow separation and recirculation zones.

Figure 9–4

Stator/rotor interaction
through the stator wakes.

Stator   Rotor

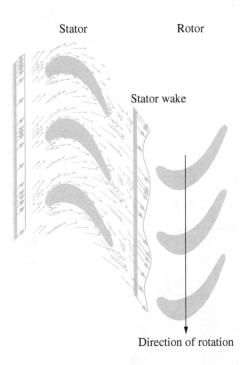

Figure 9–4

Stator/rotor interaction
through the stator wakes.

Stator wake

Direction of rotation

Forced Laminar Flow: Flat Plate

Laminar flat-plate boundary layer flows are sufficiently simple that they can be analyzed accurately using theoretical methods alone.

Problem Statement

Our objective here is to apply the fundamental mass, momentum, and energy conservation principles to a laminar boundary layer flow to obtain useful mathematical relationships to calculate the following:

- Hydrodynamic boundary layer thickness
- Thermal boundary layer thickness
- Local wall shear stress
- Total drag force
- Total heat flux
- Total heat transfer rate

These quantities are useful in the process of engineering design.

What information about the flow is required to calculate these quantities? Fortunately, the answer is quite simple. We need only determine the velocity profile through the hydrodynamic boundary layer $V(x, y)$, and the temperature boundary layer $T(x, y)$ (Figures 9–7 and 9–8). At the plate's sharp leading edge, the profile is uniform, with a discontinuity where the velocity jumps from zero at the surface to the free-stream value. As seen in Figure 9–7, the velocity gradient at the leading edge ($[\frac{\partial V}{\partial y}]_{y=0}$) is infinite at the blade leading edge. Since the fluid sticks to the plate surface (the no-slip boundary condition), a shearing action is created whereby the fluid layers are slowed below the free-stream velocity. This slowing diminishes with the perpendicular distance from the flat plate surface. Far from the plate, the local velocity gets to be equal to the free-stream

Figure 9–5

Wakes of an axial-flow
turbine stator.

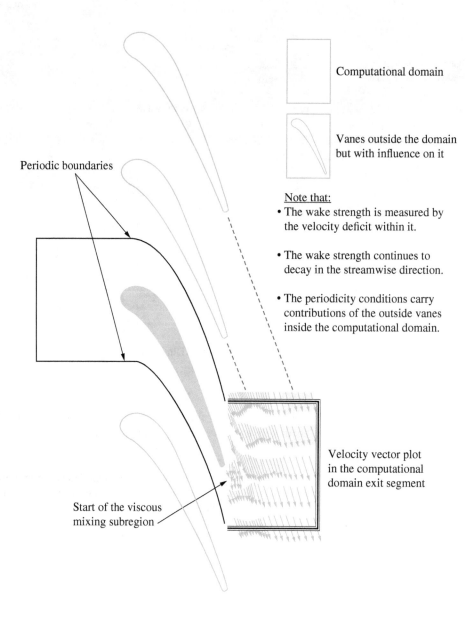

Periodic boundaries

Computational domain

Vanes outside the domain
but with influence on it

Note that:
• The wake strength is measured by
  the velocity deficit within it.

• The wake strength continues to
  decay in the streamwise direction.

• The periodicity conditions carry
  contributions of the outside vanes
  inside the computational domain.

Velocity vector plot
in the computational
domain exit segment

Start of the viscous
mixing subregion

magnitude. We formally define the hydrodynamic boundary layer thickness (δ_H) as the location where the local velocity is 99 percent of the free-stream value (Figure 9–9), i.e.,

$$\frac{V}{V_O} = 0.99$$

From this definition, we see that knowing the velocity field allows us to determine the boundary layer thickness.

In addition to providing the boundary layer thickness, the velocity field also leads to the determination of the wall shear stress as follows:

$$\tau_w = \mu \left[\frac{\partial V}{\partial y} \right]_{y=0}$$

Figure 9–6

A Venturi meter.

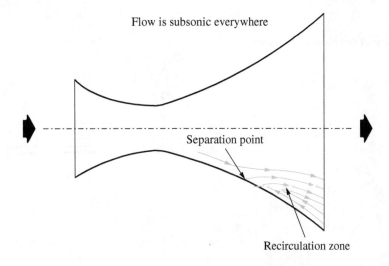

Flow is subsonic everywhere

Separation point

Recirculation zone

Figure 9–7

Boundary layer buildup over a flat plate.

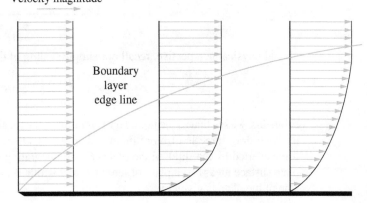

Boundary layer over a flat plate

Velocity magnitude

Boundary layer edge line

Figure 9–8

Thermal boundary layer development over a flat plate.

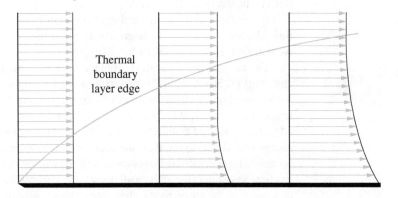

Temperature magnitude

Thermal boundary layer edge

Figure 9–9

Practical determination of the boundary layer edge.

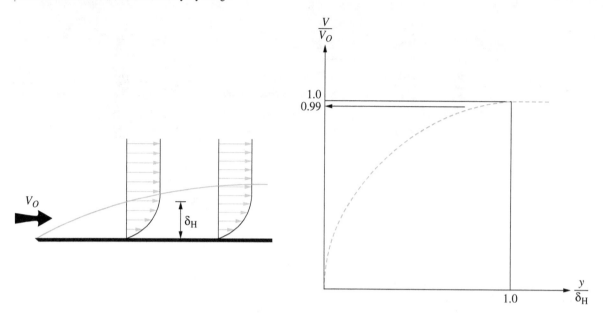

To add physical perspective, recall our simple scaling of this equation:

$$\tau_w = \frac{\mu V_O}{\delta_H}$$

From this, we see that as the boundary layer grows, the local shear stress falls. Frequently, the total drag force is of more interest than the local shear stress. We obtain the total force exerted by the fluid on the plate, F_D by integrating the wall shear stress over the plate surface area. For a plate of length L and width W, the drag force exerted on one side of the plate is given by:

$$F_D = \int_A \tau_w \, dA$$

where A is the surface area.

To keep the plate stationary, a force of equal magnitude but in the opposite direction must counteract this drag force.

In principle, the friction and drag problem is simple: Find $V(x, y)$ and apply the four preceding equations to determine all of the quantities of engineering interest. Later we will find the way to determine $V(x, y)$.

The thermal (heat transfer) problem is rather similar to the hydrodynamic (friction and drag) problem. Figure 9–8 illustrates the development of a thermal boundary layer for the situation in which the plate surface temperature T_w is greater than the free-stream temperature, T_O. Heat is thus transferred from the plate to the fluid. At the plate leading edge, we again see a uniform profile with a discontinuity at the surface. As the total energy transferred to the fluid increases with the downstream distance (x), the thermal boundary layer thickness increases as well. As a result of this thickening, the temperature gradient absolute value at the wall decreases. This can be seen by applying the graphical definition of the slope to the three temperature profiles. Thus, even without detailed knowledge of the temperature profiles, we expect the local heat flux to decrease with distance from the plate leading edge.

The thermal boundary layer thickness δ_T is defined as the perpendicular distance from the plate surface at which the temperature difference $T(x, y) - T_s$ is 99 percent of the maximum possible temperature difference $T_O - T_s$. More formally, we can write the following expression:

$$\frac{T(x, \delta_T) - T_s}{T_O - T_s} = 0.99$$

Like the hydrodynamics problem, we can determine $\delta_T(x)$. This same knowledge allows us to determine the local surface rate of heat transfer as follows:

$$\dot{Q}_x(x) = -k_f \left(\frac{\partial T(x, y)}{\partial y} \right)_{y=0}$$

where k_f is the so-called fluid thermal conductivity. Analogous to the total drag force, we obtain the total heat transfer rate from the plate to the fluid by integrating the rate of heat transfer over the plate area; thus, for one side of the plate we have:

$$\dot{Q} = \int_0^L (\dot{Q}) W \, dx$$

or

$$\dot{Q} = -k_f W \int_0^L \left(\frac{T(x, y)}{\partial y} \right)_{y=0} dx$$

We can use the preceding two equations to calculate the local and average heat transfer coefficients, $h_{\text{conv},x}$ and $\bar{h}_{\text{conv}}$, respectively, by applying the definitions of these quantities:

$$h_{\text{conv},x} = \dot{Q}(T_s - T_w)$$

and,

$$\bar{h}_{\text{conv}} = \frac{\dot{Q}}{WL(T_s - T_O)}$$

Again, to add more physical perspective, recall our simple scaling of the local heat transfer coefficient:

$$h_{\text{conv},x} = \frac{k_f}{\delta_T}$$

Here, we see that the heat transfer coefficient falls as the boundary layer grows. This behavior follows the same pattern as the wall shear stress.

Before looking at ways to obtain $V_s(x, y)$ and $T(x, y)$ we cast the problem in dimensionless terms, putting into practice what we know of the similitude principle. We know from forming the dimensionless governing equations and the dimensionless surface boundary condition that the important dimensionless parameters describing friction and drag are the friction coefficient and the Reynolds number. More explicitly, we derived the dimensionless velocity gradient at the wall related to c_f and Re (defined as the density times the incoming velocity times the characteristic length, all divided by the dynamic viscosity coefficient) as follows:

$$\left(\frac{\partial V_s^*}{\partial y^*} \right)_{y=0} = \frac{c_f \text{Re}}{2}$$

Figure 9–9 illustrates this relationship graphically where we see the dimensionless velocity ($v^* = V_s/V_O$) as a function of the dimensionless distance from the wall, $y^* = \frac{y}{L}$. A similar situation exists for the heat transfer problem. In this case, the dimensionless temperature gradient at the wall is defined to be the Nusselt number:

$$\left(\frac{\partial T^*}{\partial y^*}\right)_{y^*=0} = \text{Nu}$$

The dimensionless velocity as a function of y^* is shown in Figure 9–7. Our purpose in presenting these relationships is to show that solving the dimensionless problem cast in ordinary dimensional variables is not only possible, but is also more general. This generality is manifested in the fact that only one parameter, Re, controls the value of c_f, and that only two parameters, Re and Pr, control the value of Nu, that is:

$$c_f = f_1(\text{Re}_L)$$

and,

$$\text{Nu}_L = f_2(\text{Re}, \text{Pr})$$

where Pr is the Prandtl number defined as follows:

$$\text{Pr} = \frac{c_p \mu}{k}$$

The use of dimensionless variables takes into account all of the physical properties (ρ, k, c_p, μ), the geometric and dynamic parameters (L and V_O), and the desired engineering output (τ_w and h_L).

Solving the Problem

Two methods have traditionally been applied to determine the velocity and temperature fields within a laminar boundary layer over a flat plate. The first is known as the *similarity solution*. This method provides an exact solution to the governing boundary layer equations. Blasius was the first to apply the similarity solution to the friction and drag problem in 1908. The second method provides approximate solutions to the hydrodynamic and thermal problems by applying the integral control volume analysis. Von Karman first conceived this integral and presented results for the friction problem in 1921.

Both the exact and approximate approaches start with the equations governing the conservation of mass, momentum, and energy within a boundary layer.

The boundary layer equations are as follows:

$$\frac{\partial V_x}{\partial x} + \frac{\partial V_y}{\partial y} = 0$$

$$V \cdot \nabla V = \nu \frac{\partial^2 V_x}{\partial y^2}$$

$$V_x \frac{\partial T}{\partial x} + V_y \frac{\partial T}{\partial y} = \frac{k}{\rho c_p} \frac{\partial^2 T}{\partial y^2}$$

These equations apply to the two-dimensional steady flow of an incompressible Newtonian fluid with constant properties and negligible viscous dissipation.

Next we present an overview of the exact (similarity) solution.

Exact (Similarity) Solution: Friction and Drag

In the derivation of the conservation equations, we assumed the physical properties (ρ, μ, and c_p) to be constants. This assumption uncouples the momentum and energy equations. Thus we can first find $V_x(x, y)$ and $V_y(x, y)$ from the momentum and continuity equations, and then apply these results to solve the energy equation for $T(x, y)$.

At the heart of the similarity solution is the assumption that the velocity profile is geometrically similar at each x location when normalized by the local boundary layer thickness. Here, we see velocity profiles at two x locations. Although these two profiles are not identical when V_x is plotted as a function of y, if y is divided by the local boundary layer thickness δ_H and V_x divided by V_O, and these normalized variables are plotted, a single curve results. Mathematically, this is stated as follows:

$$\frac{V_x}{V_O} = f\left(\frac{y}{\delta_H}\right)$$

With this scaling, the continuity partial differential equation is reduced to a third-order ordinary differential equation using a transformation of variables that combines x and y from a new independent variable [i.e., $y(\frac{\rho V_O}{\mu x})^{1/2}$].

Before presenting the similarity solution for $\frac{V_x}{V_O}$, we note that the so-called similarity variable, $y(\frac{\rho V_O}{\mu x})^{1/2}$, can be manipulated to introduce a local Reynolds number that is defined as follows:

$$(\text{Re})_x = (\rho V_x x)/\mu$$

Multiplying and dividing the similarity variable by x yields:

$$y(\rho V_O \mu x)^{1/2}\left(\frac{x}{X}\right) = \left(\frac{y}{x}\right)\left(\frac{\rho V_O x}{\mu}\right)^{1/2} \equiv \text{Re}_x^{1/2}$$

Figure 9–9 presents the similarity solution showing the functional relationship between V_x/V_O and $(y/x)\text{Re}_x^{1/2}$. These results agree well with experimental measurements. This velocity profile can be used to determine how the hydrodynamic boundary layer thickness grows with the distance from the plate leading edge. We see that V_x/V_O is approximately 0.99 when $(y/x)\text{Re}_x^{1/2}$ is 5.0. Since $y \equiv \delta_H$ when $V_x/V_O = 0.99$, we conclude that:

$$(\delta_H/x)\text{Re}_x^{1/2} = 5.0$$

The x-dependency of δ_H from the preceding equation is:

$$\delta_H = 5\left(\frac{\mu}{\rho V_O}\right)x^{1/2}$$

Thus, we see that the hydrodynamic boundary layer thickness grows as the square root of x.

Figure 9–8 shows the typical development of the thermal (temperature) profile over a flat plate. Knowledge of this will make it possible to calculate the growth of the thermal boundary layer.

Example 9–1

Consider a 150-mm-long thin flat plate oriented parallel to a uniform flow. Calculate the hydrodynamic boundary layer thickness at the trailing edge of the flat plate for a flow of air and for a flow of water. The free-stream velocity for both flows is 2 m/s, and the temperature and pressure are 300 K and 1 bar, respectively.

Solution

Evaluation of Re_L requires values for the density and viscosity, which we can obtain from the tables in the Appendix. Thus:

$$Re_L = \frac{\rho V_0 L}{\mu}$$

thus:

$$Re_{L,\text{air}} = 19{,}000$$

and

$$Re_{L,\text{water}} = 351{,}700$$

Furthermore,

$$(\delta_H/L) = 5 Re_L^{-1/2}$$

$$(\delta_H/L)_{\text{air}} = 0.036$$

$$(\delta_H/L)_{\text{water}} = 0.0084$$

The dimensional boundary layer thicknesses are:

$$\delta_{H,\text{air}} = 5.4 \text{ mm}$$

$$\delta_{H,\text{water}} = 1.3 \text{ mm}$$

Figure 9–10 shows an example of a turbine engine where radial turbomachinery components are utilized. Joining this family of long streamwise chords are the mixed-flow turbomachinery components (example shown in Figure 9–11). Such a turbomachine is equipped with long flow trajectories along the blades, with the outcome being an excessive boundary layer buildup. Figure 9–11, on the other hand, shows a somewhat long-chord blade, but in a mixed-flow compressor rotor. The smallest chord, however, is shown in Figure 9–12 for a purely axial turbine rotor.

Example 9–2

Considering the drag on both sides of a flat plate, determine the force required to hold the plate stationary in a flow of air. The plate is 70 mm long.

Solution

To reinforce the use of dimensionless parameters, we first calculate the drag coefficient and then apply its definition to obtain the dimensional quantity of interest. Using the thermophysical properties and Re_L values for the air stream in Example 9–1, we have:

$$c_D = 1.328 \ Re_L^{1/2} = 0.0096$$

We also have:

$$F_{D\,\text{one-side}} = c_D(LW)\left(\frac{1}{2}\rho V_0^2\right) = 0.000237 \text{ N}$$

The same drag force is exerted on both sides, thus:

$$F_{\text{total}} = 0.000474 \text{ N}$$

Figure 9–10

An auxiliary power unit with highly large-chord blading.

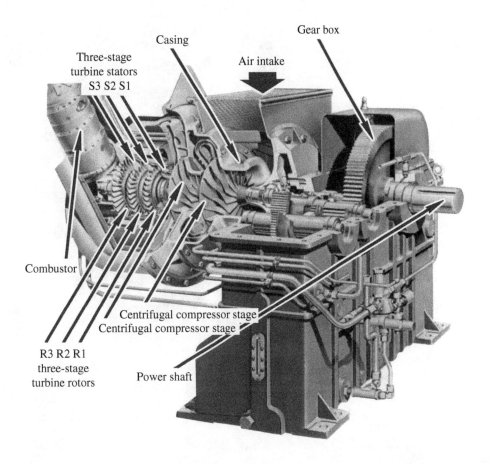

Casing

Gear box

Air intake

Three-stage turbine stators S3 S2 S1

Combustor

Centrifugal compressor stage
Centrifugal compressor stage

R3 R2 R1 three-stage turbine rotors

Power shaft

Figure 9–11

Mixed-flow compressor with large blade chords.

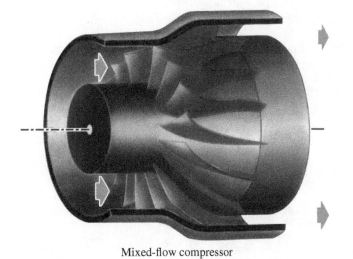

Mixed-flow compressor

Figure 9–12

An axial-flow turbine rotor with notably short chords.

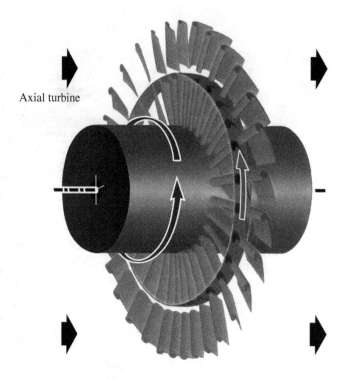

Axial turbine

Referring to Figures 9–10 through 9–12, we see that some airfoils have long chords (Figures 9–10 and 9–11), resulting in high magnitudes of the Reynolds number, while others (Figure 9–12) have significantly shorter cord lengths.

Example 9–3

A smooth, flat plate has a total length (L) of 0.75 m. The plate is to be tested in both water and air at a velocity of $U = 4.5$ m/s. The temperature of both will be 20°C, and the pressure is atmospheric. Determine if the flow at the end of the plate will be laminar or turbulent for each fluid.

Solution

If the Reynolds number $\mathrm{Re}_L \leq 0.50 \times 10^6$, the flow will be laminar, and if $\mathrm{Re}_L \geq 0.50 \times 10^6$, it will be turbulent. The kinetic viscosity ν of the air and the water are:

$$\nu_{\mathrm{air}} = 15.09 \times 10^{-6} \ \mathrm{m^2/s}$$

$$\rho_{\mathrm{air}} = 1.204 \ \mathrm{kg/m^3}$$

$$\nu_{\mathrm{water}} = 1.004 \times 10^{-6} \ \mathrm{m^2/s}$$

$$\rho_{\mathrm{water}} = 998.3 \ \mathrm{kg/m^3}$$

In water:

$$\mathrm{Re}_{L,\mathrm{water}} = \frac{UL}{\nu_{\mathrm{water}}} = 3.361 \times 10^6$$

$$\mathrm{Re}_{L,\mathrm{air}} = 0.2237 \times 10^6$$

Therefore, the flow at the end of the plate is laminar with air and turbulent with water.

9–2 **External Flows: Boundary Layer Buildup**

As a fluid moves past a solid boundary or a wall, the velocity of a typical fluid particle at the wall must be equal to the wall velocity, a fact that is in the case of stationary surfaces, termed the "no-slip" concept in fluid dynamics. This fact can be observed experimentally. As a result, a steep velocity gradient across the boundary layer adjacent to the wall materializes.

Figure 9–13 depicts a two-dimensional boundary layer flow which occurs in the direction normal to the flow. This velocity gradient occurs because of the shear stress $\tau_x(y)$ acting on the surface of the particles within the boundary layer. The action of $\tau_x(y - \frac{\Delta y}{2})$ on the bottom of the particle is to retard the fluid or decrease its velocity. On the top surface $\tau_x(y + \frac{\Delta y}{2})$ must act to move it in the flow direction if the fluid particle is to be in equilibrium, that is, summation of forces in any direction must be equal to zero. The difference in velocity between the top and the bottom of the fluid particle, Δu, is proportional to the distance Δy between the two surfaces.

$$\frac{\Delta u}{\Delta y} \text{ is proportional to } \tau_x(y)$$

If the fluid is Newtonian with a coefficient of dynamic viscosity coefficient μ, then:

$$\tau_x(y) = \mu \frac{du}{dy}$$

The viscous shear stresses in a fluid result in forces that resist the fluid motion. Energy must be added to the fluid to overcome this resistance if the fluid flow is to be maintained.

The Influence of Pressure Gradient

The force arising from the pressure differences on the fluid has, so far, been set to zero. Although this assumption is valid in some cases, it is hardly valid in many other cases. When $\frac{dp}{dx} \neq 0$ (Figure 9–14), a pressure force will exist that not only contributes to the total resistance by the fluid, but can also result in the phenomenon referred to as

Figure 9–13

Shear stress on a fluid element within the boundary layer.

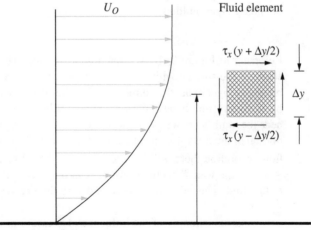

Boundary layer velocity profile

U_O

Fluid element

$\tau_x(y + \Delta y/2)$

Δy

$\tau_x(y - \Delta y/2)$

No-slip condition applies at the wall

Figure 9–14

Effect of adverse pressure gradient over a solid surface.

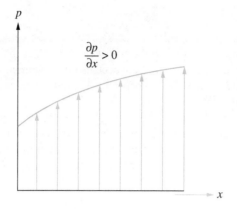

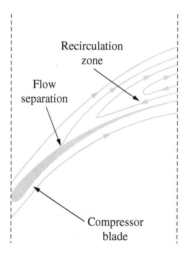

flow separation. Such a pressure gradient is commonly referred to as an *adverse* (or unfavorable) *pressure gradient*, and (as was just stated) can tremendously aggravate the boundary layer buildup.

Flow Separation

Referring to Figure 9–14, the magnitude of pressure gradient $\frac{dp}{dx}$ is dependent on the shape of the surface which, in turn, influences the variation of the velocity outside the boundary layer in the flow direction, $\frac{dU}{dx}$. The occurrence of separation can be determined by examining the shape of the u velocity profile in the boundary layer. If the flow velocity near the wall becomes zero and the flow separation takes place, the velocity gradient normal to the flow, $\frac{\partial u}{\partial y}$, must be zero at this point. Prior to flow separation, $\frac{\partial u}{\partial y} > 0$. After flow separation there will be a reverse flow, i.e., flow in the minus x direction, and $\frac{\partial u}{\partial y} \leq 0$ near the wall. This flow reversal can only occur if a force addition to τ_x is applied to the fluid. This additional force is due to the pressure gradient in the flow direction, $\frac{dp}{dx} > 0$.

A partial example of the influence of pressure gradient is the incompressible flow of a fluid through a rectangular converging-diverging channel (Figure 9–6). Therefore, an internal flow case is chosen to demonstrate the flow separation, since it is easier to describe the variation of velocity and pressure in the flow direction. The converging portion of this channel acts to increase the velocity in the region away from the walls. Such a channel is termed a *nozzle*. This produces a favorable pressure gradient in the boundary layer. At the throat section, the cross-sectional area is transitionally constant and so are the velocity U and the pressure.

Another example of favorable and adverse pressure gradients over a surface is the flow past a circular cylinder, which is initially wide in the direction normal to the flow. The flow considered over a cross section of it is practically two-dimensional. In this case $\frac{dp}{dx}$ over the forward portion of the cylinder is negative and then becomes positive (or adverse) as shown in Figure 9–15. The point at which $\frac{dp}{dx}$ is zero is located at an angle of approximately 70 degrees from the most forward point (stagnation point) on the cylinder. The stagnation point is the location where the static pressure is maximum, since the fluid velocity at this point is zero. The pressure coefficient represents the ratio of the pressure force on the cylinder to the inertia force of the fluid.

Figure 9–15

Pressure distribution over a cylindrical surface.

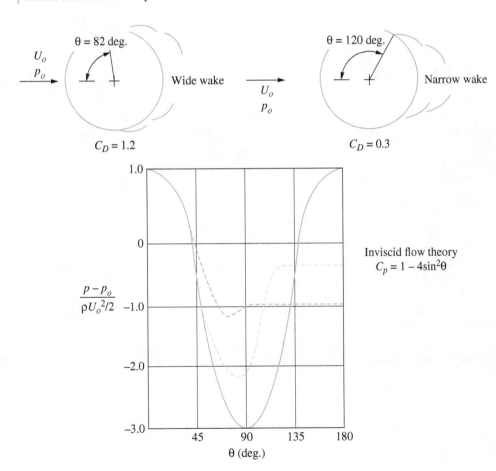

Inviscid flow theory
$$C_p = 1 - 4\sin^2\theta$$

9–3 **Potential Flow Fields**

To understand what this flow category entails, let us define a point function termed the vorticity (Ω) as follows:

$$\Omega = \nabla X \ \mathbf{V}$$

where V is the velocity vector. A flow field where Ω is zero everywhere is called *irrotational* or *potential*.

Basic Flow Equations

Inviscid flow exists outside the boundary layer and the wake (of an airfoil) in high-Reynolds-number flows around bodies (e.g., Figure 9–15 for a circular cylinder). For an airfoil (Figure 9–14), the boundary layer is rather thin (an observation that is valid perhaps exclusively to turbine airfoil cascades), and the inviscid flow provides a good approximation to the actual flow field; it is used to predict the pressure distribution on the surface (note that the static pressure across the boundary layer remains more or less constant), thereby giving a good estimate of the lift force. It will also provide us with the velocity to be used as a boundary condition in the boundary layer solution and, from that solution, we can estimate the drag and predict possible points of separation (note that this is a rather crude means of analyzing the boundary layer characteristics).

Consider a velocity field that is given by the gradient of a scalar function ϕ as follows:

$$\vec{V} = \nabla \phi$$

in which ϕ is called the *velocity potential function*. Such a velocity field is called potential or irrotational, and possesses the property that (as stated above) the vorticity is nonexisting anywhere in the flow field.

With the velocity given by the gradient of a scalar function, the differential form of the continuity equation for an incompressible potential flow is given as follows:

$$\nabla \cdot \nabla \phi = \nabla^2 \phi = 0$$

which is the well-known Laplace equation. In rectangular coordinates, we can write this equation as follows:

$$\frac{\partial^2 \phi}{\partial x^2} + \frac{\partial^2 \phi}{\partial y^2} + \frac{\partial^2 \phi}{\partial z^2} = 0$$

With the appropriate boundary conditions, this equation could be solved. However, three-dimensional problems are quite difficult, so we focus our attention on plane flows in which the velocity components u and v depend on x and y. This is acceptable for two-dimensional airfoils and other plane flows, such as the flow field around a cylinder (Figure 9–15).

Before we attempt a solution, let us define another scalar function that will aid us in our study of plane fluid flows. The continuity equation can be expressed as follows:

$$\frac{\partial u}{\partial x} + \frac{\partial v}{\partial y} = 0$$

Now, let the velocity components be defined as follows:

$$u = \frac{\partial \psi}{\partial y}$$

and

$$v = -\frac{\partial \psi}{\partial x}$$

We observe that the continuity equation is automatically satisfied; the scalar function $\psi(x, y)$ is called the *stream function*. By using the mathematical description of a streamline, $\vec{V} X d\vec{R} = 0$, we see that for a plane flow, $u\,dy - v\,dx = 0$. Substituting from these last two definitions, we get:

$$\frac{\partial \psi}{\partial y}dy + \frac{\partial \psi}{\partial x}dx = 0$$

Thus ψ is constant along a streamline (Figures 9–16 and 9–17).

The velocity vector for a plane flow has only a z-component since $\Omega = 0$, and there is no variation with z. The vorticity expression now reduces to:

$$\Omega_z = (\nabla X \vec{V}) = \frac{\partial v}{\partial x} - \frac{\partial u}{\partial y}$$

For the potential flow at hand, we demand that the vorticity be zero, meaning that:

$$\frac{\partial^2 \psi}{\partial x^2} + \frac{\partial^2 \psi}{\partial y^2} = 0$$

which is, again, Laplace equation for this plane flow.

Rather than attempt a solution of Laplace's equation for a particular flow of interest, we will use a different technique. We will identify some relatively simple functions that satisfy Laplace equation, then superimpose these simple functions to create flows of interest. It is possible to generate any plane flow desired using this technique, without having to resort to solving Laplace equation.

Before we present some simple functions, some additional observations concerning ϕ and ψ will be made.

$$u = \frac{\partial \phi}{\partial x} = \frac{\partial \psi}{\partial y}$$

and

$$v = \frac{\partial \phi}{\partial y} = -\frac{\partial \psi}{\partial x}$$

These relationships between the derivatives of ϕ and ψ are the famous Cauchy-Riemann equations from the theory of complex variables. The functions ϕ and ψ are harmonic functions since they both satisfy Laplace equation and form an analytical function $\phi + i\psi$ called the *complex velocity potential*. The theory of complex variables, with all its powerful theorems, is thus applicable to this restricted class of problems, namely plane, incompressible potential flows.

Figures 9–16 and 9–17 present the concepts of stream function and velocity potential. In these figures, we get to realize the fact that lines representing both families are perpendicular to one another.

Figure 9–16

Illustration of the stream function in a stream tube.

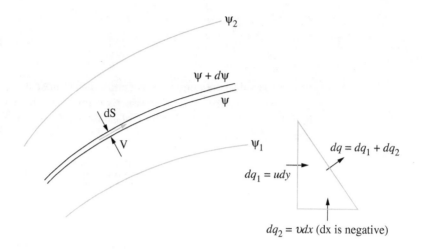

Figure 9–17

Comparison of the velocity potential and stream function.

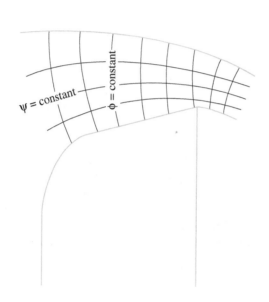

Example 9–4

Show that the difference in the stream function between any two streamlines is equal to the volumetric flow rate per unit depth between the two streamlines (q).

Solution

Consider the flow between two streamlines which are infinitesimally close to one another. The flow rate per unit depth through the elemental area is:

$$dq = dq_1 + dq_2 = u\,dy - v\,dx = \frac{\partial \psi}{\partial y}dy + \frac{\partial \psi}{\partial x}dx = d\psi$$

If this is integrated between two streamlines $\psi = \psi_1$ and $\psi = \psi_2$, the result is:

$$q = \psi_2 - \psi_1$$

which is what we need to prove.

Example 9–5   Show that the streamlines and equipotential lines of a plane, incompressible potential flow intersect one another at right angles.

Solution   If, at a point, the slope of a streamline is the negative reciprocal of the slope of an equipotential line, then the two lines are perpendicular to one another. The slope of a streamline is given by:

$$\left(\frac{dy}{dx}\right)_{\psi=\text{const}} = \frac{v}{u}$$

The slope of an equipotential line is:

$$\left(\frac{dy}{dx}\right)_{\phi=\text{const}} = \frac{\partial\phi}{\partial x} - \frac{\partial\phi}{\partial y} = -\frac{u}{v}$$

Thus we see that the slope of a streamline is the negative reciprocal of the equipotential line, which is what we need to prove.

Difference Between Turning and Rotational Flows

In illustrating the difference between a rotational flow and a simply turning flow, we will use the example of a stationary component in radial-inflow turbines called the "scroll," which is shown in Figure 9–18. The flow field in this component can be irrotational, but the flow domain is of the doubly connected type. Also referred to as the distributor, two scroll configurations with the comparison involving the general uniformity of the component at the exit station are shown in Figure 9–19. The chosen parameters in this figure are the exit mass flux and the exit flow angle.

Perhaps one of the most useful examples of a potential flow field is the inlet segment of the scroll in Figure 9–20. In this figure, the inlet passage of a radial inflow turbine scroll (or distributer) is receiving a potential flow stream. With the exception of the scroll wall, where boundary layer buildup is clearly dominant, note that a fluid particle is swirling (or rotating) around an external axis. Near the wall, however, the particle is spinning around its own axis as a result of the shear stress there.

Figure 9–21 shows a potential flow field's nondimensionalized velocity distribution in the blade-to-blade passage of the F109 (a U.S. Air Force subsonic trainer engine) first-stage stator.

By definition, a potential flow field is inviscid, for it is the steep velocity gradients near the wall in a viscous flow field that give rise to vorticity in this region in particular.

9–4   Introduction of the Velocity Potential and Stream Function

The velocity potential ϕ is defined as follows:

$$\nabla\phi = \vec{V}$$

In terms of this new function, Bernoulli's equation can be written as follows:

$$\frac{1}{2}(\nabla\phi)^2 + \frac{p}{\rho} - U = \text{constant}$$

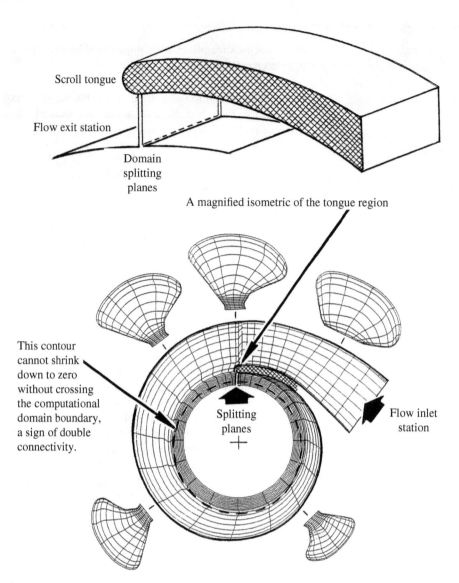

Scroll tongue

Flow exit station

Domain
splitting
planes

A magnified isometric of the tongue region

This contour
cannot shrink
down to zero
without crossing
the computational
domain boundary,
a sign of double
connectivity.

Splitting
planes

Flow inlet
station

Circulation and Vorticity

We have seen that the velocity in a potential flow is given by:

$$\vec{V} = \nabla \phi$$

Thus if $\vec{V}$ is known, ϕ can be calculated by integration:

$$\phi - \phi_A = \int_A \vec{V}.dr$$

where the subscript A denotes a typical point and r is the radius vector from some fixed origin. The difference of potential between any two points A and B is obtained by integration as follows:

$$(\phi_B - \phi_A)_I = \int_A^B \vec{V}.dt$$

Figure 9–19

Mass flux and flow angle at the scroll inlet station.

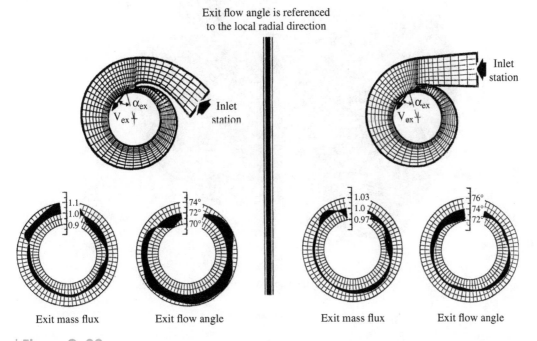

Exit flow angle is referenced
to the local radial direction

Exit mass flux   Exit flow angle   Exit mass flux   Exit flow angle

Figure 9–20

Irrotational and rotational flow zones within a volute.

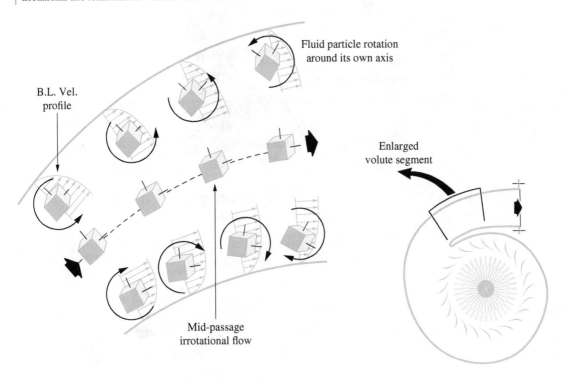

B.L. Vel.
profile

Fluid particle rotation
around its own axis

Enlarged
volute segment

Mid-passage
irrotational flow

Figure 9–21

Critical Mach number
contours in an axial
turbine stator.

F109 air force subsonic-trainer turbofan engine

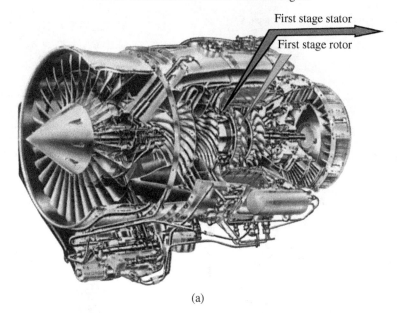

First stage stator

First stage rotor

(a)

V/V_{cr} contours in the vane-to-vane passage

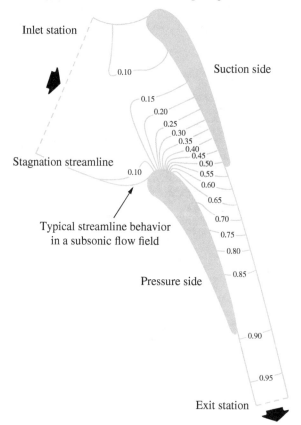

Inlet station

0.10

Suction side

0.15
0.20
0.25
0.30
0.35
0.40
0.45
0.50
0.55
0.60

Stagnation streamline

0.10

0.65

0.70

Typical streamline behavior
in a subsonic flow field

0.75

0.80

0.85

Pressure side

0.90

0.95

Exit station

(b)

The subscript I indicates the path chosen for integration. However, some alternative path II could just as well be chosen so that:

$$(\phi_B - \phi_A)_{II} = \int_A^B \vec{V}_{II} \cdot dr$$

The above defines the circulation around the airfoil. Vorticity and circulation are two different physical properties, both in magnitude and implications. Around an airfoil, the flow could possess no vorticity (meaning irrotational), but still the existence of the lifting body would create a finite magnitude of circulation. Vorticity, on the other hand, is imparted as a result of, for instance, such degrading factors as the boundary layer's velocity gradient in the vicinity of a solid surface.

9–5 Compressibility of a Working Medium: The Definition of Sonic Speed

The propagation speed of a sound wave in any fluid is called the *sonic speed*, and is defined as follows:

$$a = \sqrt{\left(\frac{\partial p}{\partial \rho}\right)_s}$$

The subscript s in this expression means constant entropy (i.e., an isentropic process in which the pressure and density change by infinitesimal magnitudes). A simple apparatus where the sonic speed can be roughly measured is that of a piston-cylinder device (Figure 9–22). Let us assume that the piston cross-sectional area is known and that the initial volume is also known, as well as the mass of the trapped air. Now, as illustrated in Figure 9–22, let us slowly add an additional weight adiabatically on the piston's outer face, causing the piston to drop through a distance Δh as frictionlessly as possible (note that we are here trying to emulate an isentropic process). As the piston, in the end, settles, let us measure the piston displacement from the initial position. Referring to the process end states by 1 and 2, we have:

$$\Delta p = \frac{\Delta w}{A_{piston}}$$

Figure 9–22

Determination of the sonic speed through a simple device.

Progression of an almost-isentropic process

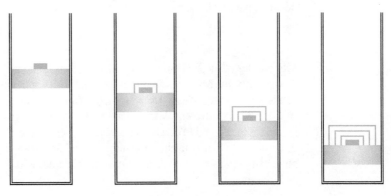

where Δw is the additional weight added to the piston's outer face. With V_1 known, we can now calculate the final volume V_2 as follows:

$$V_2 = V_1 - A_{\text{piston}}\Delta h$$

Now, with the volumes known at states 1 and 2, as well as the mass of air, we have:

$$\rho_1 = \frac{m}{V_1}$$

$$\rho_2 = \frac{m}{V_2}$$

Now, we can calculate a rather rough magnitude of the sonic speed in air at the given temperature as follows:

$$a \approx \sqrt{\left(\frac{\Delta p}{\Delta \rho}\right)}$$

Sonic Speed in Ideal Gases

With the assumptions made above, it is rather easy to prove that the path of a process, assuming the air is behaving as an ideal gas, is dictated by the following relationship:

$$\frac{p}{\rho^\gamma} = \text{constant}$$

Differentiating this relationship, and invoking the equation of state, the sonic speed expression can be rewritten as follows:

$$a = \sqrt{\gamma R T}$$

Examination of this expression reveals that the sonic speed in an ideal gas is a function of the "static" (or physically measurable) temperature only, since the specific heat ratio γ is itself a function of temperature. Again, the fact should be reemphasized that the sonic speed represents the compressibility of the working medium that is at rest.

9–6 Compressibility of the Flow Field: Definition of the Mach Number

For a fluid in motion, the term *compressibility* is defined as the tendency of the flow to undergo disproportionately high density changes in response to finite-magnitude local pressure gradients. The flow property that would profess such a flow behavior is referred to as the Mach number M. This is defined anywhere in the flow domain as the ratio between the local velocity and the speed of sound:

$$M = \frac{V}{a}$$

Magnitudes of the Mach number which are less than unity classify the flow as subsonic. Above unity, the Mach number is associated with a supersonic flow field. The importance of the sonic state, where $M = 1$, stems from the fact that it is the interface between two drastically different flow structures. For example, a subsonic flow field exclusively possesses the so-called downstream effect, whereby the fluid particle senses the existence of an obstacle that is farther downstream and accordingly adjusts its direction before it physically reaches the obstacle. No similar feature exists in supersonic flows.

Total Properties in Terms of the Mach Number

First let us define the difference between the static (or physically measurable) properties at a specific state and total properties at the same state. While static properties are physically measurable, total properties depend not only on the static properties but also on the local velocity (or Mach number).

Details of Figure 9–23 illustrate the means (definitely hypothetical) by which the total (or stagnation) flow properties can be obtained. In particular, let us focus on the exit station of a compressor stator (with the blade-to-blade passage shaped like a diffuser). The exit station in Figure 9–23, is labeled 1. In order to stagnate the flow (or get rid of the kinetic energy) we imagine introducing the flow to a fictitious infinitely long and frictionless diffuser, the objective of which is to (hypothetically) "kill" the velocity V_1 (Figure 9–23). The fictional diffuser exit (where the flow is stagnant) is labeled *stagn*.

Figure 9–23

Simulation of the stagnating process for total properties.

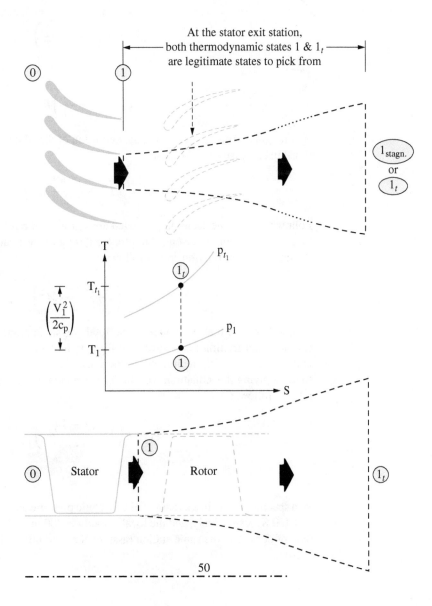

for stagnation or t for total. The properties at this particular hypothetical state are indeed the total properties at station 1. In real life, we do not have to imagine such a fictional process, for total properties can be easily computed in terms of their static counterparts and the local Mach number.

Let us now begin with the total temperature definition in terms of the Mach number by simple application of the energy-conservation equation:

$$h_t = h + \frac{V^2}{2}$$

or:

$$T_t = T + \frac{V^2}{2c_p}$$

or:

$$T_t = T\left(1 + \frac{V^2}{2c_p T}\right)$$

where:

$$c_p = \left(\frac{\gamma}{\gamma - 1}\right)R$$

Rearranging terms, and recalling the definition of the Mach number, the following relationship can easily be obtained:

$$T_t = T\left(1 + \frac{\gamma - 1}{2}M^2\right)$$

Following the same scenario presented above (about the infinitely long frictionless diffuser), and the implied isentropic process through it, we can write an equivalent total-to-static pressure relationship as follows:

$$p_t = p\left(1 + \frac{\gamma - 1}{2}M^2\right)^{\frac{\gamma}{\gamma - 1}}$$

Despite the fact that we obtained the total (or stagnation) state in a fictional device (through that frictionless isentropic diffuser), the fact remains that this particular state is thermodynamically legitimate, in the sense that the equation of state will still apply to it. Applying this equation, we can derive an expression for the total-to-static density ratio as follows:

$$\rho_t = \rho\left(1 + \frac{\gamma - 1}{2}M^2\right)^{\frac{1}{\gamma - 1}}$$

Example 9–6   At a specific station in a stream of air, the static pressure and temperature are 2.0 bars and 460 K, respectively, and the local velocity is 380 m/s. Calculate the total magnitude of density at the same station (assume a magnitude of 1.4 for the specific heat ratio).

Solution

$$a = \sqrt{\gamma RT} = 429.9 \text{ m/s}$$

$$M = \frac{V}{a} = 0.88$$

$$\rho = \frac{p}{RT} = 1.51 \text{ kg/m}^3$$

Finally:

$$\rho_t = \rho \left(1 + \frac{\gamma - 1}{2} M^2 \right)^{\frac{1}{\gamma - 1}} = 2.16 \text{ kg/m}^3$$

Example 9–7

In a hot air stream ($\gamma = 1.33$), the total and static magnitudes of pressure are 3.8 and 2.7 bars, respectively. If the static density is 1.8 kg/m³, calculate the magnitude of total enthalpy.

Solution

$$p_t = p \left(1 + \frac{\gamma - 1}{2} M^2 \right)^{\frac{\gamma}{\gamma - 1}}$$

which, upon substitution, yields:

$$M = 0.73$$

Applying the equation of state, we get:

$$T = \frac{p}{\rho R} = 522.6 \text{ K}$$

Now:

$$T_t = T \left(1 + \frac{\gamma - 1}{2} M^2 \right) = 568.6 \text{ K}$$

Using the air tables:

$$h_t = 657.7 \text{ kJ/kg}$$

Bernoulli's Equation: Accuracy and Limitations

Bernoulli's equation states that:

$$p_t = p + \rho \frac{V^2}{2}$$

On the other hand, we have:

$$\rho \frac{V^2}{2} = \frac{1}{2} M^2 a^2 = \frac{1}{2} \rho M^2 (\gamma RT) = \frac{\gamma}{2} p M^2$$

where the equation of state has already been invoked. Now let us consider the expression:

$$\frac{p_t - p}{\frac{1}{2}\rho V^2} = \frac{2}{\gamma}\left[1 + \left(\frac{\gamma - 1}{2}M^2\right)^{\frac{\gamma}{\gamma-1}}\right]$$

Expanding the right-hand side as a binomial, we get:

$$\frac{p_t - p}{\frac{1}{2}\rho V^2} = 1 + \frac{M^2}{4} + \frac{M^4}{40} + \frac{M^6}{1600} + \cdots \text{. Higher-order terms}$$

Taking, as an approximation, the first term only on the right-hand side, we get what is known as Bernoulli's equation for incompressible flows. To assess Bernoulli's equation in the general context of a compressible flow field, we obtain the following:

At $M = 0.0$, Bernoulli's equation results in zero error

At $M = 0.1$, the error is approximately 0.25 percent

At $M = 0.2$, the error is approximately 1.00 percent

At $M = 0.3$, the error is approximately 2.25 percent

At $M = 0.4$, the error is approximately 4.06 percent

At $M = 0.5$, the error is approximately 6.40 percent

At $M = 1.0$, the error is approximately 27.6 percent

At $M = 1.74$, the error is approximately 100 percent

9-7 Introduction of the Critical Mach Number

Perhaps the only undesirable characteristic of the "regular" Mach number M is that the sonic speed is not directly indicative of the local velocity in an adiabatic flow. This is true in the sense that the sonic speed itself is a function of the local temperature. To clarify this point, consider two fluid particles that are proceeding within the first stage of a compressor in a simple turbojet engine, and that through the turbine section of the same engine both proceeding at the same velocity. Because the compressor particle is exposed to a much lower temperature than its counterpart, the sonic speed at these locations will be vastly different, and so will the Mach numbers.

It would therefore be advantageous to replace the Mach number with another nondimensional velocity ratio the magnitude of which is directly indicative of the local velocity. In the following, a new nondimensional velocity ratio, the critical Mach number (or the critical velocity ratio), will be derived not only to achieve this objective but to also be a tool in classifying the flow regime as subsonic or supersonic.

Instead of the local sonic speed, let us define the so-called critical velocity V_{cr} as a velocity nondimensionalizer. To better comprehend this parameter physically, let us visualize a fictional subsonic (convergent) nozzle that admits the flow stream at the location where the magnitude of the critical velocity V_{cr} is desired. The objective of this nozzle is to "choke" the flow isentropically (i.e., to get the Mach number to the magnitude of unity). It is the flow velocity at the choking location (where $M = 1$) that is referred to as the critical velocity (V_{cr}). The name here is appropriate in the sense that most aerodynamics textbooks refer to the "throat" state of the nozzle (where $M = 1$) as the critical state.

The way to calculate a sonic speed (which is a local property) is to first compute the static temperature. Appropriately termed the *critical temperature* (T_{cr}), we can find it by simply using the total-to-static temperature expression and substituting 1.0 for the Mach number as follows:

$$T_{cr} = \frac{T_t}{1 + \frac{\gamma-1}{2}1.0^2} = \frac{2T_t}{\gamma + 1}$$

The sonic speed corresponding to this temperature is precisely the critical velocity (V_{cr}) which we are after. In order to calculate this velocity, we simply substitute T_{cr} for the static temperature (T) as follows:

$$T = T_t\left[1 - \frac{\gamma-1}{\gamma+1}\left(\frac{V}{V_{cr}}\right)^2\right]$$

As stated earlier, this particular velocity will replace the local sonic speed in nondimensionalizing the velocity, creating the "promised" critical Mach number (M_{cr}) as follows:

$$M_{cr} = \frac{V}{V_{cr}} = \frac{V}{\sqrt{\frac{2\gamma}{\gamma+1}RT_t}}$$

Figure 9–24 shows the relationship between M and M_{cr} for two common specific heat ratios, 1.4 and 1.33. Note that usage of the critical Mach number does not differ from that with the "regular" magnitude in breaking up the flow field into its subsonic and supersonic regimes for $M_{cr} = 1$ once M gets to be 1.

Figure 9–24

Relationship between the Mach and critical Mach numbers.

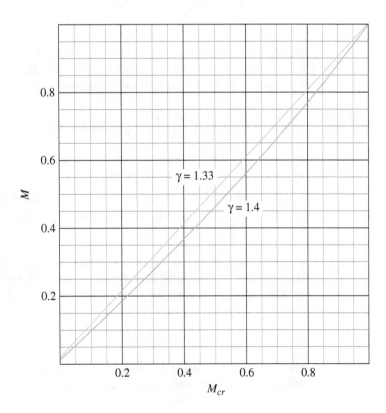

Total Properties in Terms of the Critical Mach Number

We now repeat the same procedure only using the critical Mach number instead. Let us begin with the simple total-to-static temperature relationship:

$$T_t = T + \frac{V^2}{2c_p} = T + \frac{V^2}{2\left(\frac{\gamma}{\gamma-1}R\right)}$$

which can be manipulated to yield:

$$T = T_t\left(1 - \frac{\gamma-1}{\gamma+1}M_{cr}^2\right)$$

Similar to what we did before, we have:

$$p = p_t\left(1 - \frac{\gamma-1}{\gamma+1}M_{cr}^2\right)^{\frac{\gamma}{\gamma-1}}$$

Continuity Equation in Terms of the Critical Mach Number

Let us begin with the simplest form of the continuity equation:

$$\dot{m} = \rho V A$$

which, upon algebraic manipulation and the introduction of the critical Mach number, yields:

$$\frac{\dot{m}\sqrt{T_t}}{p_t A} = \sqrt{\frac{2\gamma}{(\gamma+1)}R}\,M_{cr}\left(1 - \frac{\gamma-1}{\gamma+1}M_{cr}^2\right)^{\frac{1}{\gamma-1}}$$

Throughout the remainder of this book part, the preceding equation will be referred to as the "bulky" form of the continuity equation. The equivalent relationship, in terms of the "regular" Mach number (M) can be derived as follows:

$$\frac{\dot{m}\sqrt{T_t}}{p_t A} = \sqrt{\frac{\gamma}{R}}\,M\left(1 + \frac{\gamma-1}{2}M^2\right)^{\frac{1+\gamma}{2(1-\gamma)}}$$

Analysis of DeLaval Nozzle

Across a converging-diverging subsonic-supersonic DeLaval nozzle, the continuity equation can be applied at any arbitrary flow station as follows:

$$\rho V A = \text{constant} = \rho^* V^* A^*$$

where the asterisk signifies conditions where the Mach number, as well as the critical Mach number, attain a value of unity (i.e., at the throat section). Using the definition of the Mach number, the preceding equation can be rewritten as follows:

$$\frac{A}{A^*} = \frac{\rho^*}{\rho}\left(\frac{T^*}{T}\right)^{1/2}\frac{1}{M}$$

where M^* was equated to 1.0, which is true by definition. Since the DeLaval nozzle flow is, by definition, isentropic, all total properties remain constant across the nozzle.

Implementing this principle, we get:

$$\frac{\rho^*}{\rho} = \frac{\rho^*}{\rho_t}\frac{\rho_t}{\rho} = \left\{ \frac{1 + [(\gamma - 1)/2]M^*}{(\gamma + 1)/2} \right\}$$

and,

$$\frac{T^*}{T} = \frac{1 + [(\gamma - 1)/2]M^2}{(\gamma + 1)/2}$$

Through simple substitution, we finally obtain the following important relationship:

$$\frac{A}{A^*} = \frac{1}{M}\left\{ \frac{1 + [(\gamma - 1)/2]M^2}{(\gamma + 1)/2} \right\}^{\frac{\gamma+1}{2(\gamma-1)}}$$

This equation is valid for all Mach numbers, be they subsonic, transonic, or supersonic.

Example 9–8

We need to design a supersonic wind tunnel to operate with air (at a specific-heat ratio of 1.4) at a Mach number of 3.0. If the throat cross-sectional area is 10 cm^2, calculate the area of the test section.

Solution

Through direct substitution in the area-ratio equation (above), and using an exit (or test section) Mach number of 3.0, we get:

$$\frac{A}{A^*} = 4.23$$

which means that the test section cross-sectional area is 42.3 cm^2.

Note

The so-called compressible-flow (also termed isentropic-flow) tables are provided in the Appendix. Utilization of these tables can greatly simplify the computational approach and reduce time consumption.

Example 9–9

A supersonic wind tunnel having a square test section with a side of 15 cm is being designed to operate at a Mach number of 3.0 using air ($\gamma = 1.4$). The static magnitudes of temperature and pressure at the test section are $-20°C$ and 50 KPa, respectively. Calculate the mass flow rate.

Solution

From the tables, we find the area ratio for a Mach number of 3.0 to be 4.23. Thus the area of the throat is:

$$A^* = \frac{225}{4.23} = 53.2 \text{ cm}^2 = 0.00532 \text{ m}^2$$

Now, using the total-to-static flow relationship (misleadingly termed the isentropic relationship in some texts), we obtain the following:

$$T_t = T\left(1 + \frac{\gamma - 1}{2}M^2\right) = 708.0 \text{ K}$$

and

$$p_t = p\left(1 + \frac{\gamma - 1}{2}M^2\right)^{\frac{\gamma}{\gamma-1}} = 1.836 \text{ MPa}$$

Now, by applying the continuity equation in terms of the "regular" Mach number and total properties (stated above), we finally get a mass-flow-rate magnitude of 14.8 kg/s.

9–8 Isentropic Flow Through Varying-Area Passages

Despite the idealism that is implied in the phrase "isentropic flow" (meaning no viscosity effects, no secondary cross-flow migration, no lack of equilibrium at any point in time, etc.), such a flow regime aids in identifying important compressible flow characteristics. For instance, consideration of such a flow field helps a great deal in such cases as the area transition from a station immediately upstream to one immediately downstream from the throat section (where $M = M_{cr} = 1.0$) in a converging-diverging subsonic-supersonic nozzle, better known as a DeLaval nozzle.

To this end, the first task is to re-express the flow-governing equations in their differential forms while inserting the Mach number wherever it is appropriate. To achieve this, let us assume that the flow passage only accelerates or decelerates the flow without turning it. Now let us begin with the simplest form of the continuity equation:

$$\dot{m} = \text{constant} = \rho V A$$

Differentiating both sides, we get:

$$\frac{dA}{A} + \frac{dV}{V} + \frac{d\rho}{\rho} = 0$$

Introducing the so-called Euler's equation (since the flow field is inviscid) in its differential form:

$$V \, dV + \frac{d\rho}{\rho} = 0$$

Since the flow process is isentropic, the following equation describing the flow process path is applicable:

$$\frac{p}{\rho^\gamma} = \text{constant}$$

which, upon differentiation, yields:

$$\gamma \frac{d\rho}{\rho} - \frac{dp}{p} = 0$$

Finally, combining the equation of state with the Mach number definition, we get the following relationship:

$$M^2 = \frac{V^2}{\gamma \frac{p}{\rho}}$$

Combining the preceding expression with the rest, we get the following two relationships:

$$\frac{dA}{A} = \frac{dp}{p}\left(\frac{1 - M^2}{\gamma M^2}\right)$$

$$\frac{dV}{V} = -\left(\frac{1}{\gamma M^2}\right)\frac{dp}{p}$$

The implications of these two equations are perhaps some of the most important in the discipline of dynamics of compressible flow. Specifically speaking, the left-hand side of the first equation will determine, by its sign, the shape of the flow passage that is required to produce a subsonic or supersonic flow field. For instance, a positive sign of dA implies a diverging flow passage and vice versa. The second equation, on the other hand, spells out the fact that a flow passage that causes flow acceleration (i.e., a positive magnitude of dV) will have to simultaneously cause a decline in static pressure (i.e., a negative magnitude of dp) regardless of the Mach number value. Consequences of this, as well as the preceding set of differential equations, are discussed next. In so doing, we will distinguish subsonic from supersonic flow fields. Onset of the transitional sonic state will also be discussed.

Subsonic Flow Fields

This is obviously the simplest and most common case, for it embraces most of the real-life flow applications with which we are familiar. Let us discuss the results of applying the preceding equations to flow-accelerating passages.

In a subsonic nozzle, the condition $dV > 0$ is, by definition, fulfilled. As cited earlier, this will necessarily cause a streamwise static pressure decline. With this in mind, the first equation (above) dictates a flow passage with shrinking cross-sectional area in order for the task to be achieved. This is the familiar case of a subsonic nozzle.

The process of creating a flow-decelerating passage is conceptually similar to the process just discussed, with the exception that we now have a velocity differential dV that is negative. According to the second equation (above), the pressure differential dp will have to be positive, indicating a streamwise rise in static pressure. In this case, the first equation dictates a rising cross-flow area. This is the simple case of a diverging subsonic diffuser.

Supersonic Flow Fields

For the purpose of consistency, we begin with a flow-accelerating passage (i.e., $dV > 0$). Utilizing the second equation above, we see that this very passage will simultaneously cause a streamwise decline in static pressure. Because of its importance in this exercise, let us rewrite the second equation:

$$\frac{dA}{A} = \frac{dp}{p}\left(\frac{1 - M^2}{\gamma M^2}\right)$$

Figure 9–25

Satisfaction of the throat geometrical condition.

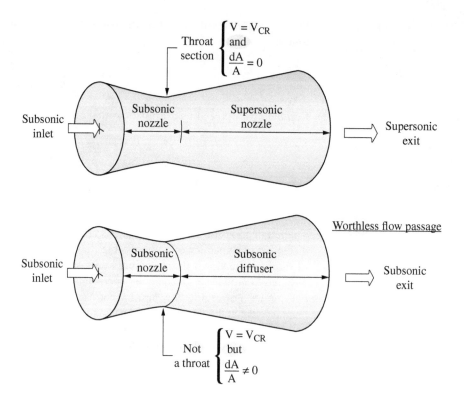

Noting that $dp < 0$, and that the term $\frac{1-M^2}{\gamma M^2}$ is negative for the supersonic nozzle at hand, we now conclude that the area differential dA will have to be positive. This simply means that, contrary to the case of a subsonic nozzle, a supersonic nozzle will have to be divergent.

Aspects concerning the case of a supersonic diffuser can be identified in the same manner. The final result is that such a flow passage will have to be convergent.

The Sonic State

This is an interesting thermophysical state whereby the flow velocity becomes identical to the sonic speed. Such a state cannot be sustained over any finite length of the flow passage. Occurrence of this transitional state would be at the minimum cross-flow area location of the famous subsonic/supersonic converging/diverging DeLaval nozzle (shown in Figure 9–25), where the minimum area section is referred to as the nozzle "throat."

An interesting feature of the throat section which allows the subsonic-to-supersonic flow transition is the following condition:

$$\frac{dA}{A} = 0$$

This means that for the throat to allow such a flow transition, the area change from converging to diverging has to be smooth (Figure 9–25). In other words, for this to occur, the cross-sectional area at this particular location has to be transitionally constant.

Example 9-10   A converging-diverging DeLaval nozzle has an area ratio of 3.0 and exhausts air into a receiver where the pressure is 1 bar. The nozzle is supplied with air at 22°C from a large chamber. At what pressure should the chamber be in order for the nozzle to operate at the design condition? Also calculate the exit velocity.

Solution   Recalling that $\frac{A_3}{A_2}$ is 3.0, we have:

$$\frac{A_3}{A_3^*} = \frac{A_3}{A_2} \frac{A_2}{A_2^*} \frac{A_2^*}{A_3^*} = 3.0$$

Using the isentropic tables:

$$M_3 = 2.64$$

$$\frac{p_3}{p_{t3}} = 0.0471$$

$$\frac{T_3}{T_{t3}} = 0.4177$$

Now:

$$p_1 = p_{t1} = \frac{p_{t1}}{p_{t3}} \frac{p_{t3}}{p_3} p_3 = 21.2 \text{ bars}$$

$$T_3 = \frac{T_3}{T_{t3}} \frac{T_{t3}}{T_{t1}} T_{t1} = 123.2 \text{ K}$$

$$V_3 = M_3 a_3 = 587.0 \text{ m/s}$$

PROBLEMS

9-1   A smooth, flat plate has a total length $L = 0.75$ m. The plate is to be tested in both water and air at a velocity $U = 4.5$ m/s. The temperature of both media is 20°C, and the pressure is the standard sea-level pressure. Determine:

(a) If the flow at the end of the plate will be laminar or turbulent for each fluid.

(b) The air velocity that is necessary to make the flows similar, that is, to have equal magnitude of the Reynolds number (based on the plate length).

9-2   (a) Calculate the total drag per unit width due to friction (D_F) on the smooth, flat plate described in Problem 9-1.

(b) Estimate the boundary layer thickness at the end of the flat plate when it is tested in both air and water.

(c) Compare the values of $\bar{c}_F$ and drag due to the friction experienced by the plate when tested in air and water at the same Reynolds number.

10

Rotating Machinery Fluid Mechanics

Chapter Outline

10–1   Classification of Turbomachinery Components   **150**

10–2   Velocity Diagrams   **152**

10–3   Sign Convention   **154**

10–4   Compressor- and Turbine-Rotor Directions of Rotation   **154**

10–5   Axial Momentum Equation   **156**

10–6   Radial Momentum Equation   **157**

10–7   Cross-Flow Area Variation   **157**

10–8   Total Pressure Variation Across Multistage Turbomachines   **158**

10–9   Variable-Geometry Stators   **159**

10–10   Design-Related Variables   **162**

10–11   Euler's Equation   **166**

10–12   Introduction of the Total Relative Properties   **166**

10–13   Incidence and Deviation Angles   **170**

10–14   Means of Assessing Turbomachinery Performance   **170**

10–15   Supersonic Stator Cascades   **180**

10–16   Sign Convention Governing Radial Turbomachines   **185**

Impeller

Figure 10–1

Nonideal Brayton cycle in a turbojet engine.

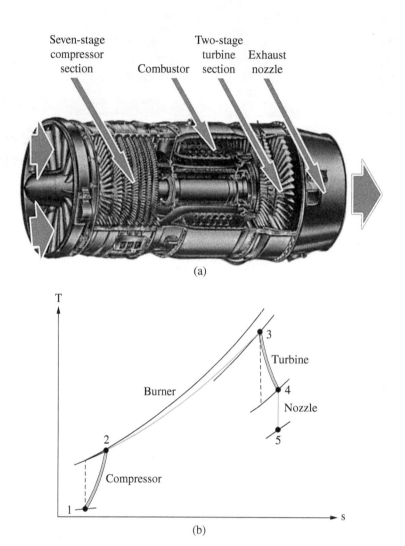

Seven-stage compressor section

Two-stage turbine section

Combustor

Exhaust nozzle

(a)

(b)

This subtopic embraces many real-life applications. Under this umbrella comes such work-absorbing turbomachinery components as compressors and pumps. On the other hand, there is a single term for work-producing components: turbines. A gas turbine is one configuration within this category, while a steam turbine is another. Figure 10–1 shows a simple turbojet engine but not very simple operating conditions. With that, we are referring to the nonreversible flow processes in each of the engine components—namely the compressor, combustor, and turbine components.

10–1 Classification of Turbomachinery Components

Among the many concepts in distinguishing turbomachines from one another comes one which is based on the z-r projection of the so-called meridional flow path. The simplest interpretation of this statement is that turbomachinery components are classified as either of the axial- or radial (centrifugal)-flow type. Figure 10–2 shows two examples of the first category, where the whole stage (stator followed by a rotor) is displayed, while the second presents a typical radial turbine and centrifugal rotors.

Shown in Figure 10–3 are two (meridional and blade-to-blade section) projections of a typical axial-flow turbine. Of these, the latter can be thought of as a constant-radius

Figure 10–2

A typical axial-flow turbine stage.

Figure 10–3

A sketch of an axial-flow turbine stage.

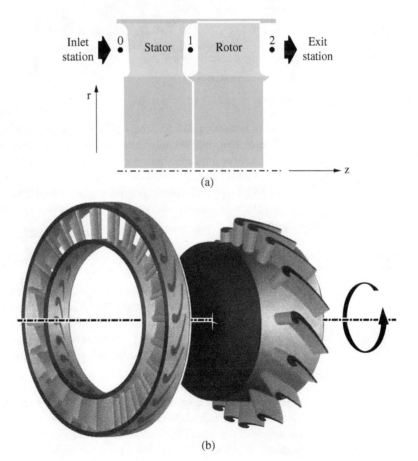

(a)

(b)

Compressor stage

Axial-flow turbomachines

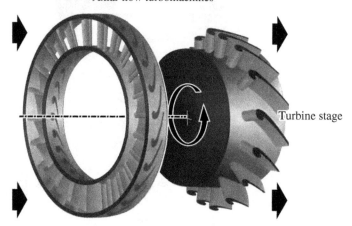

Turbine stage

section of both the stator and rotor blades. Note that the single airfoil, on the bottom, is
in reference to a single blade, while it is by now clear that there are many of these blades
forming what is commonly referred to as a blade cascade (see Figure 10–3). Figures 10–4
and 10–5 show both the compressor and turbine stages of the axial-flow type (Figure
10–4), and the centrifugal/radial type (Figure 10–5). Figure 10–6, however, shows both
the meridional (z-r) and blade-to-blade (z-θ) projections of the same axial-flow turbine
stage.

10–2   Velocity Diagrams

Referring to Figure 10–7, the velocity diagram for a rotating cascade of airfoils is simply
a vectorial relationship relating the velocity an outside stationary observer will monitor
(namely $\vec{V}$) to that monitored by a rotating observer who is "sitting" on the blade and
spinning with it, a situation that is also illustrated in Figure 10–7 (a velocity vector that
is referred to by the relative velocity vector $\vec{W}$), in terms of the rotor speed, as follows:

$$\vec{V} = \vec{W} + (\omega r)\vec{e}_\theta$$

where ω is the rotor speed (in radians/s), r is the radial coordinate, and $\vec{e}_\theta$ is the unit
vector in the tangential direction. The term $(\omega r)\vec{e}_\theta$ can be simply replaced by the vector
$\vec{U}$, which is the so-called "solid-body" rotational velocity, and is always in the r (or
vertical) direction, with a magnitude, as we should expect, that is equal to ωr.

Figure 10–5

Comparison between a radial-flow turbine and centrifugal compressor.

Radial turbine

Centrifugal compressor

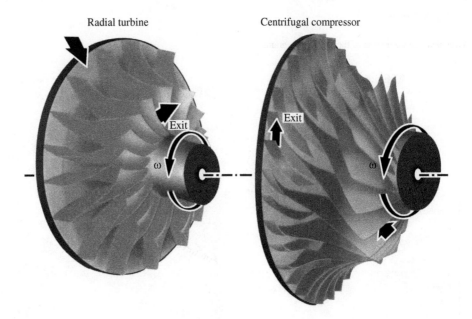

Figure 10–6

Meridional and blade-to-blade views of a turbine rotor.

Meridional flow path

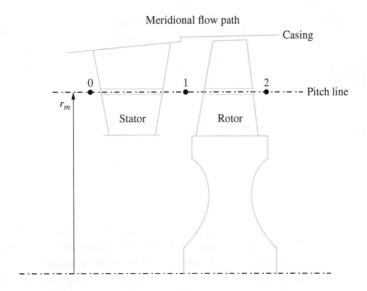

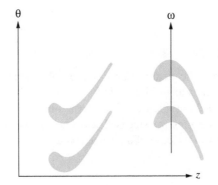

Figure 10–7

Graphical representation of the velocity diagrams.

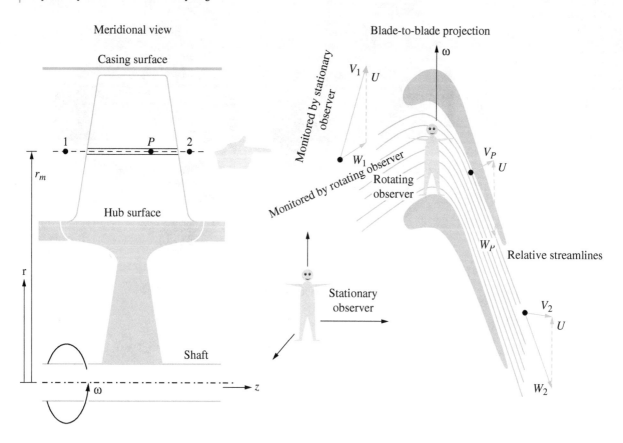

10-3 ## Sign Convention

In presenting the following analyses, it is essential to define the sign convention for both the flow and hardware and stick to it. Figure 10–8 does just that. To be brief, all airfoil (so-called metal) angles, as well as the flow angles, measured from the axial direction, are positive in the direction of rotation and negative otherwise.

10-4 ## Compressor- and Turbine-Rotor Directions of Rotation

Aside from the clear streamwise cross-flow area variation, a major distinction between compressor and turbine airfoil cascades is the direction of rotor rotation. Shown in Figure 10–9 is a side-by-side comparison between these two. In the case of a compressor cascade, the pressure side is clearly leading in the direction of rotation, while the suction side leads in the case of a turbine rotor.

Figure 10–8

Sign convention and rotor direction of rotation.

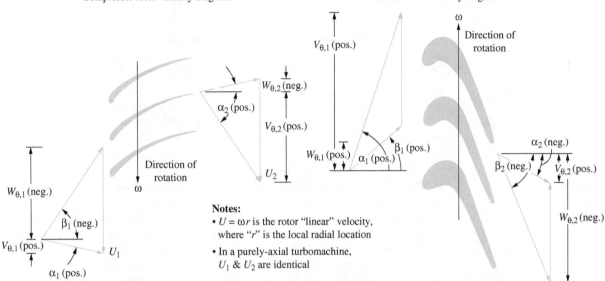

Sign convention: All angles and velocity components in the direction of rotation are positive.

Compressor rotor velocity diagrams

Turbine rotor velocity diagrams

$V_{\theta,1}$ (pos.)

$W_{\theta,2}$ (neg.)

α_2 (pos.)

$V_{\theta,2}$ (pos.)

Direction of rotation

U_2

ω

ω

Direction of rotation

$W_{\theta,1}$ (pos.) α_1 (pos.) β_1 (pos.)

α_2 (neg.)

β_2 (neg.) $V_{\theta,2}$ (pos.)

$W_{\theta,1}$ (neg.)

β_1 (neg.)

$V_{\theta,1}$ (pos.)

U_1

α_1 (pos.)

$W_{\theta,2}$ (neg.)

Notes:
- $U = \omega r$ is the rotor "linear" velocity, where "r" is the local radial location
- In a purely-axial turbomachine, U_1 & U_2 are identical

Figure 10–9

Velocity triangles in a mixed-flow compressor.

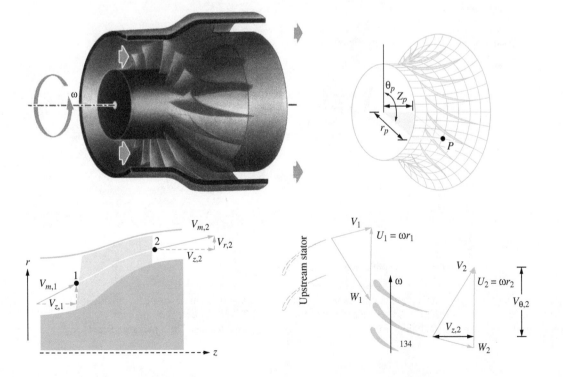

$V_{m,2}$

$V_{r,2}$

$V_{z,2}$

r

$V_{m,1}$

$V_{z,1}$

z

θ_p

Z_p

r_p

P

Upstream stator

V_1

$U_1 = \omega r_1$

W_1

ω

V_2

$U_2 = \omega r_2$

$V_{\theta,2}$

$V_{z,2}$

W_2

134

10–5 **Axial Momentum Equation**

In the general case, the axial velocity component would not really stay constant across the rotor (even though we will always assume it does in the following examples and problems, for it is usually a design intent and preference). Under such circumstances, the axial momentum equation can be expressed as follows:

$$F_z = \dot{m}(V_{z,2} - V_{z,1})$$

The axial force F_z will be absorbed, in part, by a thrust bearing in most cases. Nevertheless, part of this force can be mechanically and/or aerodynamically damaging. Referring to Figure 10–10, for an axial-flow turbine rotor, the rotating blades will indeed move axially in response to this force.

Figure 10–10 shows two different scenarios under the existence of a net axial force on the rotor blades. Movement to the left, in this figure, would close down the tip-to-casing gap. Knowing that this clearance is normally less than 0.5 mm, particularly in high-pressure turbine stages, it is perhaps obvious that this rotor displacement can very well lead to the blades rubbing against the casing, potentially causing a catastrophic mechanical failure. Referring to the other rotor-displacement scenario in the figure, the rotor displacement to the right would open up the tip clearance gap. This would encourage the pressure-to-suction side secondary flow migration over the tip (a mechanism that is referred to as *indirect tip leakage*, which is one of the most aerodynamically degrading loss mechanisms. In cases of low aspect ratio (short) blades, this rotor motion will at least partially unload not only the blade tip section, but will also render ineffective a good percentage of the rotor span (or height) near the tip. This effect will heavily influence the rotor shaft-work extraction capacity.

Figure 10–10

Cross-flow area variation in compressor and turbine rotors.

Mechanical consequences of axial momentum components across an axial-flow turbine rotor

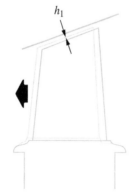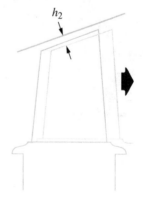

Tip clearance depth h here is exaggerated

Nominal rotor position
and tip clearance h
at room temperature

Axial force in upstream direction:
• Tip clearance closes down, and
• Rubbing problems become likely

Axial force in downstream direction:
• Tip clearance opens up, and
• Tip "leakage" becomes significant

10-6   Radial Momentum Equation

In the event where the radial velocity component varies across the rotor, the following equation becomes applicable:

$$F_r = \dot{m}(V_{r,2} - V_{r,1})$$

The radial force component F_r has little to do with the aerodynamic performance of a turbomachine. This force is normally absorbed as a journal-type load. However, depending on the lubricant flow path and its properties, this force can aggravate a cyclic shaft motion known as a shaft "whirl." This off-center shaft motion can, and historically did, result in a premature catastrophic mechanical failure, such as what reportedly happened in the Space Shuttle Main Engine turbopumps at an early design phase (Baskharone and Hensel, 1991).

10-7   Cross-Flow Area Variation

Knowing that we, in this section, are concerned with (only) subsonic-flow devices, and focusing on a work-absorbing turbomachinery component such as an air compressor, the blade-to-blade passage is diffuser-like. On the other hand, the blade-to-blade cascade unit in a turbine is shaped like a nozzle (Figure 10–11). The cross-flow blade-to-blade passage shape is crucial as far as the flow behavior is concerned. Figure 10–12 shows the typical streamline pattern in both turbine and compressor cascades, together with the suction- and pressure-side static pressure variation. In the turbine cascade's case, we see that the streamlines remain attached and fully guided by the airfoil shape, a case that is

Figure 10–11

Rotor displacement due to the change in axial momentum.

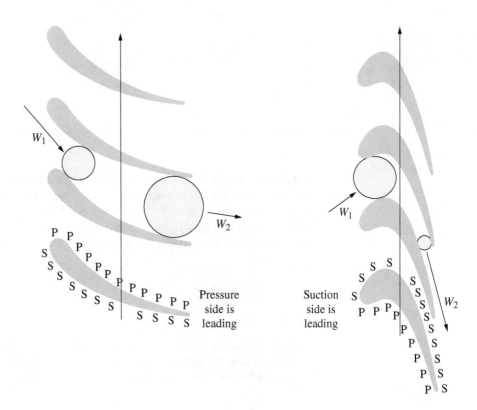

Figure 10–12

Favorable and unfavorable pressure gradients.

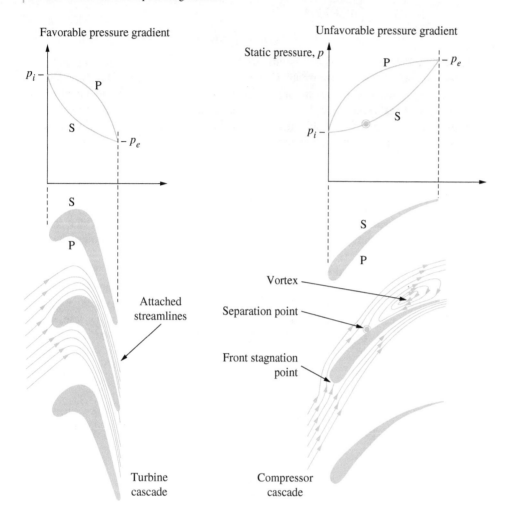

commonly referred to as one of favorable pressure gradient (i.e., steamwise-declining static pressure). As for a compressor cascade, we see that, provided that the static-to-static pressure ratio is sufficiently high, a point of flow separation might be encountered with, perhaps, no opportunity for reattachment. This case is referred to as that of an unfavorable pressure gradient, meaning the case where the streamwise pressure is rising (Figure 10–12).

10–8 Total Pressure Variation Across Multistage Turbomachines

Shown in Figure 10–13 are a two-stage compressor and a two-stage turbine. Assuming an adiabatic flow throughout, we see that the total temperature across the stators remains constant, while the total pressure suffers a decline in each case due to the irreversibilities

Figure 10–13

Multiple staging in axial-flow compressors and turbines.

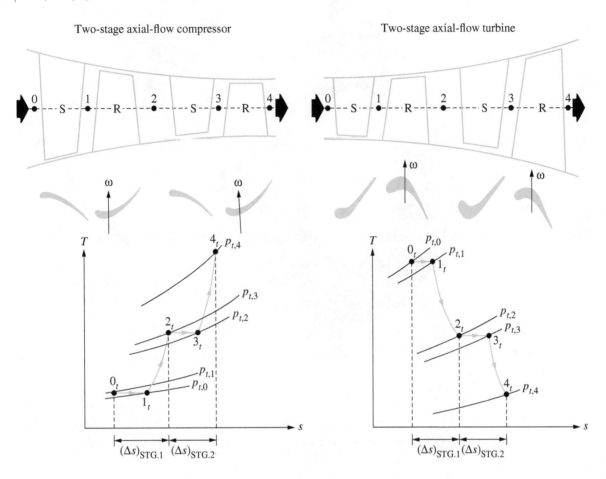

involved. Note that, as expected, the entire thermodynamic process, in each case, proceeds from left to right in accordance with the second law of thermodynamics, specifically the production of entropy principle.

10–9 Variable-Geometry Stators

Figure 10–14 shows the mechanism with which a turbine stator is opened up (in terms of the minimum cross-flow area) or closed down. This technique involves full rotation of the stator cascade, while that in Figure 10–15 (termed the articulating trailing edge) involves rotation of only the region downstream from the blade-to-blade passage throat (this term is loosely used to signify the minimum cross-flow section and not the rigorous aerodynamic throat where the Mach number is unity) down to the trailing-edge plane. Of course there are other ways of varying the stator cross-flow area, and those involve recontouring the endwalls through specially designed inserts.

Figure 10–14

Mechanism of a
rotating-vane
configuration.

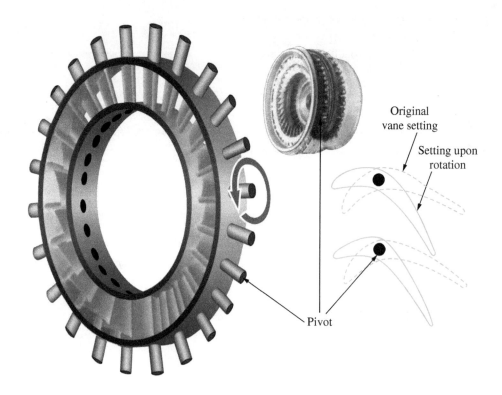

Original
vane setting

Setting upon
rotation

Pivot

Figure 10–15

Total rotation versus
articulating trailing edge.

Entire vane rotation mechanism

Articulating trailing edge mechanism

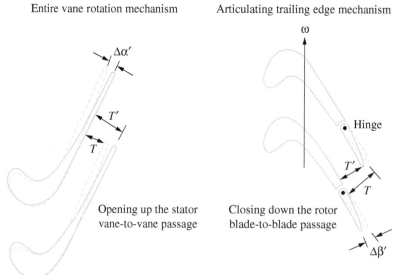

$\Delta\alpha'$

T'

T

Opening up the stator
vane-to-vane passage

ω

Hinge

T'

T

Closing down the rotor
blade-to-blade passage

$\Delta\beta'$

Example 10–1 Consider the variable-geometry axial-flow turbine stator shown in Figure 10–16, where the stator vanes (a term used to refer to stationary cascade blades) are rotatable. Initially, the vane setting was just sufficient to choke the stator flow. In this and all other setting angles, the stator-exit total temperature $T_{t,1}$ was held constant at 1150 K.

Beginning at the choking state, the stator vanes were repeatedly rotated open, with the objective of reaching a final stator-exit critical Mach number of 0.5 in increments intended to reduce the critical Mach number by 0.1 every time. For each of these vane settings, calculate the critical versus the traditional Mach number. In doing so, you may assume the following:

- Full flow "guidedness" by the stator vanes at all setting angles
- A specific heat ratio γ of 1.33

Solution In the following, we will repeatedly set the critical Mach number value. We will then proceed to calculate the "traditional" Mach number M. First, we calculate the critical velocity as follows:

$$V_{\mathrm{cr},1} = \sqrt{\frac{2\gamma}{\gamma+1} R T_{t,1}} = 613.8 \,\mathrm{m/s}$$

We can subsequently move to calculate the critical Mach number $M_{\mathrm{cr},1}$ as follows:

- Set $M_{\mathrm{cr},1} = 1.0$:

$$V_1 = M_{\mathrm{cr},1} V_{\mathrm{cr},1} = 613.8 \text{ m/s}$$

$$T_1 = T_{t,1} - \frac{V_1{}^2}{2c_p} = 987.1 \text{ K}$$

$$a_1 = \sqrt{\gamma R T_1} = 613.8 \text{ m/s}$$

$$M_1 = \frac{V_1}{a_1} = 1.0 \text{ (just as expected)}$$

- Set $M_{\mathrm{cr},1} = 0.9$:

$$V_1 = M_{\mathrm{cr},1} V_{\mathrm{cr},1} = 552.4 \text{ m/s}$$

$$T_1 = 1018.1 \text{ K}$$

$$a_1 = 623.4 \text{ m/s}$$

$$M_1 = 0.88$$

Repeating the same procedure, we obtain the following:

$$M_{\mathrm{cr},1} = 0.80 \ldots . M_1 = 0.78$$

$$M_{\mathrm{cr},1} = 0.70 \ldots . M_1 = 0.67$$

$$M_{\mathrm{cr},1} = 0.60 \ldots . M_1 = 0.57$$

$$M_{\mathrm{cr},1} = 0.50 \ldots . M_1 = 0.47$$

Figure 10–16

Stator vane rotation in a
variable-geometry stator.

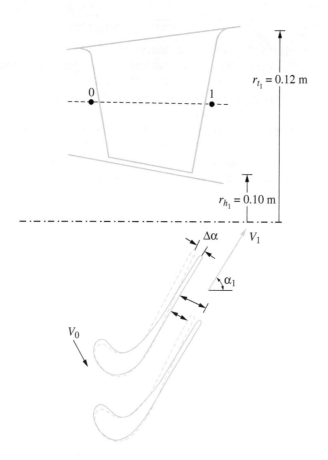

10–10 Design-Related Variables

Stage Flow Coefficient

This important variable is defined, for a symmetrical meridional flow-path projection (Figure 10–17) and for a constant axial velocity component V_z, as follows:

$$\phi = \frac{V_z}{U_m} = \frac{V_z}{\omega r_m}$$

where ω is the rotor speed (in radians/s), and r_m is the mean radius. In this definition, as stated earlier, the axial velocity component is assumed to be constant. In reality, this goes beyond being an assumption to actually being a design intent for changes in ϕ or, equivalently, V_z would physically move the rotor assembly in the axial direction, causing mechanical and/or aerodynamic problems as stated earlier.

Within a mass conservation framework, let us assume a constant axial-velocity component, together with the stage-wise density decline, create a divergent meridional flow path. To explain this fact, let us focus on the stator subdomain in Figure 10–18, assuming an isentropic flow for simplicity. In this case, the variables p_t, T_t, and V_{cr} will remain constant across the stator. Now, applying the continuity equation at a typical station within the stator region, we get:

$$\dot{m} = \rho V_z A = \rho V_z (2\pi r_m h)$$

Figure 10–17

Justification of annulus
increase over an axial
stator.

Diverging flow path of axial-flow turbines
due to the streamwise decline in "static" density

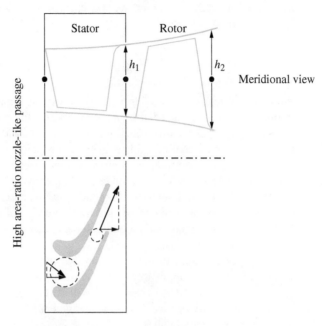

Meridional view

High area-ratio nozzle-like passage

Across the stator:
With declining density
and desire for a constant
V_z the continuity
equation requires a
divergent meridional
flow path.

Similar reasoning
applies to the rotor

Figure 10–18

A purely axial stage
versus a diverging-
endwall axial stage.

Theoretical vs. practical axial-flow turbines
and design consequences

Axial velocity grows due to the decline
in static density

Axial velocity remains constant due to
the increase in cross-flow area

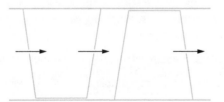

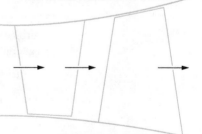

Configuration (A)
Theoretically axial flow path

Configuration (B)
Predominantly axial flow stream
contoured-endwall stage config.
is often preferable

where r_m is the mean radius and h is the annulus height (or the local stator span) in the meridional projection. With p_t and T_t being constant across the stator (isentropic flow), the total density ρ_t will also be stator-wise constant by simple application of the equation of state. At this point, we can express the static density ρ in the continuity equation (above) as follows:

$$\rho = \rho_t \left[1 - \frac{\gamma - 1}{\gamma + 1} \left(\frac{V}{V_{\text{cr}}} \right)^2 \right]^{\frac{1}{\gamma - 1}}$$

which, by reference to the rising velocity across the "nozzle-like" stator in the figure, simply means that the static density ρ will suffer a decline across the stator. Now referring, once again, to the continuity equation, the annulus height h will have to rise across the stator in order to make up for this density decline and for the continuity equation to be satisfied. As for the common misconception that the endwalls (hub and casing surfaces) of an axial-flow turbine stage should be horizontal in the meridional view, Figure 10–18 illustrates that the axial velocity component, in this case, will hardly remain constant, neither across the stator nor across the rotor, with proof of the latter being practically identical to the preceding derivation, with the exception that the so-called total-relative flow properties would be used instead.

An important point to make here is that the preceding observations and a similar derivation will be applicable to compressor stages, with the final result being that the meridional projection, in this case, will be convergent, meaning that the annulus height will gradually decline. The flow coefficient for a well-designed turbomachine is somewhere between 0.4 and 0.6.

Stage Work Coefficient

This variable is defined as follows:

$$\psi = \frac{w_s}{U_m{}^2} = \frac{V_{\theta,1} - V_{\theta,2}}{U_m}$$

where w_s is the (input or output) specific shaft work. We can rewrite the preceding expression as follows:

$$\psi = \frac{W_{\theta,2} - W_{\theta,1}}{U_m} = \frac{V_z}{U_m} (\tan \beta_2 - \tan \beta_1)$$

Normally a ψ magnitude around unity is optimum.

Stage Reaction

In order to simplify the introduction of this variable, we will limit ourselves to axial-flow turbine and compressor stages. Referring to Figure 10–19, the term *stage reaction R* is generally defined as follows:

R = [rotor static enthalpy change]/[total (static and dynamic) enthalpy change]

which, in terms of the relevant variables, can be expressed as follows:

$$R_{\text{compressor}} = \frac{0.5\left[\left(W_1{}^2 - W_2{}^2 \right) \right]}{0.5\left[\left(W_1{}^2 - W_2{}^2 \right) + \left(V_2{}^2 - V_1{}^2 \right) \right]}$$

$$R_{\text{turbine}} = \frac{0.5\left[\left(W_2{}^2 - W_1{}^2 \right) \right]}{0.5\left[\left(W_2{}^2 - W_1{}^2 \right) + \left(V_1{}^2 - V_2{}^2 \right) \right]}$$

Figure 10–19

Special stage reaction magnitudes in axial turbomachines.

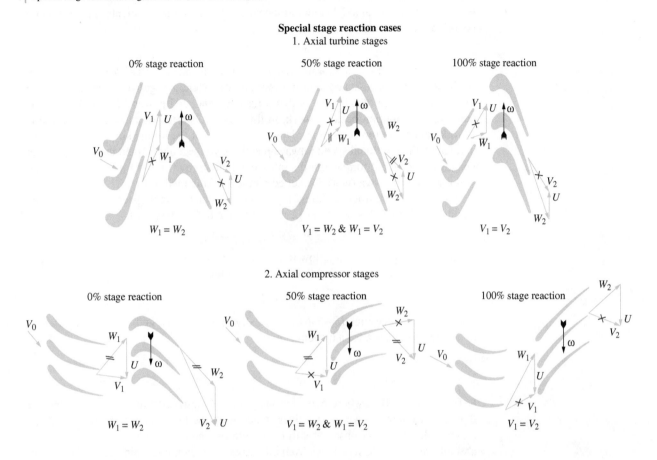

Special stage reaction cases
1. Axial turbine stages

2. Axial compressor stages

In view of the definition $h_t = h + 0.5V^2$, with the second term meaning the dynamic enthalpy, it is perhaps easy to identify that the term $0.5\Delta W^2$ stands for the static part of enthalpy change. Figure 10–19 shows both the rotor shape and velocity diagrams for both the turbine and compressor categories for the special stage-reaction magnitudes of zero, 50 percent and 100 percent. Examination of this figure reveals the following:

- For a compressor stage, the magnitude of reaction is a reflection of how "diffuser-like" the rotor blade-to-blade passage is.

- For a turbine stage, the magnitude of reaction is a reflection of how "nozzle-like" the rotor blade-to-blade passage is.

The optimum magnitude of reaction ranges from 50 percent to 60 percent on the basis of a mean-radius flow analysis. The danger, however, would come from the hub and/or tip reaction values. This is true in the sense that the velocity triangles at these points will be vastly different from those at the mean radius. At least we know that the solid-body velocity ($U = \omega r$), due to the rotor speed ω, will be a function of the radial location r.

Euler's Equation

We assume that the turbomachinery components' flow field is totally adiabatic. Referring to Figure 10–7, for an axial-flow turbine stage, the tangential momentum conservation equation can be written as follows:

$$\tau = \dot{m} r_m (V_{\theta,2} - V_{\theta,1})$$

where τ is the shaft-exerted torque, and r_m refers to the mean radius. Proper substitution (in light of the above-stated sign convention) will yield a negative magnitude for the torque. The source of this sign is that a turbine-generated power (also shaft work) is negative by definition because work, in this case, is exerted by the fluid (a turbine is a work-generating turbomachine). Now if you think of the torque simply as the power divided by the rotational speed (always positive), you can understand the source of that negative sign for the torque. On the other hand, the torque sign for a compressor stage will always be positive (work is exerted on the working medium).

Now, let us multiply the foregoing equation by ω, the rotor speed in radians/s. The result will naturally be the power generated by the turbine stage, i.e.,

$$P = \dot{m} U_m (V_{\theta,2} - V_{\theta,1})$$

Finally, dividing by the mass flow rate $\dot{m}$, we get an expression of the specific (i.e. per kilogram) shaft work delivered to the shaft as follows:

$$w_s = U_m (V_{\theta,2} - V_{\theta,1})$$

The preceding three equations have historically been lumped together in what is referred to as Euler's equation for a turbomachine.

Introduction of the Total Relative Properties

Refer to Figure 10–7, where two (stationary and rotating) observers are monitoring the flow field from their own perspectives. Aside from the flow static properties (T, p, ρ, ...etc.), the stationary observer will be monitoring the magnitude and direction of the absolute velocity V as it changes from inlet to exit. As for the spinning observer, it is the relative velocity W. As was earlier explained, it is generally the process of arithmetically adding, for instance, the static and dynamic components of static temperature (T) that gives rise to the (absolute) total temperature T_t. The dynamic temperature component in this case is $\frac{V^2}{2c_p}$, i.e., dependent on the absolute velocity magnitude. By the same token, should we add the quantity $\frac{W^2}{2c_p}$ to the static temperature (T), we will consistently arrive at the total *relative* temperature, $T_{t,r}$. Therefore, the static properties, and we are here correctly generalizing, are never a function of the frame of reference, be that stationary or rotating. In light of the above, the total relative temperature can be written as follows:

$$T_{t,r} = T + \frac{W^2}{2c_p} = \left(T_t - \frac{V^2}{2c_p} \right) + \frac{W^2}{2c_p} = T_t + \frac{W^2}{2c_p} - \frac{V^2}{2c_p}$$

While the total temperature (T_t) must drop across the turbine rotor (for shaft work to be generated), the total relative temperature (for an assumingly constant c_p) will always remain constant should the flow remain adiabatic.

The preceding relationship applies at any point in the flow field. The way to attain the total relative pressure is simply to relate it to the static magnitude at the same point, i.e.,

$$p_{t,r} = p \left(\frac{T_{t,r}}{T} \right)^{\frac{\gamma}{\gamma-1}}$$

While there is no streamwise variation (across the rotor) of the total relative temperature for an adiabatic flow, there has to be, from a second-law-of-thermodynamics viewpoint, a finite decline of the total relative pressure between the inlet and exit stations, which is indicative of the "irreversibilities" of the rotor, e.g., the presence of friction. The so-called relative critical Mach number $M_{cr,r}$ is the counterpart of the critical Mach number $M_{cr} = \frac{V}{V_{cr}}$, which was discussed earlier in this section. This new variable is defined in the same manner as M_{cr} as follows:

$$M_{cr,r} = \frac{W}{W_{cr}}$$

where W_{cr} is the relative critical velocity, defined as follows:

$$W_{cr} = \sqrt{\frac{2\gamma}{\gamma+1} RT_{t,r}}$$

Finally, a useful relationship which yields the total relative pressure will be given next. Note that the states s (for static), t (for total), and t_r (for total relative) exist on the same constant-entropy (vertical) line, and that one can always visualize an isentropic relationship going from one state to another:

$$p_{t,r} = p_t \left(\frac{T_{t,r}}{T_t} \right)^{\frac{\gamma}{\gamma-1}}$$

Figure 10–20 shows, in particular, the total relative temperature and total relative pressure variation across a compressor as well as a turbine stage. Realizing that the entropy production, by reference to the same figure, can be cast on a static-to-static, total-to-total, or total relative-to-total relative basis, it becomes logical to see, in this figure, that the total relative pressure suffers a decline across the rotor, just like the total pressure does across the stator.

| Example 10–2 | Figure 10–21a shows the off-design inlet velocity diagram for an axial-flow-turbine rotor. Use the provided loss versus incidence angle chart to calculate the total relative pressure at station X, which is just inside the airfoil leading edge, assuming the specific heat ratio γ to be 1.33. |

| Solution | Careful examination of the relative flow/airfoil situation reveals the rotor incidence angle is as follows: |

$$i_R = \beta_1 - \beta_1{}' = 30 - 50 = -20 \text{ degrees}$$

As a general rule, should the flow (or relative flow) streamlines strike the airfoil pressure side first, then the incidence angle would be positive, and vice versa. Now using the given chart, we can read off the percentage of total-relative-pressure loss as follows:

$$\frac{(\Delta p_{t,r})_{\text{incidence}}}{p_{t,r,1}} = 0.12$$

Furthermore, we have:

$$V_1 = \frac{V_z}{\cos(60°)} = 640.0 \text{ m/s}$$

$$W_1 = \frac{V_z}{\cos(30°)} = 369.5 \text{ m/s}$$

Figure 10–20

Invariance of the total relative temperature.

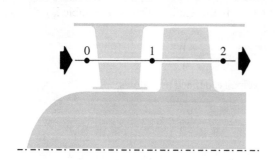
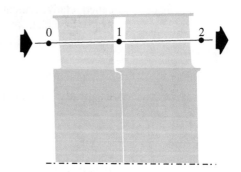

Adiabatic stator & adiabatic rotor

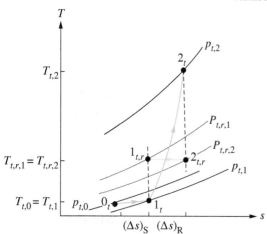
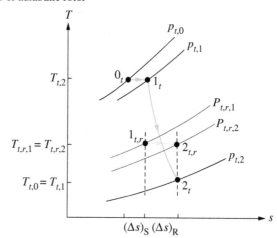

Now, we compute the rotor inlet total-relative temperature and pressure as follows:

$$T_{t,r,1} = T_{t,1} - \left(\frac{V_1^2 - W_1^2}{2c_p} \right) = 1281.9 \text{ K}$$

$$p_{t,r,1} = p_{t,1} \left(\frac{T_{t,r,1}}{T_{t,1}} \right)^{\frac{\gamma}{\gamma-1}} = 8.414 \text{ bars}$$

Now, we proceed to compute $p_{t,r,X}$ as follows:

$$0.12 = \frac{p_{t,r,1} - p_{t,r,2}}{p_{t,r,1}}$$

which yields:

$$p_{t,r,X} = 7.4 \text{ bars}$$

Figure 10–21

(a) Input variables for Example 10–2. (b) Stator-vane and rotor-blade incidence angles.

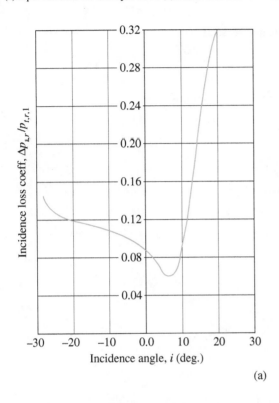

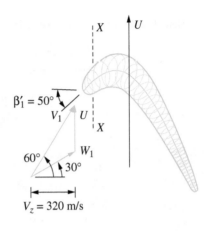

$p_{t,1} = 12$ bars
$T_{t,1} = 1400$ K

(a)

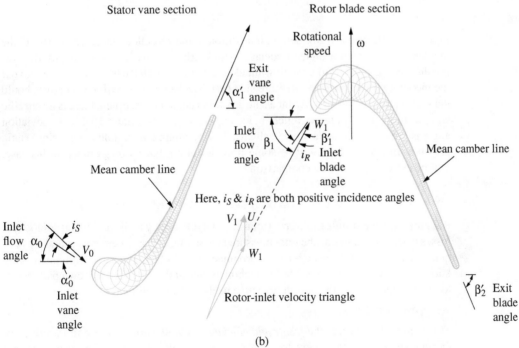

(b)

Figure 10–21
(*continued*)

(c) Incidence and deviation angles in axial-flow compressors.

Definition of the incidence and deviation angles in a compressor stage

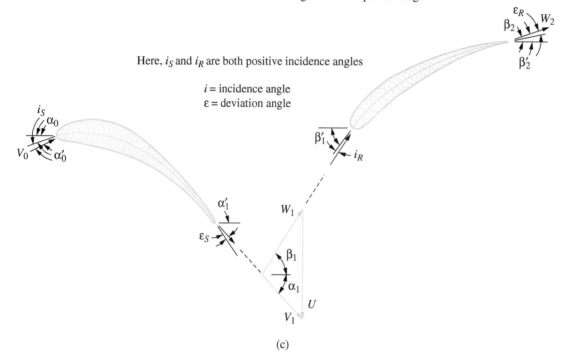

Here, i_S and i_R are both positive incidence angles

i = incidence angle
ε = deviation angle

(c)

10–13 Incidence and Deviation Angles

Figures 10–21b and 10–21c define, in addition to the so-called "mean camberline," the incidence and deviation angles associated with a turbine then compressor stages. Regarding the leading edge region, the flow incidence angle, as stated earlier, will be positive if the incoming flow meets the pressure side first. The same angle will be negative should the suction side be met first. As far as the deviation angle is concerned, there is no specific sign for it, for it will always reflect blade unguidedness. In Figure 10–20, no deviation angle is presented since the flow is always flowing along a favorable (decreasing) static pressure. In reality, there will be some flow deviation, but its magnitude, in this case, will be extremely small by comparison.

10–14 Means of Assessing Turbomachinery Performance

Ranging from the simple isentropic (total-to-total) efficiency to the individual component losses in such variables as the total pressure and kinetic energy, there are several ways of quantifying the performance level of turbomachines. In fact, and aside from universally known performance assessors, turbomachinists can, and do, define their own parameters to do the same should the design objective itself be nontraditional.

Total-to-Total Efficiency

This is normally termed the *isentropic efficiency*, a well-known phrase in introductory thermodynamics texts, except that the properties at the end states, in such texts, are usually left to simply imply static properties. In our case, the efficiency is defined on the basis of total properties at both the inlet and exit stations, as follows:

$$(\eta_{t-t})_T = \frac{\Delta h_{t,\text{actual}}}{\Delta h_{t,\text{ideal}}} \dots \text{for a turbine}$$

$$(\eta_{t-t})_C = \frac{\Delta h_{t,\text{ideal}}}{\Delta h_{t,\text{actual}}} \dots \text{for a compressor}$$

Under an adiabatic-flow assumption, the turbine efficiency is further simplified and graphically represented in Figure 10–22. In the case of a compressor, it is the ideal

Figure 10-22

Total-to-total versus total-to-static efficiencies.

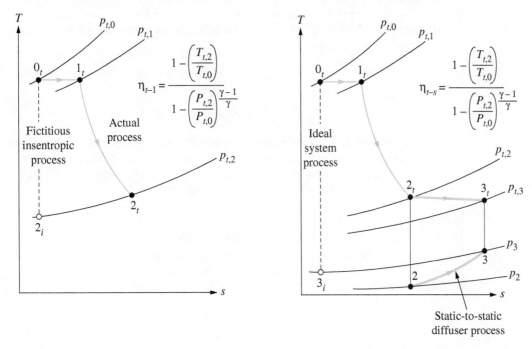

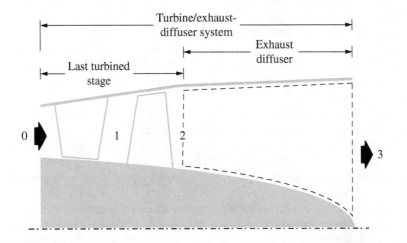

(isentropic-path) shaft work that constitutes the numerator, a magnitude with which the compressor would yield the same total-to-total pressure ratio.

Referring to the same figure, a beginner may make a usually undetectable mistake by simply stating the phrase "*assuming* the ideal exit pressure to be equal to the actual exit pressure, ..." The fact of the matter is that the actual (entropy-producing) process will exist naturally first. As a matter of definition, we next construct a *fictional* isentropic process between the same inlet and exit total pressures for the purpose of comparison. There is, accordingly, no room here for any assumption, as this is simply a matter of implementing a definition.

Total-to-Static Efficiency

Restricting the discussion to the stage exit station of an entire turbine or compressor section, the mere existence of a finite nonzero swirl angle is indicative of inefficiency.

For one thing, a nonzero swirl-velocity component simply means a waste of what would have, otherwise, been a shaft work contributor in light of Euler's equation of angular-momentum conservation. Moreover, the nonzero swirl angle at such location will give rise to higher magnitudes of friction-related losses, for the streamlines over the endwalls will, in this case, be elongated as the fluid particles will have to undergo several tangential trips from one station to the other, and thus longer surfaces for the boundary layer to grow on.

In any turbomachinery stage, it is both desirable and nearly possible to produce a zero exit swirl velocity, at least under the design-point operation mode. Although the nonzero exit swirl angle is itself a loss indicator, a separately defined variable termed the *total-to-static efficiency* is used to characterize, in particular, the turbine performance in turboshaft (e.g., turboprop) engines. Such engines typically end with exhaust diffusers which, by definition, are unfavorable-pressure-gradient passages. The static pressure recovery in this case is highly sensitive to the diffuser's inlet swirl angle. Note that the turbine-exit static pressure, in this case, is usually less than the ambient pressure, which would cause a highly damaging flow reversal to reach the turbine section should the diffuser-wise static pressure rise be offset by a total pressure loss.

Referring to Figure 10–22, the total-to-static efficiency in the case of a turboprop engine is expressed as follows:

$$\eta_{t-s} = \frac{1 - \left(\frac{T_{t-2}}{T_{t-1}}\right)}{1 - \left(\frac{p_{s,2}}{p_{t,1}}\right)^{\frac{\gamma-1}{\gamma}}} \ldots \text{ for a turbine}$$

A similar expression of the same variable can be obtained for a compressor from the expression for the total-to-total efficiency by simply replacing $p_{t,\text{exit}}$ with p_{exit}. Comparing the total-to-static efficiency expressions to their total-to-total counterparts, we get to realize that the former will always be smaller than the latter.

Kinetic Energy Loss Coefficient

This is a popular loss coefficient, for it applies to bladed and unbladed turbomachinery components. Although the definition applies to rotating cascades (as based on the relative velocity W), it is almost exclusively a primary performance assessor for stators. The coefficient (represented as $\bar{e}$ in Figure 10–23) is defined as follows:

$$\bar{e} = 1 - \frac{V_{\text{actual}}^2}{V_{\text{ideal}}^2}$$

Figure 10–23

Definition of kinetic
energy loss coefficient
and total pressure loss
coefficient.

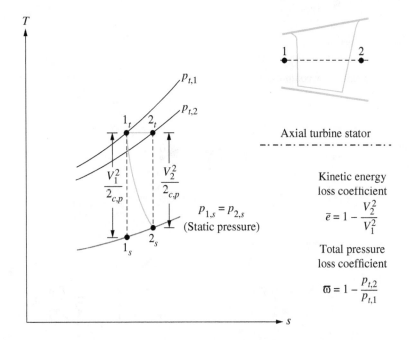

where V_{actual} is the actual exit velocity and V_{ideal} is the ideal velocity that is associated with a component-wise isentropic flow process.

Total Pressure Loss Coefficient

Based on its name alone, one can detect that this variable applies to stationary components, although this is not necessarily the case. The reason is that the total pressure will certainly change across the rotor subdomain should any shaft work interaction take place. The popularity of this coefficient is due to the fact that it is an explicit statement of the real-life decline in total pressure, one of the most physically meaningful properties and the most traceable on the T-s diagram (Figure 10–23) because of irreversibilities (e.g., friction effects) within the component at hand. This coefficient (denoted $\bar{\omega}$) is defined as follows:

$$\bar{\omega} = \frac{\Delta p_t}{p_{t,\text{in}}} = 1 - \frac{p_{t,\text{ex}}}{p_{t,\text{in}}}$$

The same concept applies to a rotor subdomain once the total relative pressures are used instead. Note that such a definition will exclusively apply to axial-flow turbomachinery components. This is due to the fact that the mere change in radius across the so-called master streamline, in radial (or centrifugal) turbomachines, will cause the total relative pressure to change.

Entropy-Based Efficiency

As the phrase might imply, this variable is based on the second law of thermodynamics. Unfortunately, this variable is unpopular in turbomachinery applications despite its unique versatility, for it is as applicable to rotors as it is to stators. In fact, this variable even applies to the entirety of multistage compressor and turbine sections, as shown in Figure 10–24. Moreover, the variable remains applicable to any flow process that is

Figure 10–24

Accumulative entropy production.

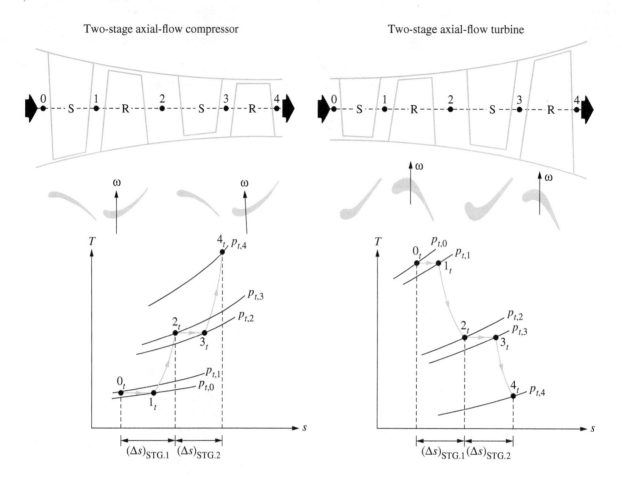

Two-stage axial-flow compressor

Two-stage axial-flow turbine

defined (or electively viewed) on a total-to-total, total-to-static, or static-to-static basis. Let us begin with the loss coefficient $\bar{q}$ defined as follows:

$$\bar{q} = 1 - e^{\frac{-\Delta s}{R}}$$

where Δs is the sum of all specific-entropy productions (Figure 10–24) for all components under consideration.

An efficiency-like parameter η_s is also common in the gas turbine industry. This is defined as follows:

$$\eta_s = e^{\frac{-\Delta s}{c_p}}$$

Note that the magnitude of η_s will always lie between zero and 100 percent, with the latter corresponding to the ideal case of zero entropy production (i.e., a perfectly isentropic flow process).

Figure 10–25

Input variables for
Example 10–3.

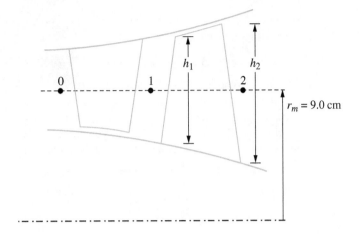

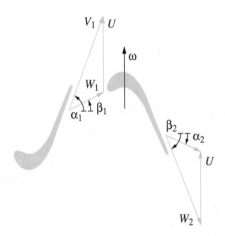

Example 10–3

Figure 10–25 shows an axial-flow turbine stage, together with its major dimensions.
The stage is operating under the following conditions:

- Rotor-inlet total pressure = 12.5 bars
- Rotor-inlet total temperature = 1285.0 K
- Rotor speed = 46,000 r/min
- Rotor-inlet critical Mach number = 0.86
- Rotor-inlet swirl angle = 66.0°
- Rotor total-to-total efficiency = 89 percent
- Mass flow rate = 13.4 kg/s
- Stage reaction = 50 percent
- Adiabatic flow throughout the stage
- Constant axial velocity component

Assuming a specific heat ratio γ of 1.33, calculate the rotor-exit annulus height h_2 by applying the continuity equation:

(a) In the stationary frame of reference

(b) In the rotating frame of reference

Solution   Since the stage reaction is 50 percent, it then suffices to compute the rotor-inlet velocity-triangle variables, as the exit velocity triangle is simply a mirror image of the former (this is the reason such a stage is often termed *symmetric*).

$$V_{cr,1} = \sqrt{\left(\frac{2\gamma}{\gamma + 1}\right) RT_{t,1}} = 648.9 \text{ m/s}$$

$$V_1 = M_{cr,1} V_{cr,1} = 558.0 \text{ m/s}$$

$$V_{z,1} = V_1 \cos \alpha_1 = 227.0 \text{ m/s}$$

$$V_{\theta,1} = V_1 \sin \alpha_1 = 509.8 \text{ m/s}$$

$$U_m = \omega r_m = 433.5 \text{ m/s}$$

$$W_{\theta,1} = V_{\theta,1} - U_m = 76.3 \text{ m/s}$$

$$W_1 = \sqrt{W_{\theta,1}{}^2 + V_z{}^2} = 239.5 \text{ m/s}$$

$$\beta_1 = \tan^{-1} \frac{W_{\theta,1}}{V_z} = 18.6°$$

$$\rho_1 = \frac{p_{t,1}}{RT_{t,1}} \left[1 - \left(\frac{\gamma - 1}{\gamma + 1}\right) M_{cr,1}{}^2\right]^{\frac{1}{\gamma - 1}} = 2.42 \text{ kg/m}^3$$

Let us now apply the continuity equation at the stator exit station, in order to calculate h_1:

$$h_1 = \frac{\dot{m}}{\rho_1 V_z 2\pi r_m} = 4.3 \text{ cm}$$

Moving to the rotor exit station, we have:

$$V_2 = W_1 = 239.5 \text{ m/s (50\%-reaction stage)}$$

$$W_2 = V_1 = 558.0 \text{ m/s}$$

$$\alpha_2 = -\beta_1 = -18.6°$$

$$\beta_2 = -\alpha_1 = -66.0°$$

$$V_{\theta,2} = -W_{\theta,1} - 76.3 \text{ m/s}$$

Now we apply Euler's equation as follows:

$$T_{t,2} = T_{t,1} - \frac{U_m}{c_p}(V_{\theta,1} - V_{\theta,2}) = 1065.3 \text{ K}$$

Also:

$$V_{cr,2} = \sqrt{\left(\frac{2\gamma}{\gamma+1}\right)RT_{t,2}} = 590.8 \text{ m/s}$$

$$M_{cr,2} = \frac{V_2}{V_{cr,2}} = 0.405$$

$$T_{t,r,2} = T_{t,r,1} = T_{t,1} + \frac{W_1^2 - V_1^2}{2c_p} = 1175.2 \text{ K (adiabatic flow in}$$
$$\text{an axial-flow stage)}$$

$$W_{cr,2} = \sqrt{\left(\frac{2\gamma}{\gamma+1}\right)RT_{t,r,2}} = 620.5 \text{ m/s}$$

$$M_{cr,r,2} = \frac{W_2}{W_{cr,2}} = 0.899$$

Applying the definition of total-to-total efficiency, we get

$$\eta_{t-t} = 0.89 = \frac{1 - (T_{t,2}/T_{t,1})}{1 - (p_{t,2}/p_{t,1})^{\frac{\gamma-1}{\gamma}}}$$

which yields:

$$p_{t,2} = 5.29 \text{ bars}$$

$$p_{t,r,2} = p_{t,2}\left(\frac{T_{t,r,2}}{T_{t,2}}\right)^{\frac{\gamma}{\gamma-1}} = 7.86 \text{ bars}$$

Part a:

Let us now compute the rotor-exit annulus height, h_2, by applying the continuity equation in the stationary frame of reference using, of course, the absolute thermophysical properties [note that $A_2 = (2\pi r_m h_2)\cos\alpha_2$].

Substituting in what was earlier termed the "bulky" version of the continuity equation (in terms of total conditions and critical Mach number), we get:

$$h_2 = 6.48 \text{ cm}$$

Part b:

We now apply the continuity equation in the rotating frame of reference, using the relative flow thermophysical properties this time around, as follows:

$$\frac{\dot{m}}{\sqrt{T_{t,r,2}}}p_{t,r,2}(2\pi r_m h_2)\cos\beta_2 = \sqrt{\frac{2\gamma}{(\gamma+1)R}}M_{cr,r,2}\left[1 - \frac{\gamma-1}{\gamma+1}M_{cr,r,2}^2\right]^{\frac{1}{\gamma-1}}$$

which yields:

$$h_2 = 6.48 \text{ cm} \ldots. \text{ which is precisely the same result as in Part a.}$$

Figure 10–26

A single-stage
compressor.

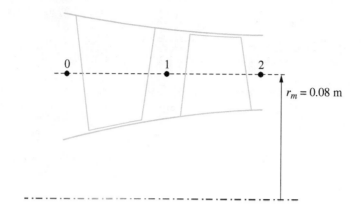

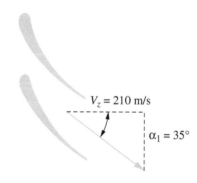

$V_z = 210$ m/s

$\alpha_1 = 35°$

$r_m = 0.08$ m

Example 10–4

Figure 10–26 shows the rotor-inlet velocity triangle in an axial-flow compressor rotor. The compressor-stage reaction is 50 percent at the mean radius. The inlet absolute total temperature $(T_{t,1})$ is 500 K, and the inlet absolute flow angle (α_1) is 15°. Assuming a constant axial velocity component, an isentropic stator, and a specific heat ratio γ of 1.4:

(a) Sketch the rotor-exit velocity triangle.

(b) At the rotor exit station, calculate T_t and T.

Solution

Part a:
Noting that the stage is symmetric (i.e., has a 50 percent reaction), and referring to the rotor-inlet velocity triangle, we have:

$$W_2 = V_1$$
$$V_2 = W_1$$
$$\alpha_2 = -\beta_1 = +30°$$
$$\beta_2 = -\alpha_1 = -15°$$

With these parameters, construction of the rotor-exit velocity triangle (shown on the same figure) is straightforward.

Part b:
In order to calculate the required variables, note that $T_{t,r,2} = T_{t,r,1}$ for an adiabatic rotor flow, as we proceed as follows:

$$V_z = 250.0 \cos 30° = 216.5 \text{ m/s}$$

$$V_1 = \frac{V_z}{\cos 15°} = 224.1 \text{ m/s}$$

$$U_m = V_{\theta,1} - W_{\theta,1} = V_1 \sin(15°) - W_1 \sin(-30°)$$

$$= 183.0 \text{ m/s (in the direction of } \omega)$$

$$T_{t,r,2} = T_{t,r,1} = T_{t,1} - \frac{V_1^2 - W_1^2}{2c_p} = 506.1 \text{ K}$$

Combination of the energy-conservation equation and Euler's equation gives:

$$w_s = U_m(V_{\theta,2} - V_{\theta,1}) = c_p(T_{t,2} - T_{t,1})$$

Application of this expression enables us to proceed toward computing both the rotor-exit total relative and static temperatures as follows:

$$T_{t,2} = T_{t,1} + \frac{U_m(V_{\theta,2} - V_{\theta,1})}{c_p} = 512.2 \text{ K}$$

$$T_2 = T_{t,2} - \frac{V_2^2}{2c_p} = 475.0 \text{ K}$$

Example 10–5

Shown in Figure 10–25 is the power turbine stage in an auxiliary power unit (APU), and the stator-exit mean-radius absolute velocity vector. The rotor speed is 17,900 r/min, and the specific shaft work at the mean radius is 75,000 J/kg. Assuming an average specific heat ratio γ of 1.33, calculate:

(a) The stage-exit absolute and relative flow angles (α_2 and β_2)

(b) The stage reaction

(c) The rotor incidence angle, knowing that the airfoil (or metal) inlet angle is 45.0°. Also sketch the rotor-blade airfoil at the mean radius.

Solution

Part a:
Let us assume that $V_{\theta,2}$ is positive, an assumption which can later be verified. Let us now proceed with computing the rotor-exit absolute and relative flow angles (i.e., α_2 and β_2, respectively):

$$U_m = \omega r_m = 150.0 \text{ m/s}$$

$$V_{\theta,1} = V_z \tan \alpha_1 = 321.3 \text{ m/s}$$

$$V_{\theta,2} = V_{\theta,1} - \frac{w_s}{U_m} = -178.7 \text{ m/s}$$

This negates the assumption we made earlier. Now:

$$\alpha_2 = \tan^{-1}\frac{V_{\theta,2}}{V_z} = -38.5°$$

$$\beta_2 = \tan^{-1}\frac{W_{\theta,2}}{V_z} = -55.6°$$

Part b:

$$V_1 = \sqrt{V_{\theta,1}^2 + V_z^2} = 392.3 \text{ m/s}$$

$$V_2 = \sqrt{V_{\theta,2}^2 + V_z^2} = 287.3 \text{ m/s}$$

$$W_1 = \sqrt{(V_{\theta,1} - U_m)^2 + V_z^2} = 282.8 \text{ m/s}$$

$$W_2 = \sqrt{(V_{\theta,2} - U_m)^2 + V_z^2} = 398.3 \text{ m/s}$$

The stage reaction (R) can now be calculated as follows:

$$R = \frac{(W_2^2 - W_1^2)}{(V_1^2 - V_2^2) + (W_2^2 - W_1^2)} = 52.4\%$$

Part c:

In order to calculate the rotor-blade incidence angle (i_R), we first calculate the rotor-inlet relative flow angle β_1 as follows:

$$\beta_1 = \tan^{-1}\frac{W_{\theta,1}}{V_z} = 37.3°$$

Now, with the rotor-blade inlet (metal) angle β_1' being 45° (which is greater than β_1), we conclude (by a simple sketch of the blade-inlet/flow interaction) that the blade incidence angle i_R is negative.

$$i_R = \beta_1 - \beta_1' = -7.7°$$

Based on the relative flow angles (above), the mean-radius rotor-blade airfoil can be easily sketched (Figure 10–25).

10–15   Supersonic Stator Cascades

One way of maximizing the turbine-stage power output is to achieve a substantially high stator-exit kinetic energy. To this end, and besides contouring the endwalls (Figure 10–30), the stator blade-to-blade passage can be shaped like a converging-diverging DeLaval nozzle, giving rise to a supersonic exit stream. Figure 10–30 shows an example of such a stator passage design. The same figure shows the two means of creating the minimum-area throat section across which the flow stream, working at the design intent, proceeds to be supersonic. These include a simple endwall contouring as well as a special blade-to-blade cross section that resembles a converging-diverging DeLaval nozzle. Note the smooth area transition across the throat section, which is necessary for the subsonic-to-supersonic flow transition to occur.

Figure 10–27

Input variables for
problem 10–3.

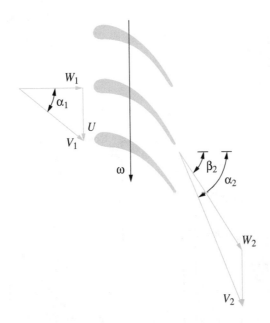

An obvious drawback of this stator design is the inability to produce a supersonic exit flow stream under some off-design operation modes. The different inlet total properties under such operation modes may very well require a different (smaller) throat area. In a fixed-geometry stator, the inability to produce sonic conditions at the minimum-area location will result in nothing but a Venturi-meter type of blade-to-blade flow passage, with the diverging segment acting as a flow decelerator (subsonic diffuser). This would materialize at obviously far (maybe even near) off-design operation modes. The stator exit flow under such circumstances would possess a significantly low Mach number, defeating the initial purpose, which is maximizing the stator exit velocity.

Example 10–6

Consider the adiabatic subsonic-supersonic stator shown in Figure 10–31. At the design point, the stator operating conditions are as follows:

- Inlet total pressure $(p_{t,0}) = 12.3$ bars
- Inlet total temperature $(T_{t,0}) = 1452$ K
- Shaft speed $(N) = 46{,}000$ r/min
- Mass flow rate $(\dot{m}) = 6.44$ kg/s
- Stator exit critical Mach number $(M_{cr,1}) = 2.35$
- Stator exit flow angle $(\alpha_1) = 79°$

The stator mean radius r_m is 0.325 m. Furthermore, the following simplifications apply:

- Constant axial velocity V_z component across the stator
- Isentropic flow between the throat section and the exit station
- An average specific heat ratio $\gamma = 1.33$

Figure 10–28

Input variables for problem 10–5.

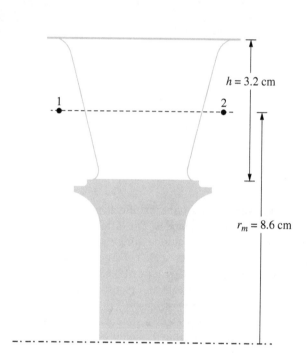

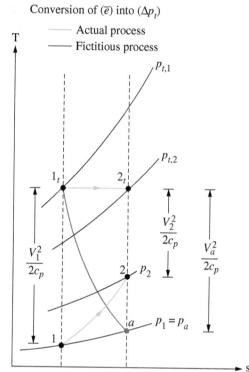

Conversion of $(\bar{e})$ into (Δp_t)

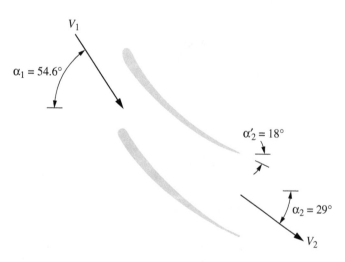

Figure 10–29

Input variables for
problem 10–6.

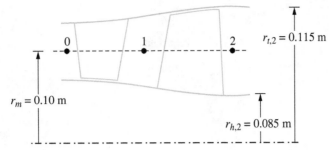

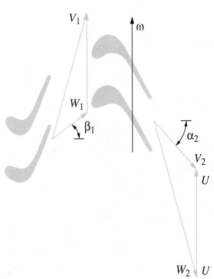

(a) Calculate the total pressure loss percentage.

(b) Determine whether the stator will produce an exit supersonic flow.

(c) Repeat item b (above) upon changing the shaft speed to 28,000 r/min.

Solution

Let us add the superscript * to identify the thermophysical properties at the throat
section

Part a:
Applying the "bulky" form of the continuity equation at the stator exit station, we
get:

$$p_{t,1} = 11.4 \text{ bars}$$

Now, the total pressure loss percentage is:

$$\frac{p_{t,0} - p_{t,1}}{p_{t,0}} = 7.3 \text{ percent}$$

Figure 10–30

The supersonic
stator-discharge option.

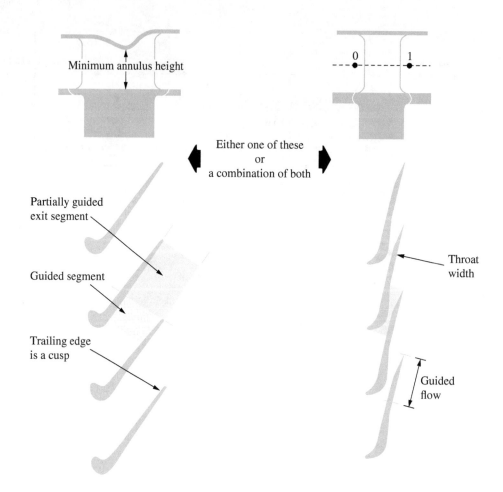

Minimum annulus height

Either one of these
or
a combination of both

Partially guided
exit segment

Guided segment

Trailing edge
is a cusp

Throat
width

Guided
flow

Part b:
At the stator exit station, we have:

$$V_1 = M_{cr,1} V_{cr,1} = 1618.3 \text{ m/s}$$

$$V_{z,1} = V_1 \cos \alpha_1 = 308.8 \text{ m/s}$$

$$V_{\theta,1} = V_1 \sin \alpha_1 = 1588.0 \text{ m/s}$$

$$U_m = \omega r_m = 1565.6 \text{ m/s}$$

$$W_{\theta,1} = V_{\theta,1} - U_m = 23.0 \text{ m/s}$$

$$W_1 = \sqrt{W_{\theta,1}{}^2 + V_z{}^2} = 309.7 \text{ m/s}$$

$$T_{t,r,1} = T_{t,1} + \frac{W_1{}^2 - V_1{}^2}{2c_p} = 386.5 \text{ K}$$

$$W_{cr,1} = \sqrt{\left(\frac{2\gamma}{\gamma+1}\right) R T_{t,r,1}} = 355.3 \text{ m/s}$$

$$\frac{W_1}{W_{cr,1}} = 0.87$$

Figure 10–31

Numerical results.

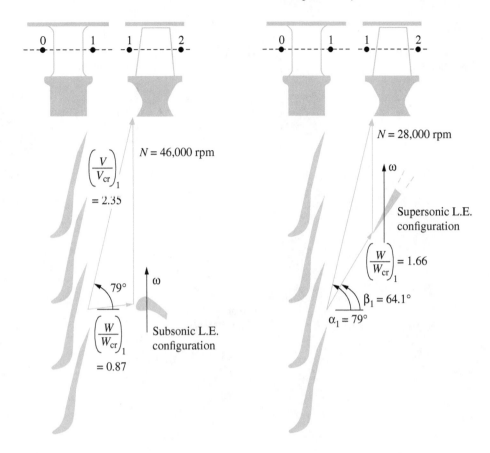

In words, this result simply means that an observer that is attached to the rotor blade and spinning with it will register an incoming flow stream that is subsonic. Note that an outside stationary observer will register an "absolute" Mach number that is above unity, actually 2.35.

Part c:
Repeating the entire procedure for the now-lower magnitude of shaft speed, the final result is:

$$\frac{W_1}{W_{cr,1}} = 1.66$$

10–16   Sign Convention Governing Radial Turbomachines

So far, we have dealt exclusively with axial-flow turbomachines, where both the rotor inlet and exit segments are in the axial direction (by reference to their meridional projections). Now we deviate slightly from this configuration, focusing on radial turbomachinery components. Referring to the meridional view in Figure 10–33, we find that while the rotor inlet segment is axial, the exit segment is radial by reference to the meridional flow

Figure 10–32

Input variables for
problem 10-7.

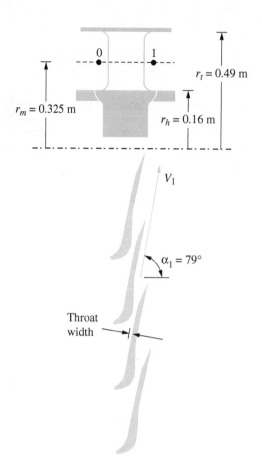

path. In the latter case, therefore, the angles (both blade and flow) will be measured from
the radial direction, as shown on the extreme left in Figure 10–33. Again, flow and blade
angles in the direction of rotation will be positive and, otherwise, negative.

It follows that tangential components of both the velocity vector (i.e., V_θ), and
relative velocity vector (meaning W_θ) will be positive in the direction of rotation, and
negative otherwise. Also note that the "arm" of $V_{\theta,2}$ is different (in this case longer) than
that of $V_{\theta,1}$. In this case, the shaft work version of Euler's equation for the centrifugal
compressor configuration in Figure 10–33 can be rewritten as follows:

$$w_s = U_2 V_{\theta,2} - U_1 V_{\theta,1} = \omega (r_2 V_{\theta,2} - r_1 V_{\theta,1})$$

Figure 10–33

Meridional projections of
axial and centrifugal
compressors.

Meridional flow path

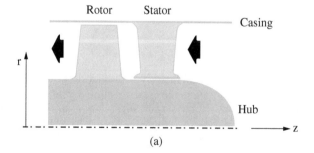

(a)

Figure 10–33
(*continued*)

Meridional projections of axial and centrifugal compressors.

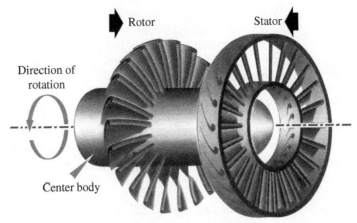

Axial-flow compressor stage

(b)

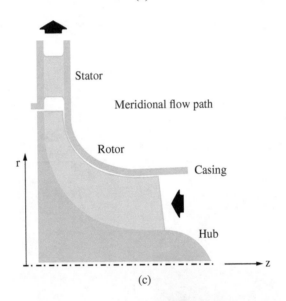

(c)

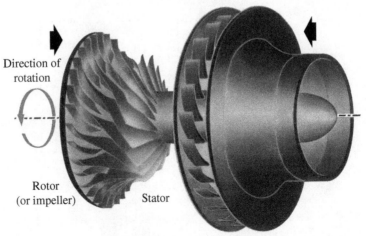

Centrifugal compressor stage

(d)

Worth noting is the fact that the inlet velocity vector $\vec{V}_1$ usually possesses no swirl action, meaning that $V_{\theta,1}$ is more or less zero.

Finally note, by reference to Figure 10–33, that a radial turbine rotor has diametrically opposed features as compared to a centrifugal compressor rotor. In other words, it is the inlet rotor segment here that is radial, while the exit segment is axial.

Example 10–7

Figure 10–34a shows the meridional views of a centrifugal compressor and its major variables. These variables have the following magnitudes.

- Mean inlet radius (r_1) = 4.6 cm
- Exit radius (r_2) = 10.5 cm
- Inlet blade height (h_1) = 5.2 cm
- Exit blade height (h_2) = 1.8 cm

Figure 10–34

(a) A centrifugal impeller rotor.

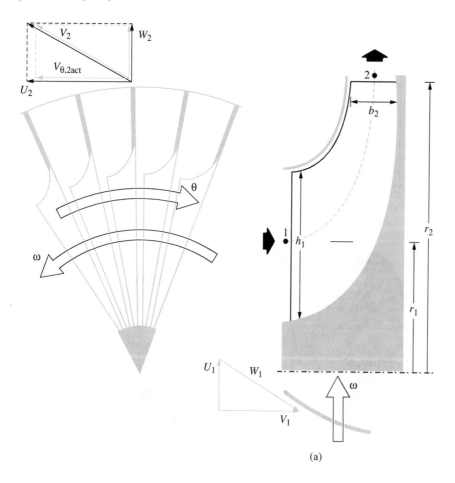
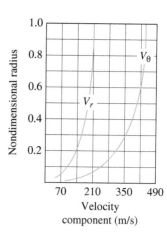

(a)

Figure 10–34
(*continued*)

(b) Results of the numerical example.

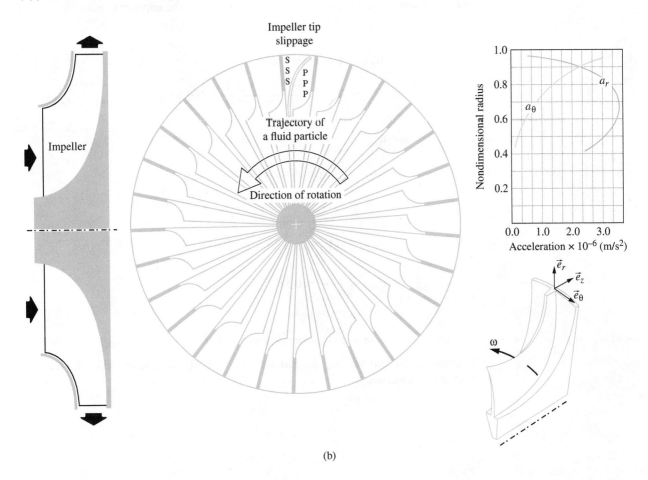

(b)

The stage design point gives rise to the following data:

- Rotational speed (N) = 41,600 r/min
- Inlet total temperature ($T_{t,1}$) = 490 K
- Inlet total pressure ($p_{t,1}$) = 1.88 bars
- Inlet swirl angle (α_1) = 0
- Exit total pressure ($p_{t,2}$) = 6.02 bars
- Exit total temperature ($T_{t,2}$) = 683 K
- Radial blading with no deviation angle (i.e., $\beta_2 = 0$)
- Slip factor (σ_s) = 92.6 percent
- Choked flow at the impeller inlet station
- Sonic flow at the impeller exit station

Assuming a parabolic dependency of the velocity components (V_r and V_θ) on the radius (Figure 10–34a), calculate the components of both the Coriolis and centripetal accelerations at key radial locations of your choice. Use the results to explain the tangential shifting of a fluid particle which is traversing the impeller subdomain.

Solution

$$W_{\theta,1} = U_1 = 209.1 \text{ m/s}$$

$$T_{t,r,1} = T_{t,1} - \frac{\left(W_{\theta,1}{}^2 - V_{\theta,1}{}^2\right)}{2c_p} = 496.8 \text{ K}$$

$$W_{cr,1} = \sqrt{\frac{2\gamma}{\gamma+1} RT_{t,r,1}} = 407.8 \text{ m/s}$$

$$W_1 = W_{cr,1} = 407.8 \text{ m/s}$$

$$V_1 = V_{cr,1} = V_{z,1} = \sqrt{W_1{}^2 - U_1{}^2} = 350.2 \text{ m/s}$$

$$V_{cr,1} = \sqrt{\left(\frac{2\gamma}{\gamma+1}\right) RT_{t,1}} = 405.1 \text{ m/s}$$

$$M_{cr,1} = \frac{V_1}{V_{cr,1}} = 0.88$$

$$\rho_1 = 0.976 \text{ kg/m}^3$$

$$\dot{m} = \rho_1 V_{z,1}(2\pi r_1 h_1) = 5.14 \text{ kg/s}$$

Referring to the exit velocity diagram in Figure 10–34a, we can calculate the following exit variables:

$$V_{cr,2} = 478.2 \text{ m/s}$$

$$V_2 = V_{cr,2} = 478.2 \text{ m/s}$$

$$(V_{\theta,2})_{id} = U_2 457.4 \text{ m/s}$$

$$(V_{\theta,2})_{act} = 423.6 \text{ m/s}$$

$$V_{r2} = \sqrt{V_2{}^2 - V_{\theta,2}{}^2} = 222.0 \text{ m/s}$$

Parabolic Interpolation of V_r and V_θ

Let us define the nondimensional coordinate ($\bar{r}$) as follows:

$$\bar{r} = \frac{r - r_1}{r_2 - r_1}$$

Using the impeller exit magnitudes of V_r and V_θ as boundary conditions, we get the following parabolic relationships:

$$V_r = 222.0\bar{r}^2$$

$$V_\theta = 423.6\bar{r}^2$$

Noting that the rotational speed ω is opposite to the θ direction, we can express the combined centripetal and Coriolis acceleration components as:

$$(a)_{net} = [(\omega^2 r) - 2(\omega W_\theta)]e_r + [2\omega W_r]e_\theta$$

Noting that $W_\theta = V_\theta - \omega r$, we can now calculate the net acceleration components at the following five radial locations:

$$(a_r)_{\bar{r}=0.2} = 3.14 \times 10^6 \text{ m/s}^2 \text{ and } (a_\theta)_{\bar{r}=0.2} = 0.08 \times 10^6 \text{ m/s}^2$$

$$(a_r)_{\bar{r}=0.4} = 3.38 \times 10^6 \text{ m/s}^2 \text{ and } (a_\theta)_{\bar{r}=0.4} = 0.31 \times 10^6 \text{ m/s}^2$$

$$(a_r)_{\bar{r}=0.6} = 3.30 \times 10^6 \text{ m/s}^2 \text{ and } (a_\theta)_{\bar{r}=0.6} = 0.70 \times 10^6 \text{ m/s}^2$$

$$(a_r)_{\bar{r}=0.8} = 2.95 \times 10^6 \text{ m/s}^2 \text{ and } (a_\theta)_{\bar{r}=0.8} = 1.24 \times 10^6 \text{ m/s}^2$$

$$(a_r)_{\bar{r}=1.0} = 2.29 \times 10^6 \text{ m/s}^2 \text{ and } (a_\theta)_{\bar{r}=1.0} = 1.94 \times 10^6 \text{ m/s}^2$$

Now, the following conclusions can be drawn:

- The radial acceleration component continues to decline as the impeller exit station is approached. This is due to the decline in the Coriolis acceleration radial component. As a result, a fluid particle in this region will continually give rise to the radial momentum.
- Over the same exit subregion, the tangential component of Coriolis acceleration continually grows, causing the tangential shift in the fluid particle that is sketched and labeled "particle trajectory" in Figure 10–34b.
- Combination of the two behavioral characteristics (above) gives rise to the "slip" phenomenon in centrifugal compressors, which manifests itself close to and at the blade tip.

Example 10–8

Figure 10–35 shows the rotor subdomain of an axial-flow turbine rotor and its major dimensions. The rotor operating conditions are as follows:

- Inlet total pressure = 13.2 bars
- Inlet total temperature = 1405 K
- Inlet critical Mach number = 0.92
- Exit total pressure = 5.4 bars
- Exit total temperature = 1162 K
- $M_{cr,2}$ is small enough to justify the equality of static and total densities at the rotor exit station
- Mass flow rate = 11.4 kg/s
- Shaft speed = 48,300 r/min

In addition, the following geometrical items are also applicable:

- Number of blades (N_b) = 29
- Trailing edge thickness ($t_{t.e.}$) = 2.5 mm
- Mean-radius axial chord length (C_z) = 2.5 cm
- Mean-radius true chord (C) = 4.6 cm
- Mean camber-line length (L) = 9.2 cm

Figure 10–35

Example 10–8 input variables.

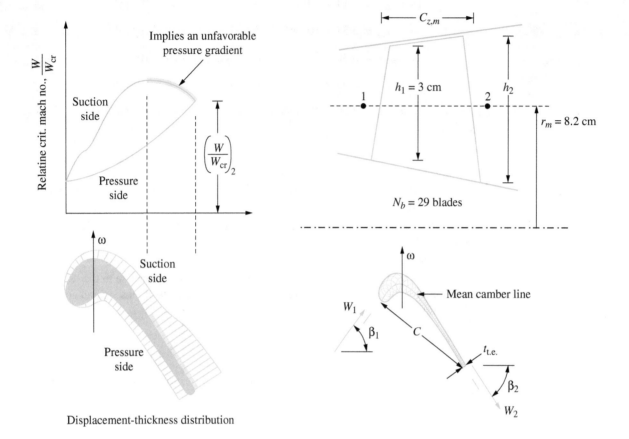

Displacement-thickness distribution

Segment 1

(a) Beginning with the assumption of an unchoked rotor, calculate the blade-exit annulus height (h_2) that will ensure a rotor-wise constant magnitude of the axial velocity component (V_z).

(b) Now verify whether the rotor is actually choked.

(c) In the event of rotor choking, recompute all variables which are pertinent to the exit velocity triangle.

Segment 2

Consider the following expression for calculating the kinetic energy loss coefficient ($\bar{e}_R$):

$$\bar{e}_R = \left(\frac{\theta_{tot}}{S_m \cos \beta_2 - t_{t.e.} - \delta^*_{tot}} \right) \left(\frac{A_{3D}}{A_{2D}} \right) \left(\frac{Re}{Re_{ref}} \right)^{-0.2}$$

where:

- $\bar{e}_R$ is the kinetic energy loss coefficient.
- θ_{tot} refers to the combined (pressure plus suction sides) momentum thickness.
- δ_{tot} refers to the combined displacement thicknesses.
- S_m refers to the mean-radius blade-to-blade linear spacing.
- β_2 is the rotor-exit relative flow angle.
- $t_{\text{t.e.}}$ is the trailing edge thickness.
- A_{2D} is the blade surface area that is in contact with the flow stream.
- A_{3D} is A_{2D} plus the blade-to-blade hub and casing surface areas.
- Re is the Reynolds number based on the true chord and the exit relative velocity (i.e., C and W_2, respectively).
- Re_{ref} is a reference Reynolds number and is equal to 7.57×10^6.

Now apply the $\bar{e}_R$ expression to calculate the relative pressure loss (i.e., $p_{t,r,1} - p_{t,r,2}$) over the rotor.

You may also assume that the suction side is the same as that of the pressure side multiplied by a factor of 3.5.

Clarification   Referring to Figure 10–35, the reason for this multiplier is that the boundary layer grows much faster on the suction side, by comparison. Consequently, both the displacement and momentum thicknesses are typically much larger than over the blade pressure side. Part of the reason behind such a substantial difference is the strong likelihood of the suction side being exposed to flow deceleration (known as *diffusion*) which causes a local rise in static pressure. This constitutes an adverse (or unfavorable) pressure gradient, a factor that aggravates the suction-side boundary layer buildup.

Comment   The importance of this problem has to do with some of the most serious rotor-choking consequences. As seen in Figure 10–35, there can really be major changes in the rotor-exit velocity triangle as a result of choking, as opposed to "blindly" treating the rotor as unchoked. In reality, turbomachinists (including the author himself) are not immune from making the mistake of ignoring the process of verifying whether the passage (stationary or rotating) is choked. This is particularly the case in situations where the need to compute the critical Mach number (absolute or relative) is nonexistent.

Noteworthy, as well, is a late step in the problem solution where a loss assesor, namely the kinetic energy loss coefficient ($\bar{e}_R$), is converted into a total pressure loss coefficient ($\bar{\omega}$). Because the object in this problem is a rotor, a new definition of $\bar{e}$ will now be cast in terms of relative thermophysical properties. The conversion from $\bar{e}$ into $\bar{\omega}$ (the reader will notice) is made through an intermediate step, where the entropy production Δs will be executed.

Segment 1
Part 1a:

$$V_{cr,1} = \sqrt{\left(\frac{2\gamma}{\gamma+1}\right) RT_{t,1}} = 678.5 \text{ m/s}$$

$$V_1 = M_{cr,1} V_{cr,1} = 624.2 \text{ m/s}$$

$$\rho_1 = \left(\frac{p_{t,1}}{RT_{t,1}}\right)\left(1 - \frac{\gamma-1}{\gamma+1} M_{cr,1}{}^2\right)^{\frac{1}{\gamma-1}} = 2.22 \text{ kg/m}^3$$

$$\dot{m} = \rho_1 V_{z,1} A_1 = \rho_1 V_{z,1}(2\pi r_m h_1)$$

which yields:

$$V_{z,1} = 332.2 \text{ m/s}$$

Now we proceed toward eventually calculating the rotor-exit annulus height (h_2) as follows:

$$\alpha_1 = \cos^{-1}\left(\frac{V_{z,1}}{V_1}\right) = 57.8°$$

$$V_{z,1} = V_1 \sin\alpha_1 = 528.4 \text{ m/s}$$

$$U = \omega r_m = 414.8 \text{ m/s}$$

$$W_{\theta,1} = V_{\theta,1} - U = 113.6 \text{ m/s}$$

$$W_1 = \sqrt{W_{\theta,1}{}^2 + V_{z,1}{}^2} = 351.1 \text{ m/s}$$

$$\beta_1 = \tan^{-1}\left(\frac{W_{\theta,1}}{V_{z,1}}\right) = 18.9°$$

$$T_{t,r,1} = T_{t,1} - \left(\frac{V_1{}^2 - W_1{}^2}{2c_p}\right) = 1289.9 \text{ K}$$

$$p_{t,r,1} = p_{t,1}\left(\frac{T_{t,r,1}}{T_{t,1}}\right)^{\frac{\gamma}{\gamma-1}} = 9.35 \text{ bars}$$

Now we set $V_{z,2} = V_{z,1} = 332.2$ m/s, and the exit density equal to the exit total magnitude as an approximation:

$$\rho_2 \approx \rho_{t,2} = \frac{p_{t,2}}{RT_{t,2}} = 1.62 \text{ kg/m}^3$$

These two simplifications enable us to compute the rotor-exit annulus height by applying the continuity equation as follows:

$$h_2 = \frac{\dot{m}}{\rho_2 V_{z,2}(2\pi r_m)} = 4.11 \text{ cm}$$

Part 1b:

In order to verify the choking status of the rotor passage, we have to calculate the relative critical Mach number at the passage exit station. To this end, we proceed as follows:

$$V_{\theta,2} = V_{\theta,1} - \frac{c_p(T_{t,1} - T_{t,2})}{U} = -149.2 \text{ m/s}$$

$$\alpha_2 = \tan^{-1}\left(\frac{V_{\theta,2}}{V_{z,2}}\right) = -24.2°$$

$$V_2 = \sqrt{V_\theta^2 + V_z^2} = 364.2 \text{ m/s}$$

$$W_{\theta,2} = V_{\theta,2} - U = -564.0 \text{ m/s}$$

$$W_2 = \sqrt{W_{\theta,2}^2 + V_{z,2}^2} = 654.6 \text{ m/s}$$

$$\beta_2 = \tan^{-1}\left(\frac{W_{\theta,2}}{V_{z,2}}\right) = -59.5°$$

$$T_{t,r,2} = T_{t,r,1} = 1289.9 \text{ K}$$

$$W_{\text{cr},2} = \sqrt{\frac{2\gamma}{\gamma + 1}RT_{t,r,2}} = 650.1 \text{ m/s}$$

$$\frac{W_2}{W_{\text{cr},2}} = 1.01 \text{ (impossible for a convergent nozzle)}$$

which means that the rotor blade-to-blade passage is choked.

Part 1c:

Due to the newly discovered rotor choking status, we have to implement the following corrective actions:

- Set the rotor exit relative velocity (W_2) equal to the relative critical velocity ($W_{\text{cr},2}$).
- Re-apply the continuity equation at the rotor exit station and in the rotating frame of reference, so that the actual magnitude of the exit relative flow angle (β_2) can be obtained.
- Make all other changes in the variables comprising the rotor-exit velocity triangle. Note that the axial velocity component (V_z) will no longer remain constant across the rotor.

We now proceed to implement these changes, beginning with the calculation of the rotor-exit relative properties:

$$p_{t,r,2} = p_{t,2}\left(\frac{T_{t,r,2}}{T_{t,2}}\right)^{\frac{\gamma}{\gamma-1}} = 8.23 \text{ bars}$$

$$T_{t,r,2} = 1289.9 \text{ K (computed earlier)}$$

Applying the "bulky" version of the continuity equation in the rotating frame of reference at the rotor exit station, we get:

$$\beta_2 = 53.7°$$

Referring to Figure 10–35, the sign of β_2 is indeed negative, i.e.,

$$\beta_2 = -53.7°$$

This value is certainly different from the previously computed value of $-59.5°$, with the difference being the recognition (at this point) of the rotor passage choking status. Pursuing this corrective procedure, we have:

$$W_{\theta,2} = W_2 \sin \beta_2 = -523.9 \text{ m/s}$$

$$T_2 = T_{t,r,2} - \frac{W_2{}^2}{2c_p} = 1107.2 \text{ K}$$

$$V_{\theta,2} = W_{\theta,2} + U = -109.1 \text{ m/s}$$

$$V_{z,2} = W_{z,2} = W_2 \cos \beta_2 = 384.9 \text{ m/s}$$

$$V_2 = \sqrt{V_{\theta,2}{}^2 + V_{z,2}{}^2} = 400.1 \text{ m/s}$$

$$\alpha_2 = \tan^{-1} \frac{V_{\theta,2}}{V_{z,2}} = -15.8°$$

$$T_{t,2} = T_2 + \frac{V_2{}^2}{2c_p} = 1176.4 \text{ K (to be compared to the previously}$$
$$\text{computed magnitude of 1162.0 K)}$$

With the velocity components recalculated, we can easily generate the choking-altered rotor-exit velocity triangle, which is shown in Figure 10–36.

Figure 10–36

Rotor choking effect on the exit velocity triangle.

— Velocity triangles with the initial assumption of unchoked rotor

— Effect of rotor choking on the exit velocity triangle

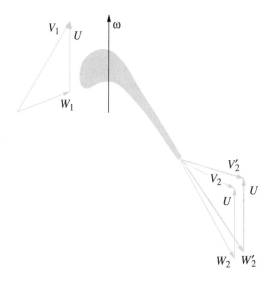

Figure 10–37

Dynamic viscosity coefficient for air at high temperatures.

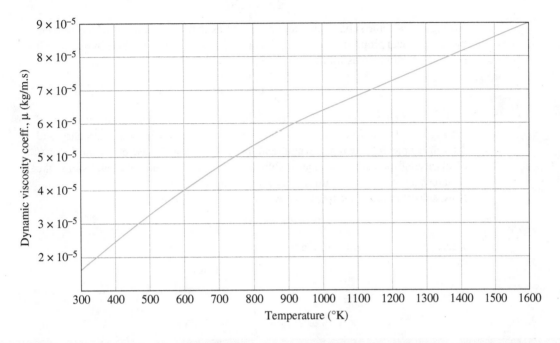

Segment 2

A good first step is to list the different magnitudes of variables which are relevant to the calculation of the kinetic energy loss coefficient $\bar{e}_R$ as follows:

- $h_{av} = \frac{1}{2}(h_1 + h_2) = 3.56$ cm
- $S_m = \frac{2\pi r_m}{N_b} = 1.78$ cm (blade-to-blade linear spacing)
- $\beta_2 = -59.7°$
- $t_{t.e.} = 2.5$ mm
- $T_2 = 1107.2$ K
- $\mu_2 = 6.7 \times 10^{-5}$ kg/(m·s) (from Figure 10–37)
- $\rho_2 \approx \rho_{t,2} = 1.62$ kg/m^3
- $\nu_2 = \frac{\mu_2}{\rho_2} = 4.14 \times 10^{-5} = 4.14 \times 10^{-5}$ m^2/s
- $W_2 = 650.1$ m/s
- Mean camber-line length $(L) = 9.2$ cm
- True chord $(C) = 4.6$ cm
- $\delta^*_{press} = 0.057 L \left(\frac{W_2 L}{\nu_2}\right) = 0.493$ mm
- $\delta_{tot} = \delta^*_{press} + \delta^*_{suc} = \delta^*_{press} + 3.5\delta^*_{press} = 2.22$ mm
- $\theta_{press} = 0.022 L \left(\frac{W_2 L}{\nu_2}\right) = 0.190$ mm
- $\theta_{tot} = 0.857$ mm
- $A_{2D} \approx 2L h_{av} = 0.00655$ m^2
- $A_{3D} \approx A_{2D} + 2S_m c_{z,m} = 0.00744$ m^2
- $Re = \frac{W_2 C}{\nu_2} = 7.22 \times 10^{-5}$

Upon substituting these variables for $\bar{e}_R$ in the provided expression, and reading off the dynamic viscosity coefficient from Figure 10–37, we obtain:

$$\bar{e}_R = 0.268$$

Noting that a rotor subdomain can be viewed as that of a stator, provided relative thermophysical properties are used, we can legitimately adapt the $\bar{e}$ definition to suit our rotor subdomain as follows:

$$\bar{e}_R = 1 - \frac{W_{act}^2}{W_{id}^2} \text{ for a rotor subdomain}$$

In the computational procedure (above), the dynamic viscosity coefficient was read off Figure 10–37. Note that the static states are common in both stationary and rotating frames of reference when dealing with a rotor. However, in the interest of clarifying the aspects of thermodynamics behind the computational steps, the variables involved in computing $\bar{e}_{rotor}$ are displayed.

Now, substituting in $\bar{e}_R$ expression, we get:

$$W_{act} = 300.4 \text{ m/s}$$

where the following substitution was made:

$$W_{id} = W_1 = 351.1 \text{ m/s}$$

Also:

$$T_{id} = T_{t,r,1} - \frac{W_1^2}{2c_p} = 1236.6 \text{ K}$$

$$T_{act} = T_{t,r,1} - \frac{W_{act}^2}{2c_p} = 1250.9 \text{ K}$$

By selecting the most convenient states (namely 1_{id} and 2_{act}) in this case, we can calculate the rotor-generated entropy production as follows:

$$\Delta s_R = c_p \ln\left(\frac{T_{act}}{T_{id}}\right) - R \ln\left(\frac{p_{act}}{p_{id}}\right) = 13.29 \text{ J/(kg} \cdot \text{K)}$$

In applying this expression, we are free to select any two (inlet and exit) states. In other words, the rotor flow process (when we get to this point) can be viewed from any of the total-to-total, total-to-total relative, or total relative-to-static standpoints. In order to calculate $p_{t,r,2}$ (the next step), let us select the two states $1_{t,r}$ and $2_{t,r}$ (both being total relative states) for expressing Δs_{rotor}:

$$\Delta s_R = 13.29 \text{ J/(kg} \cdot \text{K)} = c_p \ln\left(\frac{T_{t,r,2}}{T_{t,r,1}}\right) - R \ln\left(\frac{p_{t,r,2}}{p_{t,r,1}}\right)$$

Recalling that we are dealing with an adiabatic process, we know that:

$$T_{t,r,2} = T_{t,r,1}$$

which, together with the Δs_R expression, yields:

$$p_{t,r,2} = 8.93 \text{ bars}$$

which is less than $p_{t,r,1}$, as would be anticipated. Finally, we can calculate the total relative pressure loss, by reference to Figure 10–38, as follows:

$$\frac{\Delta p_{t,r}}{p_{t,r,1}} = 4.53 \text{ percent}$$

Figure 10–38

Axial distribution of the static, total, and total relative pressures.

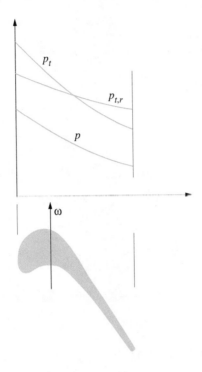

Figure 10–39

Input variables for problem 10-8.

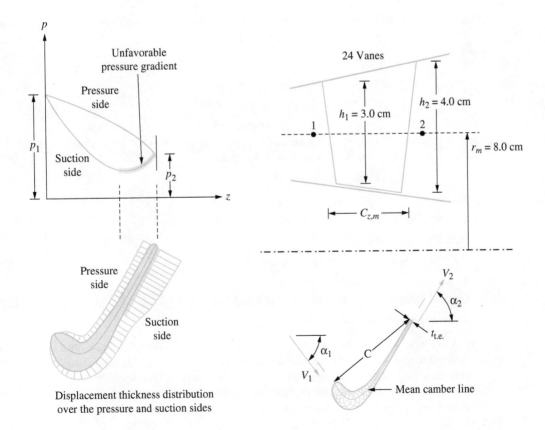

Displacement thickness distribution over the pressure and suction sides

Figure 10–40

Example 10–9 input data.

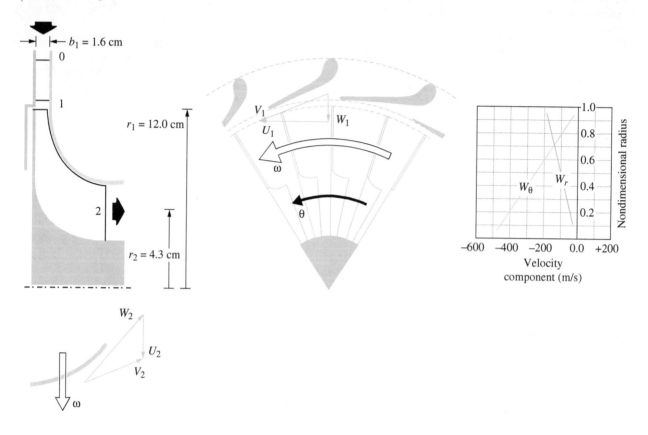

Example 10–9
Figure 10–40 shows different views of a radial inflow turbine stage, the stage inlet and exit velocity diagrams, and the stage major dimensions, with the stator/rotor radial-gap length being negligible. The stage operating conditions are as follows:

- Rotational speed = 46,500 r/min
- Mass flow rate = 5.04 kg/s
- Inlet total pressure $(p_{t,1}) = 11.8$ bars
- Inlet total temperature = 1265.0 K
- Both of the stator and rotor passages are choked
- Rotor-inlet incidence angle $(i_R) = 0$
- Specific shaft work produced $(w_s) = 404.56$ kJ/kg
- Total-to-total pressure ratio = 4.85

By assuming a linear variation of W_r and W_θ with radius across the rotor, calculate the combined magnitudes of centripetal and acceleration components at several radial locations. Using the resulting tangential component of these two components, sketch the relative trajectory of a fluid particle as it traverses the rotor subdomain. Also comment on the radial variation of the radial acceleration component.

Solution   The inlet and exit solid-body rotational velocities are calculated first:

$$\omega = 4869.5 \text{ rad/s}$$

$$U_1 = \omega r_1 = 584.3 \text{ m/s}$$

$$U_2 = \omega r_2 = 209.4 \text{ m/s}$$

Applying the "bulky" version of the continuity equation at the stator exit radius, we get:

$$\alpha_1 = 71.5°$$

The rotor-inlet absolute velocity components can now be calculated:

$$V_{\theta,1} = U_1 = 584.3 \text{ m/s (Note that } \beta_1 \text{ is zero.)}$$

$$V_{r,1} = \frac{V_{\theta,1}}{\tan \alpha_1} = -195.5 \text{ m/s (negative sign inserted)}$$

The exit total properties can be computed as follows:

$$p_{t,2} = 2.43 \text{ bars}$$

$$T_{t,2} = T_{t,1} - \frac{w_s}{c_p} = 915.2 \text{ K}$$

The rest of the stage-exit thermophysical properties can be calculated as follows:

$$V_{\theta,2} = \left(\frac{U_1}{U_2}\right) V_{\theta,1} - \frac{c_p}{U_2}(T_{t,1} - T_{t,2}) = 301.8 \text{ m/s}$$

$$W_{\theta,2} = V_{\theta,2} - U_2 = -511.2 \text{ m/s}$$

$$T_{t,r,2} = T_{t,2} + \frac{W_{\theta,2}{}^2 - V_{\theta,2}{}^2}{2c_p} = 988.8 \text{ K}$$

$$W_{cr,2} = \sqrt{\left(\frac{2\gamma}{\gamma+1}\right)RT_{t,r,2}} = 569.2 \text{ m/s}$$

$$W_2 = W_{cr,2} = 569.2 \text{ (rotor is choked)}$$

$$\beta_2 = \sin^{-1}\left(\frac{W_{\theta,2}}{W_2}\right) = -63.9°$$

$$V_{z,2} = W_{z,2} = 250.3 \text{ m/s}$$

$$\alpha_2 = \tan^{-1}\left(\frac{V_{\theta,2}}{V_{z,2}}\right) = -50.3°$$

Noting that $W_r = V_r$, and using the inlet and exit magnitudes as boundary conditions, we can similarly establish the following linear relationship:

$$W_\theta = -796.6 + 6638.5r \text{ (where the radius } r \text{ is in meters)}$$

This is graphically represented in Figure 10–41, along with the inlet and exit velocity diagrams.

With the preceding two expressions, let us calculate both relative velocity components at six equidistant radial locations between the rotor inlet and exit radii (r_1 and r_2), beginning at the rotor inlet radius as follows:

at $r = 0.1200$ m, $W_\theta = 0.0000$ m/s and $W_r = -195.5$ m/s

at $r = 0.1046$ m, $W_\theta = -102.2$ m/s and $W_r = -156.4$ m/s

at $r = 0.0892$ m, $W_\theta = -204.4$ m/s and $W_r = -117.3$ m/s

at $r = 0.0738$ m, $W_\theta = -306.7$ m/s and $W_r = -78.20$ m/s

at $r = 0.0584$ m, $W_\theta = -408.9$ m/s and $W_r = -39.10$ m/s

at $r = 0.0430$ m, $W_\theta = -511.1$ m/s and $W_r = 0.0000$ m/s

At this point, we have to define our relative (spinning) frame of reference, which we will use in the acceleration calculations (to follow). The reference frame here is defined in Figure 10–41 in the form of unit vectors in the r, θ, and z directions. Using this frame of reference, the Coriolis acceleration vector can be written in the form of the following vector product:

$$\vec{a}_C = 2\omega X \vec{W} = [(0.0)\vec{e}_r + (0.0)\vec{e}_\theta + (2\omega)\vec{e}_z] \times [(W_r)\vec{e}_r + (W_\theta)\vec{e}_\theta + (W_z)\vec{e}_z]$$

The preceding vector product is usually expressed in the form of a determinant that, upon expansion, yields the following nonzero components:

$$\vec{a}_c = (-2\omega W_\theta)\vec{e}_r + (2\omega W_r)\vec{e}_\theta$$

The centripetal acceleration, however, can consistently be written as follows:

$$\vec{a}_{\text{cent}} = (\omega^2 r)\vec{e}_r$$

Utilizing the foregoing dependencies of W_r and W_θ on the local radius, as well as the different acceleration components, we are now in a position to compute the net acceleration components acting on a typical fluid particle. To this end, let us choose the same six radial locations that we defined earlier in computing the relative velocity components. The net acceleration components (in m/s^2) at these radii are as follows:

at $r = 0.1200$ m, $a_r = +2.85 \times 10^6$ and $a_\theta = -1.90 \times 10^6$

at $r = 0.1046$ m, $a_r = +3.48 \times 10^6$ and $a_\theta = -1.52 \times 10^6$

at $r = 0.0892$ m, $a_r = +4.11 \times 10^6$ and $a_\theta = -1.14 \times 10^6$

at $r = 0.0738$ m, $a_r = +4.74 \times 10^6$ and $a_\theta = -0.76 \times 10^6$

at $r = 0.0584$ m, $a_r = +5.36 \times 10^6$ and $a_\theta = -0.38 \times 10^6$

at $r = 0.0430$ m, $a_r = +6.00 \times 10^6$ and $a_\theta = 0.0$

Using the tangential-acceleration-component magnitudes (above), a trajectory sketch of a typical fluid particle, within the rotor subdomain, should look like that in Figure 10–41.

Figure 10–41

Final results.

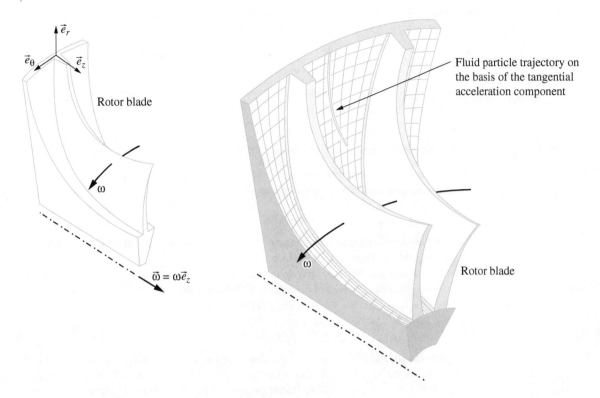

Rotor blade

$\vec{e}_r$

$\vec{e}_\theta$ $\vec{e}_z$

ω

$\vec{\omega} = \omega \vec{e}_z$

Fluid particle trajectory on the basis of the tangential acceleration component

ω

Rotor blade

PROBLEMS

10–1 Consider a single-stage axial-flow turbine, with a mean radius r_m of 0.1 m. The turbine operating conditions are as follows:

- Mass flow rate = 4.3 kg/s
- Stage-inlet total temperature = 1400 K
- Stage-inlet total pressure = 12.5 bars
- Stator flow process is isentropic
- Rotor speed = 34,000 r/min
- Constant axial velocity component
- Stator-exit critical Mach number = 0.85
- Stator-exit absolute flow angle $\alpha_1 = 68.0°$
- Rotor-exit relative Mach number $M_{cr,r2} = 0.82$
- Rotor-relative total pressure ratio (which simply indicates a rotor-subdomain loss) $\frac{p_{t,r,2}}{p_{t,r,1}} = 0.822$

Considering the preceding data, calculate:

(a) The torque (τ) delivered by the turbine stage

(b) The stage total-to-total efficiency (η_{t-t})

10–2   Consider an adiabatic compressor stage, with a mean radius of 0.08 m. The stage operating conditions are as follows:

- Rotor-inlet absolute total pressure = 1.85 bars
- Rotor-inlet absolute total temperature = 480 K
- Rotor speed = 58,000 r/min
- Stage-wise axial velocity component is constant
- Stator-exit absolute flow angle $\alpha_1 = +35°$
- Rotor-exit absolute total pressure $p_{t,2} = 3.1$ bars
- Rotor-exit relative flow angle β_2 is negative
- Rotor-exit static temperature $(T_2) = 496.0$ K

Assuming an average specific heat ratio of 1.4, calculate:
(a) The change in total relative pressure across the rotor
(b) The stage reaction
(c) The rotor-exit static pressure (p_2)

10–3   The turbine-rotor blade cascade in Figure 10–27 is switched to operate as a compressor, by transmitting (to the blades) a torque which gives rise to the direction of rotation indicated in the figure (blade pressure side is leading in the direction of rotation). If V_z is maintained constant across the rotor, indicate which of the following statements are true, and which are false:
(a) Reaction is greater than 100 percent . . . . (TRUE or FALSE)
(b) The change in Δh_t will be negative . . . . (TRUE or FALSE)
(c) α_2 will be greater than β_2 . . . . (TRUE or FALSE)

10–4   Air enters the first stage of an axial-flow compressor at a total pressure and temperature of 2.0 bars and 350 K, respectively. The axial velocity component (constant throughout the stage) is 200 m/s, and the stator flow process is assumed isentropic. In addition, the stage total-to-total pressure ratio is 1.75, the total-to-total efficiency is 82 percent, the rotor speed is 32,000 r/min, the rotor-exit relative flow angle (β_2) is zero, and the average specific heat ratio (γ) is 1.4. Calculate the following variables:
(a) Rotor inlet and exit absolute flow angles
(b) Rotor inlet and exit total relative temperatures
(c) Stage reaction

10–5   The meridional projection of an axial-flow compressor stator is shown in Figure 10–28 along with the major dimensions. The stator, which has a purely axial meridional flow path, is operating under the following conditions:

- Mass flow rate = 8.59 kg/s
- Inlet total pressure = 5.1 bars
- Inlet total temperature = 507.0 K
- Airfoil (or metal) exit angle $(\alpha_2') = 18°$
- Flow deviation (or underturning) angle $(\epsilon) = 11.0°$
- Kinetic energy loss coefficient $(\bar{e}) = 0.075$

   Assuming an adiabatic flow, and a constant specific heat ratio of 1.4, calculate the streamwise rise or decline in the following variables:
(a) The total pressure (p_t)
(b) The static pressure (p)

(c) The static density (ρ)

(d) The axial velocity component (V_z)

10-6 The turbine stage shown in Figure 10–29 is a symmetric (50 percent reaction) stage, and is running at 45,000 r/min. The stage mean radius is 0.1 m, and the axial velocity component is constant throughout the stage. The flow conditions at station (1) are as follows:

- Stator-exit total pressure = 7.4 bars
- Stator-exit total temperature = 1450 K
- Stator-exit critical Mach number ($M_{cr,1}$) = 0.85
- Stator-exit absolute flow angle (α_1) = 72°

Assuming an adiabatic flow process, and an average specific heat ratio (γ) of 1.33, calculate:

(a) The static enthalpy drop across the rotor

(b) The rotor-exit absolute critical Mach number ($M_{cr,2}$)

(c) The power (P) produced by the turbine stage

10-7 Consider the subsonic/supersonic stator shown in Figure 10–32. At the design point, the stator operating conditions are as follows:

- Inlet total pressure = 12.3 bars
- Inlet total temperature = 1452 K
- Shaft speed = 46,000 r/min
- Mass flow rate = 6.44 kg/s
- Stator-exit critical Mach number = 2.35
- Stator-exit (absolute) flow angle = 79 deg.

Furthermore, the following simplifications apply:

- Constant axial velocity component
- Isentropic flow between the throat and the exit station
- Specific heat ratio = 1.32

a) Calculate the total pressure loss percentage.

b) Determine whether the rotor will receive a supersonic flow stream.

10-8 Following the steps presented in Example 10–8, and considering the axial-flow turbine stator in Figure 10–39, calculate the stator kinetic energy loss coefficient.

The stator vanes have the same geometry and vane count as that of the rotor. Also the inlet conditions (inlet total properties and mass flow rate) all remain the same. Finally, the following two conditions apply:

- Stator-exit critical Mach number = 0.93
- Stator-exit flow angle = 68 deg
- Inlet critical Mach number = 0.35

11

Variable-Geometry
Turbomachinery Stages

Chapter Outline

11–1   Definition of a Variable-Geometry Turbomachine   **208**

11–2   Examples of Variable-Geometry Turbomachines   **208**

Impeller

11–1 Definition of a Variable-Geometry Turbomachine

For a gas turbine engine that operates over a wide spectrum of engine operation modes, there are several means of improving its off-design performance by abandoning the fixed-geometry option in the design of bladed components. Turbomachines, in this engine category, are collectively referred to as "variable-geometry" turbomachines.

11–2 Examples of Variable-Geometry Turbomachines

Examples of Variable-Geometry turbomachines are depicted (for both axial-flow and radial-inflow turbines) in Figures 11–1 through 11–4. As seen in these figures, the change in geometry in such turbomachines corresponds to a process where the stator exit cross-flow area is either decreased or increased by rotating the entire airfoil, rotating its exit segment, or changing the shape or position of one endwall (or sidewall). The common objective here is to efficiently respond to a change in the mass flow rate and/or shaft speed. Figure 11–5 shows an example of the efficiency gain as a result of using variable-versus fixed-geometry radial turbines in a process where the shaft speed is reduced during operation.

PROBLEMS

11–1 The turbine stage in Figure 11–6 is symmetric (i.e., has a 50 percent reaction) and is running at 45,000 r/min. The stage mean radius is 0.1 m, and the axial-velocity component is assumed to be constant throughout the entire stage. The flow conditions at station 1 are as follows:

- The stator-exit total pressure = 7.4 bars
- The stator-exit total temperature = 1450 K
- The stator-exit critical Mach number = 0.85
- The stator-exit absolute flow angle = 72°

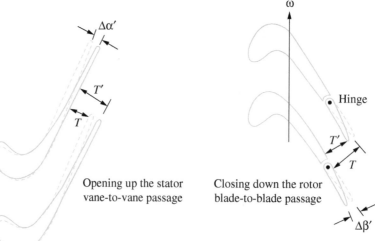

Figure 11–1

Rotating airfoil versus
rotating trailing edge.

Entire vane rotation mechanism

Articulating trailing edge mechanism

Opening up the stator
vane-to-vane passage

Closing down the rotor
blade-to-blade passage

Figure 11–2

Articulating trailing edge
in a radial turbine stator.

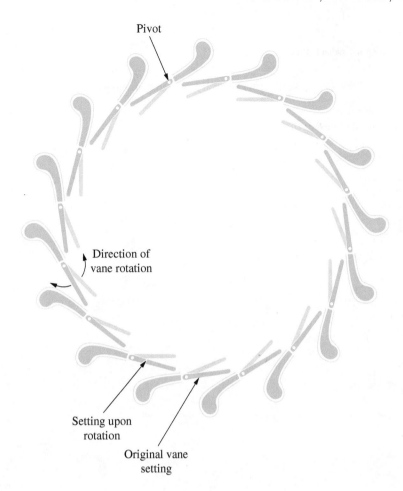

Pivot

Direction of
vane rotation

Setting upon
rotation

Original vane
setting

Assuming an adiabatic flow process and an average specific heat ratio γ of 1.33, calculate:

(a) The static enthalpy drop across the rotor

(b) The rotor-exit absolute critical Mach number

(c) The power P produced by the turbine stage

11–2   Shown in Figure 11–7 is the power turbine stage in an auxiliary power unit (APU) and the stator-exit mean-radius absolute velocity vector. The rotor speed is 17,900 r/min, and the specific shaft work, at the mean radius, is 75,000 J/kg. Assuming γ to be 1.33, calculate:

(a) The stage-exit flow angles (absolute and relative) α_2 and β_2

(b) The stage reaction

11–3   Figure 11–8 shows a single-stage turbine with a mean radius of 0.1 m. The turbine operating conditions are as follows:

- The mass flow rate $= 4.3$ kg/s

- The turbine-inlet total temperature $= 1400$ K

- The turbine-inlet total pressure $= 12.5$ bars

- The stator flow process is assumingly isentropic

- The rotor speed $= 34,000$ r/min

Figure 11–3

Purely translating insert in a radial turbine.

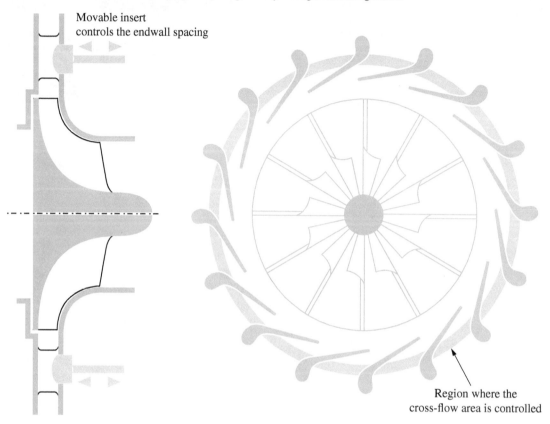

Varying the stator geometry through translating inserts

Movable insert
controls the endwall spacing

Region where the
cross-flow area is controlled

- The axial velocity component is constant throughout the stage
- The stator-exit critical Mach number $= 0.85$
- The stator-exit absolute flow angle $= 68°$
- The rotor-exit relative critical Mach number $= 0.82$
- The rotor relative-to-total pressure ratio $= 0.822$

With this information, calculate:

(a) The torque τ delivered by the rotor
(b) The stage total-to-total efficiency

11–4   Figure 11–9 shows an adiabatic compressor stage with a mean radius of 0.08 m. The stage operating conditions are as follows:

- The rotor-inlet absolute total pressure $= 1.85$ bars
- The rotor-inlet absolute total temperature $= 480$ K
- The rotor speed $= 58,000$ r/min
- The axial velocity component is constant throughout the stage.

Figure 11–4

Purely rotating insert in a variable-geometry radial stator.

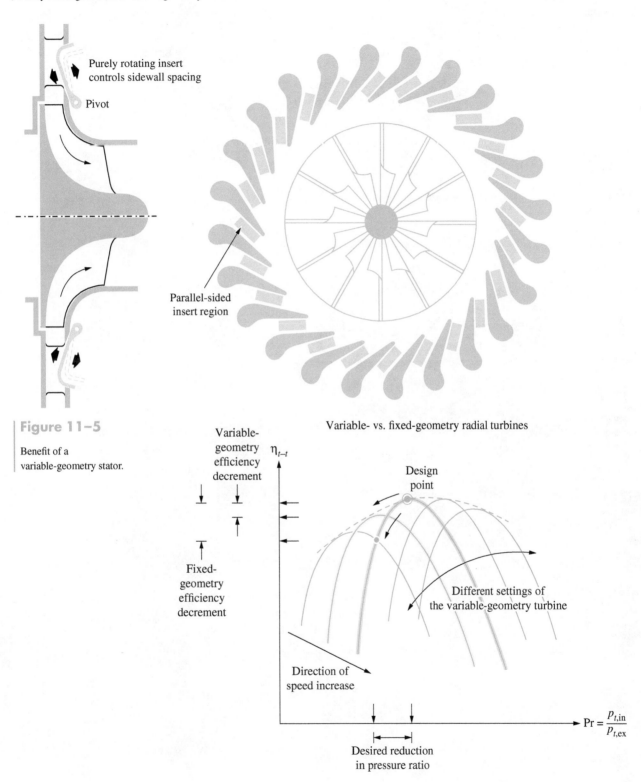

Purely rotating insert
controls sidewall spacing

Pivot

Parallel-sided
insert region

Variable- vs. fixed-geometry radial turbines

Figure 11–5

Benefit of a
variable-geometry stator.

Variable-
geometry
efficiency
decrement

η_{t-t}

Fixed-
geometry
efficiency
decrement

Design
point

Different settings of
the variable-geometry turbine

Direction of
speed increase

$Pr = \dfrac{p_{t,in}}{p_{t,ex}}$

Desired reduction
in pressure ratio

Figure 11–6

Input variables for problem 11–1.

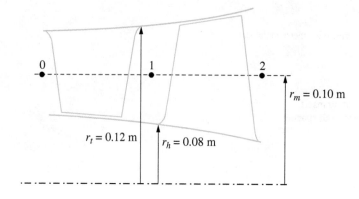

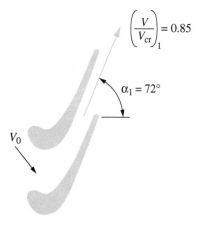

Figure 11–7

Input variables for problem 11–2.

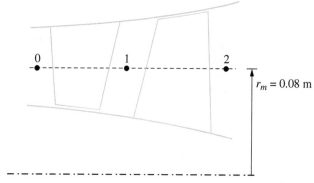

Figure 11–8

Figure 11–8

Input variables for
problem 11–3.

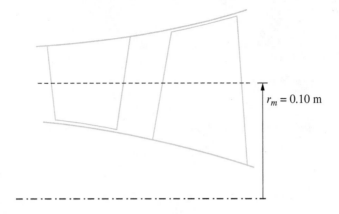

$r_m = 0.10$ m

Figure 11–9

Input variables for
problem 11–4.

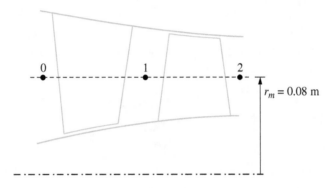

$r_m = 0.08$ m

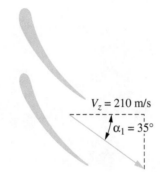

$V_z = 210$ m/s

$\alpha_1 = 35°$

- The stator-exit absolute flow angle = $+35°$
- The rotor-exit absolute total pressure = 3.1 bars
- The rotor-exit relative flow angle (β_2) is negative
- The rotor-exit static temperature = $496°$

Assuming an average specific heat ratio (γ) of 1.4, calculate:
(a) The change in total relative pressure across the rotor
(b) The stage reaction
(c) The rotor-exit static pressure

12

Normal Shocks

Chapter Outline

12–1   Introduction **216**

12–2   Shock Analysis **216**

12–3   Normal Shock Tables **219**

12–4   Shocks in Nozzles **220**

Impeller

Figure 12–1

Upstream versus
downstream flow
conditions across shock.

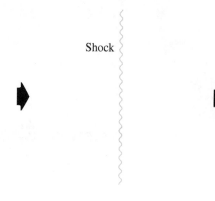

Figure 12–1

Upstream versus
downstream flow
conditions across shock.

Shock

12–1 Introduction

Up to this point, we have only considered continuous (both tangentially and normally) flow fields, that is, flow systems in which state changes occur continuously and in which the process can be identified, represented, and plotted (e.g., on the p-v diagram). This chapter covers a tangential discontinuity in practically all of the thermophysical properties. The term "tangential" here refers to the fact that the phenomenon occurs at a certain (tangential) distance of the flow stream (Figure 12–1). Recall that infinitesimal pressure disturbances are called sound waves, and these travel at the local speed of sound, which is determined by the flowing medium and the local state.

Due to the complex interactions involved, an analysis of the changes within the shock wave is beyond the scope of this book. Thus, we will deal only with the properties that exist on each side of the discontinuity. We will discover that the phenomenon manifests itself only when the case is that of supersonic flow, and that it is basically a compression process. We will apply the basic concepts of gas dynamics to analyze a shock wave in an arbitrary fluid, and then develop working equations for an ideal (or perfect) gas. This procedure naturally leads to compilation tables which will greatly simplify a typical problem solution.

12–2 Shock Analysis

Figure 12–2 shows a standing normal shock in the diverging section of a DeLaval nozzle. First, we establish a control volume (shown in Figure 12–2) which includes the shock region and an infinitesimal amount of fluid on each side of the shock. In this manner, we deal only with changes that occur across the shock. It is important to recognize that since the shock wave is extremely thin, a control volume in the form just described is, in turn, extremely thin in the flow direction. This permits the following simplifications to be made:

- The cross-flow areas on both sides of the shock are identical.
- There is but negligible surface contact with the nozzle wall, meaning a frictionless flow process.

Continuity Equation
Per unit cross-sectional area:

$$\rho_1 V_1 = \rho_2 V_2$$

Figure 12–2

Analysis of a standing
normal shock.

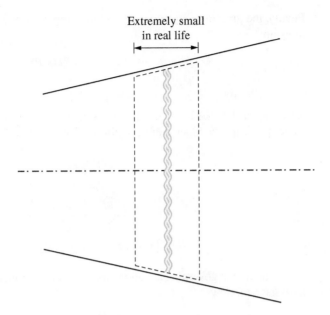

Energy Equation

$$h_t = h_1 + \frac{V_1{}^2}{2} = h_2 + \frac{V_2{}^2}{2} = \text{constant}$$

Referring to Figure 12–3, we can write this equation as shown next.

Momentum Equation

$$F_x = \dot{m}(V_2 - V_1) = A(p_1 - p_2)$$
$$p_1 - p_2 = \rho_2 V_2{}^2 - \rho_1 V_1{}^2$$

Figure 12–3

Expanding the
"infinitesimally short"
normal shock.

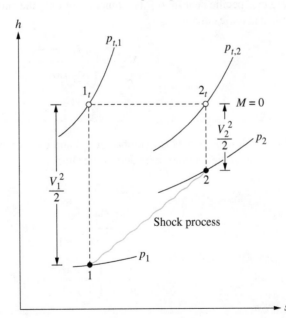

Finally, the preceding equations can all be lumped together to give the following relationship:

$$p_1 + \rho_1 V_1{}^2 = p_2 + \rho_2 V_2{}^2$$

Governing Equations for an Ideal Gas

Referring back to the basic aerodynamics equations, including the equation of state, the sonic speed, and the Mach number definitions, we can develop the following relationship:

$$\rho_1 V_1 = \rho_2 V_2$$

$$V = Ma = M\sqrt{\gamma RT}$$

$$\frac{p_1 M_1}{\sqrt{T_2}} = \frac{p_2 M_2}{\sqrt{T_2}}$$

$$p_1\left[1 + \gamma M_1{}^2\right] = p_2\left[1 + \gamma M_2{}^2\right]$$

At this point, the governing equations for a standing normal shock can be written for an ideal gas as follows:

$$\frac{p_1 M_1}{\sqrt{T_1}} = \frac{p_2 M_2}{\sqrt{T_2}}$$

Note that the Mach number at station 2 is a function of that at station 1, as well as the specific heat ratio (assumed constant for a small change of static temperature across the shock). A trivial, probably stupid, solution of this equation is to equate M_2 to M_1, meaning no shock whatsoever. To solve this equation for a real-life result, we algebraically manipulate it as follows:

$$p_1\left(1 + \frac{\gamma - 1}{2}M_1{}^2\right) = p_2\left(1 + \frac{\gamma - 1}{2}M_2{}^2\right)$$

where the specific heat ratio γ is a function of only the static temperature. The solution to this quadratic equation is:

$$M_2{}^2 = \frac{M_1{}^2 + \frac{2}{\gamma - 1}}{\left(\frac{2\gamma}{\gamma - 1}\right)M_1{}^2 - 1}$$

Example 12–1

Helium is flowing at a Mach number of 1.8 and enters a normal shock. Determine the static-to-static pressure ratio across the shock.

Solution

$$M_2{}^2 = \frac{M_1{}^2 + \frac{2}{\gamma - 1}}{\left(\frac{2\gamma}{\gamma - 1}\right)M_1{}^2 - 1} = 0.641$$

$$\frac{p_2}{p_1} = \frac{1 + \gamma M_1{}^2}{1 + \gamma M_2{}^2} = 3.80$$

Normal Shock Tables

We have found that for any given fluid with a specific set of conditions entering a normal shock, there is one and only one set of conditions that can result just downstream from the shock (Figure 12–4). An iterative-type solution results for a fluid which cannot be treated as an ideal gas, whereas the case of an ideal gas produces an explicit solution. The latter case opens the door to further simplifications since the preceding equation yields the exit Mach number for any inlet Mach number, and we can now eliminate the Mach number at the exit station from all previous equations.

For example, the static temperature relationship can be solved for the temperature ratio:

$$\frac{T_2}{T_1} = \frac{1 + \frac{\gamma-1}{2}M_1^2}{1 + \frac{\gamma-1}{2}M_2^2}$$

or:

$$\frac{T_2}{T_1} = \frac{1 + \frac{\gamma-1}{2}M_1^2}{1 + \frac{\gamma-1}{2}M_2^2}(\gamma+1)^2\left(\frac{2}{\gamma-1}M_1^2\right)$$

Similarly, the static pressure ratio can be re-expressed as follows:

$$\frac{p_2}{p_1} = \frac{2\gamma}{\gamma+1}M_1^2 - \left(\frac{\gamma-1}{\gamma+1}\right)$$

Combining the two preceding equations and invoking the equation of state, we can derive an expression for the static density ratio:

$$\frac{\rho_2}{\rho_1} = \frac{(\gamma+1)M_1^2}{(\gamma-1)M_2^2} + 2$$

Figure 12–4

Shock pattern in a subsonic/supersonic DeLaval nozzle.

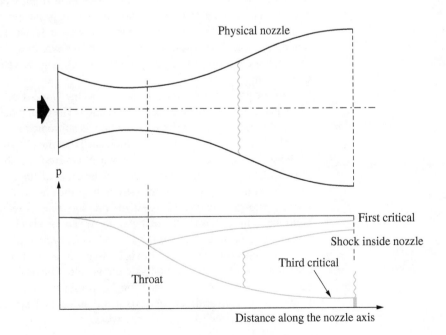

Now we consider the total-to-total pressure ratio:

$$\frac{p_{t2}}{p_{t1}} = \frac{p_2}{p_1}\left(\frac{1 + \frac{\gamma-1}{2}M_2{}^2}{1 + \frac{\gamma-1}{2}M_1{}^2}\right)$$

In fact, we can combine the preceding three equations to obtain an explicit relation for the specific entropy as a function of the inlet Mach number and the specific heat ratio γ.

12-4 Shocks in Nozzles

We have already discussed the isentropic flow operation in a converging-diverging DeLaval nozzle. Recall that this type of nozzle is physically identified by its area ratio or by the ratio of exit-to-throat area, with the latter being the section at which the Mach number gets to be unity. Furthermore, its flow conditions are determined by the operating pressure ratio or the ratio between the receiver pressure and that at the inlet station. There are two critical pressure ratios which are significant. For any pressure ratio above the first critical, the nozzle is not chocked and is allowing a subsonic flow field (typical Venturi operation). The first critical represents flow which is subsonic in both the converging and diverging segments, but is chocked with a Mach number of unity at the throat section. The reason, in this case, is the lack of smooth transition in the cross-flow area at the minimum area location where the nozzle "aerodynamic" throat was to exist. The third critical represents operation at the design conditions with supersonic flow beginning just downstream from the throat section. The first and third criticals represent the only operating points characterizing the nozzle operation and are as follows:

- Isentropic flow throughout the nozzle
- Mach number of unity at the throat section
- Exit pressure that is equal to the receiver pressure

Remember that, with subsonic flow at the exit station, the exit pressure must be equal to the receiver pressure. Imposing a pressure ratio slightly below that of the first critical presents a problem in that there is no way that an isentropic flow can meet the boundary condition of pressure equilibrium at the exit station. However, there is nothing to prevent a nonisentropic flow adjustment from occurring within the nozzle. This internal adjustment takes the form of a standing normal shock, which we now know involves entropy production.

As the pressure ratio is decreased below the first critical point, a normal shock takes place just downstream from the throat section. The remainder of the nozzle is now acting as a diffuser, since after the shock the flow is subsonic and the cross-flow area is rising. The shock will locate itself in a position such that the pressure changes that occur ahead of the shock, across the shock, and downstream from the shock will produce a pressure that exactly matches the exit (or receiver) pressure. In other words, the operating pressure ratio determines the location and strength of the shock. An example of this operation mode is shown in Figure 12–4. As the pressure ratio is reduced further, the shock will continue to move toward the exit station. When the shock is located at the exit station, this condition is referred to as the *second critical operation mode*.

If the operating pressure ratio is between the second and third criticals, then a compression process takes place outside the nozzle. This is called *overexpansion*, meaning that the flow has been expanded too far within the nozzle. If the receiver pressure is below the third critical, then an expansion process takes place outside the nozzle. This condition is referred to as *underexpansion*.

Figure 12-5

A normal shock at the
throat section of a
subsonic nozzle.

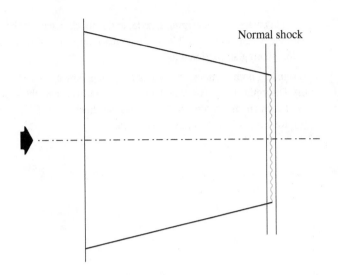

Normal shock

Example 12-2

Oxygen enters the converging section of the nozzle shown in Figure 12–5, and a
normal shock occurs at the exit station. The entering Mach number is 2.8, and the
area ratio (A_1/A_2) is 1.7. Calculate the overall static temperature ratio (T_3/T_1).

Solution

Let us proceed as follows:

$$\frac{A_2}{A_2^*} = \frac{A_2}{A_1}\frac{A_1}{A_1^*}\frac{A_1^*}{A_2} = 2.06$$

Thus:

$$M_2 \approx 2.23$$

And we get the following from the shock tables:

$$M_3 = 0.543$$

and

$$T_3/T_2 = 1.884$$

Now:

$$\frac{T_3}{T_1} = \frac{T_3}{T_2}\frac{T_{t2}}{T_{t1}}\frac{T_{t1}}{T_1} = 2.43$$

PROBLEMS

12-1

Air is treated as an ideal gas. Before it enters a shock, the conditions are:

- $M_1 = 2.0$
- $p_1 = 1.3$ bars
- $T_1 = 480$ K

Determine the conditions after the shock and the shock-wise entropy change.

12–2 Oxygen enters the converging nozzle, and a shock occurs at the exit station (Figure 12–5). The entering Mach number is 2.8, and the area ratio is 1.7. Calculate the overall static temperature ratio, ignoring all frictional losses.

12–3 A standing normal shock occurs in air which is flowing at a Mach number of 1.8.
(*a*) Determine the pressure and temperature ratios across the shock.
(*b*) Calculate the entropy change across the shock.
(*c*) Repeat part (b) for a flow at $M = 2.8$.

13

Oblique Shocks

Chapter Outline

13–1   Oblique Shock Tables and Charts   **227**
13–2   Boundary Condition of Flow Direction   **227**

Impeller

Let us consider a standing normal shock. Recall that the velocity drops as the fluid passes through the shock, and thus the normal velocity component suffers an instant decline across the shock. Also recall that for this type of shock, $V_{1,n}$ must always be supersonic, and that the normal velocity component downstream from the shock is always subsonic.

Let us now superimpose on the entire flow field a tangential velocity component which is perpendicular to the normal velocity component. This is equivalent to running along the shock front with a constant tangential velocity component. The resulting picture is depicted in Figure 13–1. As before, we realize that velocity superposition does not affect the static states of the fluid.

We would normally view this picture in a slightly different way if we concentrate on the resulting velocity, rather than on its components. Now, the following conclusions can be drawn:

- The shock is no longer normal to the approaching flow.
- The flow has been deflected away from the normal.

The velocity component downstream from the shock could be supersonic should the tangential velocity component there be sufficiently large.

We define the shock angle θ as the acute angle between the approaching flow stream and the shock front. The deflection angle is the angle through which the flow direction has been altered.

Referring to Figure 13–2, viewing the oblique shock in the preceding manner (as a combination of normal plus tangential velocity components) permits us to use the normal-shock governing equations and tables to solve oblique shock problems for an ideal gas.

$$V_{1,n} = V_1 \sin \theta$$

Since the sonic speed, for an ideal gas, is a function of temperature only, then

$$a_{1,n} = a_1$$

Dividing the preceding two equations, we get:

$$M_{1,n} = M_1 \sin \theta$$

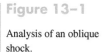

Figure 13–1

Analysis of an oblique shock.

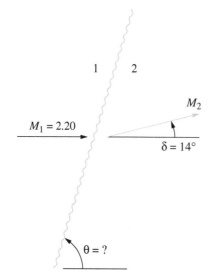

Figure 13–2

Tangential velocity
component remains
constant.

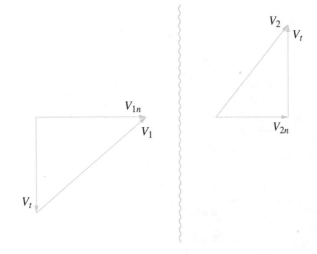

Therefore, if we know the approaching Mach number and the shock angle, the normal-shock tables can be utilized by using the normal Mach number. This procedure can be used to obtain the static pressure and temperature changes across the shock, since these are unchanged by superposition of the tangential velocity component on the original normal shock picture.

Let us now investigate the range of possible shock angles that may exist for a given Mach number. We know that for a shock to exist:

$$M_1 \sin \theta < 1$$

$$\theta_{\min} = \sin^{-1} \frac{1}{M_1}$$

Note that this is a limiting condition whereby no shock would really occur. For this reason, the shock front is called the Mach wave rather than the shock wave. The maximum magnitude that θ can achieve is obviously 90°. This is another limiting case and represents a normal shock.

Notice that as the shock angle θ decreases from 90° to the Mach angle μ, $M_{1,n}$ decreases from M_1 to unity. Since the strength of the shock is dependent upon the normal Mach number, we have the means to produce a shock of any strength equal to or less than the normal shock.

Example 13–1

Figure 13–3 shows the details of an oblique shock. Calculate the flow conditions after the shock.

Solution

$$a_1 = \sqrt{\gamma R T_1} = 472.4 \text{ m/s}$$

$$V_1 = M_1 a_1 = 758.3 \text{ m/s}$$

$$M_{1,n} = M_1 \sin \theta = 1.39$$

$$V_{1,n} = M_{1,n} a_1 = 656.8 \text{ m/s}$$

$$V_\theta = V_1 \cos \theta = 379.1 \text{ m/s}$$

Figure 13–3

Example input variables.

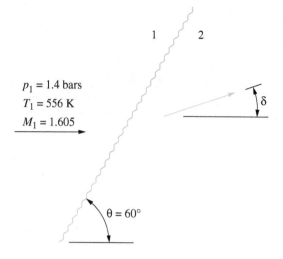

Using information from the normal shock tables at $M_{1,n} = 1.39$, we find $M_{2,n} = 0.744$, $T_2/T_1 = 1.248$, $p_2/p_1 = 2.088$, and $p_{t,2}/p_{t,1} = 0.961$. Note that the static temperatures and pressures are the same whether we are talking about the normal or the oblique shock. Now, using this information, we have:

$$p_2 = (p_2/p_1)/p_1 = 2.84 \text{ bar}$$
$$T_2 = (T_2/T_1)T_1 = 948.9 \text{ K}$$
$$a_2 = \sqrt{(\gamma R T_2)} = 527.9 \text{ m/s}$$
$$V_{2,n} = M_{2,n}a_2 = 392.8 \text{ m/s}$$
$$V_{2,\theta} = V_{1,\theta} = 379.1 \text{ m/s}$$
$$V_2 = \sqrt{\left(V_{2,n}^2 + V_{2,\theta}^2\right)} = 545.8$$
$$M_2 = V_2/a_2 = 1.034$$

Note that, although the normal component is subsonic after the shock, the velocity after the shock is supersonic in this case. Now we calculate the deflection angle:

$$\tan\beta = \frac{1244}{1289} = 0.965$$
$$\beta = 44°$$

Now:

$$90 - \theta = \beta - \delta$$

which yields:

$$\delta = 14°$$

Let us now calculate the total pressures:

$$p_{t,1} = \frac{p_{t,1}}{p_1}p_1 = 5.83 \text{ bars}$$

$$p_{t,2} = \frac{p_{t,2}}{p_2}p_2 = 5.59 \text{ bars}$$

Figure 13–4

Weak versus strong shocks.

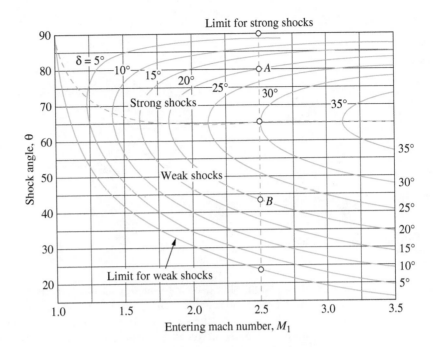

13-1 Oblique Shock Tables and Charts

At this point, we have a relationship relating the shock angle to the deflection angle and the entering Mach number. Our objective in obtaining this relationship was to solve problems in which the shock angle θ is the unknown, but we found that an explicit solution for $\theta = f(M, \delta, \gamma)$ was not possible. The next best thing is to plot this relationship. This can be achieved in several ways, but it is perhaps most instructive to look at a plot of the shock angle θ versus the entering Mach number M_1 for various deflection angles δ. This plot is shown in Figure 13–4.

13-2 Boundary Condition of Flow Direction

We have seen that one of the characteristics of an oblique shock is that the flow direction is changed. Consider a supersonic flow over a wedge-shaped object as shown in Figure 13–5 and the body of revolution in Figure 13–6. For example, this configuration may represent the leading edge of a supersonic airfoil.

In this case, the flow is forced to change direction to meet the boundary condition of flow tangency along the wall, and this can only be done through the mechanism

Figure 13–5

Shock at the tip of a conical body of revolution.

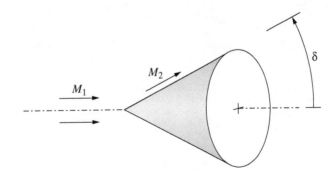

Figure 13-6

Shocks for a blunt body
of revolution.

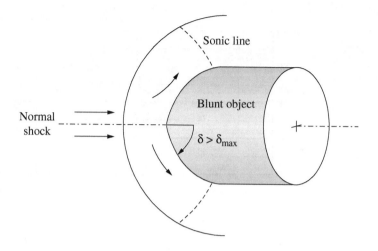

of an oblique shock. Now for any given Mach number and deflection angle, there are two possible shock angles. Thus, the question naturally arises as to which solution will actually take place, the strong or the weak one. Here is where the surrounding pressure must be considered. Recall that the strong shock occurs at the higher shock angle and results in a large pressure change. In order for this solution to occur, a physical situation must exist which creates the necessary pressure differential. It is conceivable that such a case might exist in an internal flow situation. However, for an external flow, such as around an airfoil, there is no means available to support the greater pressure difference required by the strong shock. Thus, in external flow problems (flow around objects) we always find the weak shock situation.

PROBLEMS

13-1 Observation of an oblique shock in air reveals that a Mach 2.2 flow at 550 K and 2 bar is deflected by 14°. Calculate the conditions after the shock. Assume that the weak shock prevails.

13-2 A supersonic Pitot tube indicates a total pressure of 2 bars and a static pressure of zero. Determine the free-stream velocity if the temperature of the air is 250 K.

13-3 A converging-diverging nozzle with an area ratio of 5.9 is fed by air from a chamber with a total pressure of 6.8 bars. Exhaust is to the atmosphere at 1 bar. Show that this nozzle is operating between the second and third critical modes, and determine the conditions after the first shock.

14

Prandtl-Meyer Flow

Chapter Outline

14–1   Argument for Isentropic Turning Flows **230**

14–2   Pressure and Entropy Changes Versus Deflection Angles for Weak Oblique Shocks **231**

14–3   Isentropic Turns from Infinitesimal Shocks **231**

14–4   Analysis of Prandtl-Meyer Flow **235**

14–5   Prandtl-Meyer Tables **238**

Impeller

14–1 Argument for Isentropic Turning Flows

Throughout this chapter, we will assume that the working medium is an ideal gas, as this will enable us to develop some precise relations. We begin by recalling the outcome of the continuity equation:

$$\rho_1 V_1 = \rho_2 V_2$$

As for the outcome of the momentum analysis, we have:

$$p_1 - p_2 = \rho V_{av}(V_2 - V_1)$$

Choosing $\rho_1 V_1$ for the term ρV, we get:

$$p_1 - p_2 = \rho_1 V_1 (V_2 - V_1) = \rho_1 V_1{}^2 \left(\frac{V_2}{V_1} - 1 \right)$$

Invoking the ideal gas equation of state, we get:

$$p_2 - p_1 = \left(\frac{p_1}{RT_1} \right) V_1{}^2 \left(1 - \frac{V_2}{V_1} \right)$$

In addition, the following well-known relationship applies:

$$V_1{}^2 = M^2 a_1{}^2 = \gamma M_1{}^2 R T_1$$

which yields the following relationship:

$$\frac{p_2 - p_1}{p_1} = \gamma M_1{}^2 \left(1 - \frac{V_2}{V_1} \right)$$

Furthermore, the density ratio across the shock can be expressed as follows:

$$\frac{\rho_2}{\rho_1} = \frac{(\gamma + 1) M_1{}^2}{(\gamma - 1) M_1{}^2 + 2}$$

As for the velocity ratio, we have (through the use of the continuity equation) the following:

$$\frac{V_1}{V_2} = \frac{\rho_2}{\rho_1} = \frac{(\gamma + 1) M_1{}^2}{(\gamma - 1) M_1{}^2 + 2}$$

Manipulating the last two equations, we get:

$$\frac{p_2 - p_1}{p_1} = \gamma M_1{}^2 \left[1 - \frac{(\gamma - 1)^2}{M_1} + 2(\gamma + 1) M_1{}^2 \right]$$

Finally we arrive at the following relationship:

$$\frac{p_2 - p_1}{p_1} = \frac{2\gamma}{\gamma + 1} (M_1{}^2 - 1)$$

This relationship shows that the pressure rise across a normal shock is directly proportional to the quantity $(M_1{}^2 - 1)$.

14–2 Pressure and Entropy Changes Versus Deflection Angles for Weak Oblique Shocks

The relationships developed for normal shocks equally apply to the "normal component" of an oblique shock. Now:

$$M_{1,n} = M_1 \sin\theta$$

We can, therefore, derive the following relationship:

$$\frac{p_2 - p_1}{p_1} = \frac{2\gamma}{\gamma + 1}(M_1^2 \sin^2\theta - 1)$$

We now proceed to relate the quantity $(M_1^2 \sin^2\theta - 1)$ to the deflection angle for the case of a weak oblique shock:

$$\frac{\tan(\theta - \delta)}{\tan\theta} = \frac{(\gamma - 1)M_1^2 \sin^2\theta + 2}{(\gamma + 1)M_1^2 \sin^2\theta}$$

or:

$$\frac{(\gamma + 1)\tan(\theta - \delta)}{2\tan\theta} = \frac{\gamma - 1}{2} + \frac{1}{M_1^2 \sin^2\theta}$$

With the use of appropriate trigonometric identities, one can put the preceding equation in the following form:

$$M_1^2 \sin^2\theta - 1 = \frac{\gamma + 1}{2} M_1^2 \frac{\sin\theta \sin\delta}{\cos(\theta - \delta)}$$

Now we restrict ourselves to the consideration of very weak shocks, in which the deflection angle δ is very small. Thus:

$$\cos(\theta - \delta) \approx \cos\theta$$

$$\sin\delta \approx \delta$$

which paves the way to the following relationship:

$$M_1^2 \sin^2\theta - 1 = \left[\frac{\gamma + 1}{2} M_1^2 \tan\theta\right]\delta$$

Note that for a very weak oblique shock, the shock angle θ approaches $\theta_{\min}$ or the Mach angle μ. Utilizing the above-derived relationships, the following relationship can be attained:

$$\frac{p_2 - p_1}{p_1} = \frac{2\gamma}{2}\left(\frac{\gamma + 1}{2} M_1^2 \tan\mu_1\right)\delta$$

14–3 Isentropic Turns from Infinitesimal Shocks

We have laid the groundwork to show a remarkable phenomenon. Figure 14–1 shows a finite turn divided into n equal segments of δ each. The total turning angle will be indicated by δ_T, and thus:

$$\delta_T = n\delta$$

Figure 14–1

Breaking up a sharp turn
into several small turns.

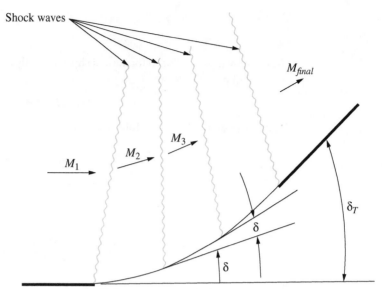

Finite turn composed of many small turns

Each segment of the turn causes a shock wave to form with an appropriate change in the Mach number, pressure, temperature, entropy, and other properties. As we increase the number of segments n, δ becomes very small, which means that each shock will become a very weak oblique shock, and the results of the previous section become applicable. Thus, for each segment, we may write the following relationship:

$$\Delta p' = \delta$$

$$\Delta s' = \delta^3$$

where $\Delta p'$ and $\Delta s'$ are the pressure and entropy changes across each segment. Now, for the total turn we have:

Total $\Delta p = \Sigma \Delta p'$ and is proportional to $n\delta$

Total $\Delta s = \Sigma \Delta s'$ and is proportional to $n\delta^3$

We are now in a position to express δ as $\frac{\delta_T}{n}$.

As n tends to infinity, we get the following expression:

$$\text{Total } \Delta p = Limit\left[\frac{\delta_T}{n}\right], \text{ which is proportional to } \delta_T.$$

$$\text{Total } \Delta s = Limit\left[\frac{\delta_T}{n}\right]^3, \text{ and it tends to zero.}$$

In the limit as n goes to infinity, we conclude that:

- The wall makes a smooth turn through the angle δ_T.
- The shock waves approach Mach waves.
- The Mach number continually changes.
- There is a finite pressure change.
- There is no entropy change.

Figure 14–2

Compression shock
pattern past a smooth turn.

Prandtl-Meyer compression

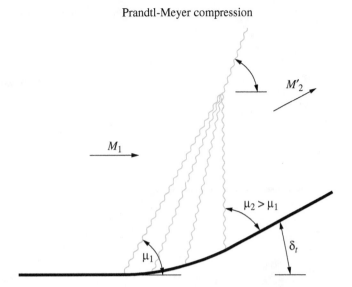

The final result is shown in Figure 14–2. Note that as the turn progresses, the Mach number decreases, and thus the Mach waves are at ever-increasing angles. Thus, we observe an envelope of Mach lines that forms a short distance from the wall. The Mach waves coalesce to form an oblique shock inclined at the proper angle θ corresponding to the initial Mach number and the overall deflection angle δ_T.

We return to the flow in the neighborhood of the wall as this is interesting. Here we have an infinite number of infinitesimal compression waves. We have achieved a decrease in Mach number and an increase in pressure without any change in entropy. Since we are dealing with an adiabatic flow, an isentropic process indicates that there are no losses, that is, the process is reversible.

The reverse process (an infinite number of infinitesimal expansion waves) is shown in Figure 14–3. Here we have a smooth turn in the other direction from that previously

Figure 14–3

Prandtl-Meyer expansion
process past a smooth
turn.

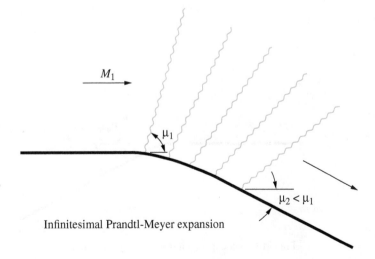

Infinitesimal Prandtl-Meyer expansion

Figure 14–4

Prandtl-Meyer expansion
past a sharp turn.

General Prandtl-Meyer expansion

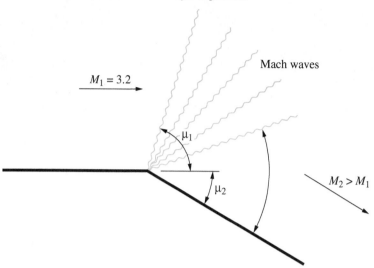

discussed. In this case, the turn progresses, and the Mach number increases. Thus, the Mach angles are decreasing, and the Mach waves will never interact.

If the corner were sharp, all of the "expansion waves" would emanate from the corner (Figure 14–4). This is called a "centered expansion fan."

All of the preceding isentropic flows can be lumped together as the Prandtl-Meyer flow. At a smooth concave wall, we have a Prandtl-Meyer compression. Flows of this type are not too important since the boundary layer and other real-life effects interfere with the isentropic region near the wall. At a smooth convex wall or at a sharp convex turn (Figure 14–5), we have Prandtl-Meyer expansions. These expansions are quite prevalent in a supersonic flow, as the examples given later in this chapter will show. Incidentally, we have now discovered the second means by which the flow direction of a supersonic stream may be changed.

Figure 14–5

Prandtl-Meyer expansion
with a unity upstream
Mach number.

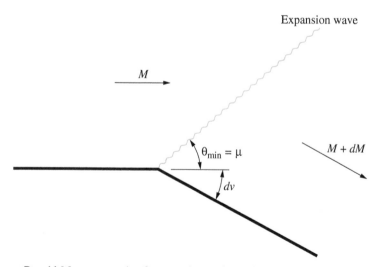

Prandtl-Meyer expansion from a unity mach number

Figure 14–6

Illustration of an
incremental change in
velocity.

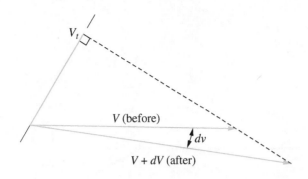

Figure 14–6

Illustration of an
incremental change in
velocity.

14–4 Analysis of Prandtl-Meyer Flow

We have already established that the flow is isentropic through a Prandtl-Meyer compression or expansion. If we know the final Mach number, we can use the isentropic flow relationships to compute the final thermodynamic state for any given set of initial conditions. Thus, our objective in this section is to relate the changes in Mach number to the turning angle in a Prandtl-Meyer flow.

Figure 14–6 shows a single Mach wave caused by turning the flow through an infinitesimal angle $d\nu$. It is more convenient to measure ν positive in the direction shown, which corresponds to an expansion wave. The pressure difference across the wave front causes a momentum change and, therefore, a velocity change perpendicular to the wave front. There is no mechanism by which the tangential velocity component can be changed. In this respect, the situation is similar to that of an oblique shock. A detail of this velocity relationship is shown in Figure 14–7, where V represents the magnitude of velocity before the expansion wave and $V + dV$ is the magnitude after the wave. In both cases, the tangential component of velocity is V_t. From the velocity triangle in Figure 14–8 we see that:

$$V_t = V \cos \nu$$

and

$$V_t = (V + dV) \cos(\nu + d\nu)$$

Equating these, we obtain:

$$V \cos \mu = (V + dV) \cos(\mu + d\nu)$$

Figure 14–7

Further illustration of an
incremental change in
velocity.

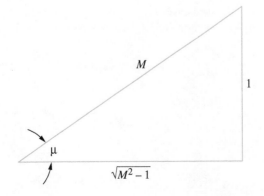

Figure 14–8

Input variables for
Example 14–1.

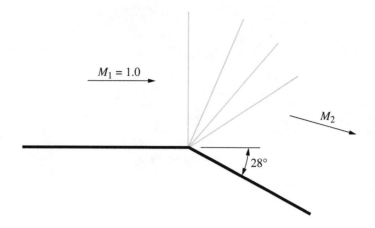

Figure 14–8

Input variables for
Example 14–1.

Expanding $\cos(\mu + d\nu)$, this becomes:

$$V \cos \mu = (V + dV)(\cos \mu \cos d\nu - \sin \mu \sin d\nu)$$

But $d\nu$ is a very small angle, therefore:

$$\cos d\nu \approx 1$$

and

$$\sin d\nu = d\nu$$

and the preceding equation becomes:

$$V \cos \mu = (V + dV)(\cos \mu - d\nu \sin \mu)$$

The final result is:

$$d\nu = \cot \mu \frac{dV}{V}$$

Now the cotangent of μ can easily be obtained in terms of the Mach number. We know that $\sin \mu = \frac{1}{M}$. From the triangle in Figure 14–7, we see that:

$$\cot \mu = \sqrt{M^2 - 1}$$

Through simple substitution, we get:

$$d\nu = \sqrt{M^2 - 1} \frac{dV}{V}$$

Recall that our objective is to obtain a relationship between the Mach number and the turning angle $d\nu$. Thus, we seek a means of expressing dV/V as a function of the Mach number. In order to obtain an explicit expression, we will assume the fluid to be an ideal gas. First:

$$V = Ma = M\sqrt{\gamma R T}$$

Therefore:

$$dV = dM\sqrt{\gamma R T} = \frac{M}{2}\sqrt{\frac{\gamma R}{T}} dT$$

Thus:

$$\frac{dV}{V} = \frac{dM}{M} + \frac{dT}{2T}$$

Now, knowing that:

$$T_t = T\left[1 + \frac{\gamma - 1}{2}M^2\right]$$

Then:

$$dT_t = dT\left[1 + \frac{\gamma - 1}{2}M^2\right] + T(\gamma - 1)MdM$$

But since there is no shaft work or heat transfer to or from the fluid as it passes through the expansion wave, the total enthalpy h_t remains constant. For our ideal gas, this means that the total temperature T_t remains (more or less) constant, or:

$$dT_t = 0$$

Solving the preceding two equations, we get:

$$\frac{dT}{T} = \frac{(\gamma - 1)MdM}{\left[1 + \frac{\gamma - 1}{2M^2}\right]}$$

or:

$$\frac{dV}{V} = \frac{1}{\left[1 + \frac{\gamma - 1}{2}M^2\right]}\frac{dM}{M}$$

We can now accomplish our objective by simple substitution, which yields the following relationship:

$$dv = \frac{(M^2 - 1)^{1/2}}{\left[1 + \frac{\gamma - 1}{2}M^2\right]}\frac{dM}{M}$$

This is a significant relationship, for it states that:

$$dv = f(M, \gamma)$$

For a reasonable range of static temperature, γ is fixed, and the preceding equation can be integrated to yield:

$$v + \text{constant} = \left(\frac{\gamma + 1}{\gamma - 1}\right)^{1/2}\tan^{-1}\left[\frac{\gamma - 1}{\gamma + 1}(M^2 - 1)\right]^{1/2} - \tan^{-1}(M^2 - 1)^{1/2}$$
$$- \tan^{-1}(M^2 - 1)^{1/2}$$

If we set $v = 0$ when $M = 1$, then the constant will be zero, and we end up with the following relationship:

$$v = \left(\frac{\gamma + 1}{\gamma - 1}\right)^{1/2}\tan^{-1}\left[\frac{\gamma - 1}{\gamma + 1}(M^2 - 1)\right]^{1/2} - \tan^{-1}(M^2 - 1)^{1/2}$$

Establishing the constant as zero in the preceding manner attaches a special significance to the angle v. This is the angle, measured from the flow direction where $M = 1$, through which the flow has been turned (by an isentropic process) to reach the indicated Mach number. The preceding expression is referred to as the *Prandtl-Meyer function*.

14-5 **Prandtl-Meyer Tables**

The preceding equation is the basis for solving all problems involving Prandtl-Meyer expansion or compression. If the Mach number is known, it is relatively easy to solve for the turning angle. However, in a typical problem the turning angle might be prescribed and no explicit solution is available for the Mach number. This function (ν) has been included as a column of the isentropic table. The following example illustrates how rapidly problems of this type are solved.

Example 14–1

The wall in Figure 14–8 turns at an angle of 28°, with a sharp corner. The fluid, which is initially at $M = 1$, must follow the wall, and in so doing it executes a Prandtl-Meyer expansion at the corner. Recall that ν represents the angle (measured from the flow direction where $M = 1$) through which the flow has turned.

Solution

Since M_1 is unity, $\nu_2 = 28°$.

From the isentropic table, we find that this Prandtl-Meyer function corresponds to $M_2 \approx 2.06$.

Figure 14–9

Input variables for problem 14–2.

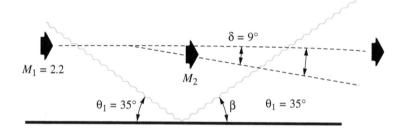

Figure 14–10

Input data for problem 14–3.

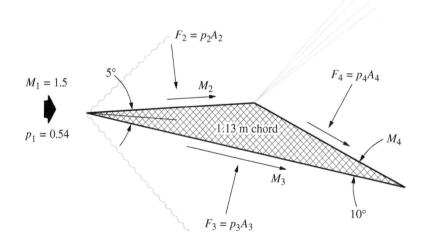

PROBLEMS

14-1   Calculate the lift force per unit span of a flat-plate airfoil with a chord of 2 m when flying at $M = 1.8$ and an angle of attack of 5°. Ambient air pressure is 0.4 bar.

14-2   Air at a Mach number of 2.2 passes through an oblique shock at an angle of 35°. The shock runs into a physical boundary as shown in Figure 14–9. Find the angle of reflection and compare the strengths of the two shock waves.

14-3   A supersonic airfoil shape is shown in Figure 14–10, which is composed of an isosceles triangle with 10° equal angles and a 1.13-m chord. When it is operating at a 5° angle of attack, the air flow appears as shown in Figure 14–10. Find the pressures on the various surfaces and the lift and drag forces when flying at $M = 1.5$ through air with a pressure of 0.54 bar.

15

Internal Flows: Friction, Pressure Drop, and Heat Transfer

Chapter Outline

15–1   Assumptions   **242**

15–2   Mass Conservation   **242**

15–3   Pressure Drop and Wall Shear Stress   **242**

15–4   Friction Factors   **244**

15–5   Mechanical Energy Conservation   **244**

15–6   Energy Conservation   **245**

Impeller

15–1   Assumptions

For the analyses that follow, we invoke the following assumptions:

- The flow is steady.
- The fluid is incompressible (i.e., the Mach number is much less than 0.4).
- The flow entrance effects are negligible (i.e., the flow is fully developed).
- The flow is through a straight duct with a constant circular cross-sectional area.
- Uniform pressure exists over the flow cross-sectional area.

15–2   Mass Conservation

Consider the cylindrical control volume shown in Figure 15–1. Mass enters at station 1 and exits at station 2 through the circular cross-sectional area. In general:

$$\dot{m}_1 = \dot{m}_2$$

This equation can be restated using the velocity distributions at stations 1 and 2 as follows:

$$\left[\rho \int_A V\, dA \right]_1 = \left[\rho \int_A V\, dA \right]_2$$

In general, the velocity distribution depends on both the radial and axial coordinates. However, our assumption of fully developed flow implies that the velocity distribution depends only on the radial coordinate.

We can express the mass conservation principle in terms of the average velocities as follows:

$$[\rho V_{\mathrm{av}} A_1]_1 = [\rho V_{\mathrm{av}} A_2]_2$$

or

$$\dot{m} = \rho V_{\mathrm{av}} \pi R^2 = \text{constant}$$

where R is the tube radius.

15–3   Pressure Drop and Wall Shear Stress

To generalize, the duct is now inclined at an arbitrary angle (θ) to explicitly include the gravitational body forces. This inclined control volume is shown in Figure 15–2 along

Figure 15–1

Effect of wall friction.

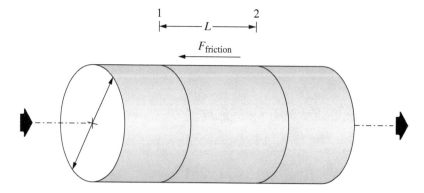

Figure 15–2

Forces acting on a control volume.

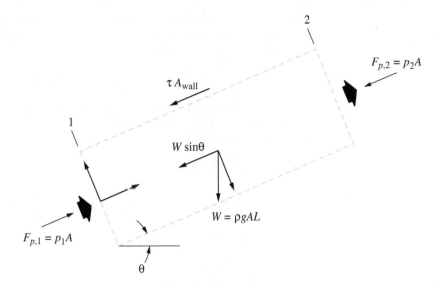

with all the forces acting on it. Among these forces are the pressure forces at stations 1 and 2, both acting inward by definition. We assume that the pressure forces acting on the cylinder sides are symmetrical and, thus, do not introduce any net force. The viscous force acts on the side wall of the cylindrical control surface opposing the flow. Again, because the flow is fully developed, the wall shear stress doesn't depend on x and, hence, is constant. Thus, the viscous force is simply the product of the wall shear stress τ_{wall} and the side-wall area A_{wall}, as shown in Figure 15–2. The body force is just the weight of the fluid within the control volume, and acts vertically downward. The component $W \sin \theta$ acts in the negative x-direction, opposing the flow stream.

The flow momentum entering the control volume in the x-direction can be expressed as follows:

$$x\text{-momentum at station } 1 = \int_{A, x-\sec} \rho V_x^2 dA_{x-\sec}$$

Similarly, the flow momentum exiting in the x-direction is:

$$x\text{-momentum at station } 2 = \int_{A, x-\sec} \rho V_x^2 dA_{x-\sec}$$

With all the x-direction forces and momentum flows defined, we are now prepared to apply the momentum conservation principle as follows:

$$(\text{momentum flow in}) - (\text{momentum flow out}) + \Sigma F_x = 0$$

By simple substitution, we get:

$$p_1 A - p_2 A - \tau_{\text{wall}} A_{\text{wall}} - \rho g \sin \theta A L = \int_A \rho V_x^2 dA - \int_A \rho V_x^2 dA$$

We immediately notice that the right-hand side of this equation is zero. With that, let us rearrange the terms in this equation as follows:

$$(p_1 - p_2) - \rho g L \sin \theta = \tau_{\text{wall}} \frac{A_{\text{wall}}}{A} = 4 \tau_{\text{wall}} \frac{L}{D}$$

Since the wall shear stress is constant (i.e., independent of x), we conclude that the pressure distribution along the length of a tube must be linear. Allowing L to be a variable (i.e., $L \equiv x$), we obtain the following:

$$p(x) = p_1 - 4\frac{\tau_w}{D}x$$

At this point, we have transformed the problem of calculating the pressure drop to that of calculating the wall shear stress τ_w. To determine the wall shear stress, however, requires knowledge of the details of the flow structure. Specifically speaking, for a Newtonian fluid, knowledge of the velocity distribution allows us to evaluate the shear stress using the relationship:

$$\tau_w = -\mu \left[\frac{\partial V_x}{\partial r}\right]_{r=R}$$

where R is the tube radius.

15-4 Friction Factors

Analogous to the dimensionless friction and drag coefficients defined for external flows, friction factors are defined for internal flows. The Darcy friction factor is a dimensionless wall shear stress defined as follows:

$$F_D = \frac{4\tau_w}{1/2\rho V_{av}^2}$$

15-5 Mechanical Energy Conservation

Applying the mechanical energy equation to the control volume in Figure 15–2 for an incompressible, nominally isothermal flow yields the following:

$$p_1 - p_2 + 1/2\rho\left(\alpha_1 V_{av,1}^2 - \alpha_2 V_{av,2}^2\right) + \rho g(z_1 - z_2) = \rho g h_L$$

where α_1 and α_2 are the so-called kinetic energy correction factors, h_L is the head loss, and z is the elevation above a specified datum. Since we are dealing with a fully developed flow, mass conservation requires that $\alpha_1 = \alpha_2$, and $V_{av,1} = V_{av,2}$. Furthermore, we note that the elevation change $z_2 - z_1$ relates to the length of the control volume as:

$$z_2 - z_1 = L \sin \theta$$

Now, we can get the following simplified equation:

$$p_1 - p_2 - \rho g L \sin \theta = \rho g h_L$$

Comparing this equation with the momentum conservation equation, we see that $\rho g h_L$ must be equal to $\frac{4\tau_w L}{D}$. The head loss, therefore, is related to the wall shear stress as follows:

$$h_L = 4\frac{\tau_w L}{\rho g D}$$

Now the problem reduces to determining the wall shear stress.

The head loss can also be related to the Darcy friction factor f_D by combining the two preceding equations as follows:

$$h_L = f_D \frac{L}{D}\frac{V_{av}^2}{2g}$$

This equation is known as the Darcy-Weisbach equation. For a horizontal pipe, we get the following relationship:

$$p_1 - p_2 = f_D \frac{L}{D} \frac{1}{2} \rho V_{av}^2$$

15-6   Energy Conservation

To apply the energy conservation principle, we distinguish between a nominally isothermal flow where the temperature variations are quite small and the nonisothermal flows where the temperature variations in both the radial and axial directions are too much to ignore. The former case applies to flows in which the only heating of the fluid is a result of frictional effects, whereas the latter applies to flows where the fluid is deliberately heated or cooled by applying a heat flux to the tube wall or fixing the tube wall temperature above or below that of the entering fluid.

Nominally Isothermal Flow

We can write the energy equation for the control volume in Figure 15–2 as follows:

$$p_1 - p_2 + \rho g(z_1 - z_2) = \rho \left[u_2 - u_1 + \frac{\dot{Q}_{cv,net,out}}{\dot{m}} \right]$$

This equation is a result of cancelling the kinetic energy terms and deleting the work term. Since the flow is nominally isothermal, the conservation of energy principle is not particularly useful to our analysis. We are now in a position to obtain the following relationship:

$$g h_L = u_2 - u_1 + \frac{\dot{Q}_{cv,net,out}}{\dot{m}}$$

Once again, we have cancelled the kinetic energy terms. We see that if the flow is adiabatic, the head loss is directly proportional to the increase in the internal energy of the fluid. Furthermore, for an ideal gas (and many liquids), the thermal energy change is proportional to the temperature change. Adding the momentum conservation principle to these observations, we conclude that:

$$(T_2 - T_1) \text{ is proportional to } \tau_w$$

That is, the fluid friction at the wall produces a small temperature rise in a fluid that is flowing through a tube, a conclusion that is consistent with our intuition.

Nonisothermal Flow

In this subsection, we apply integral energy conservation analyses, first, to the flow through a tube in which a uniform heat flux is applied through the tube wall and, second, to a flow through a tube in which the wall temperature is maintained at a fixed value above (or below) the temperature of the entering fluid. In both cases, our objective is to provide mathematical relationships that describe the following:

- Bulk mean temperature as a function of the axial location x
- The tube wall temperature as a function of the axial location x
- The local heat flux as a function of the axial location x
- The total heat transferred to the fluid over the distance $x = 0$ to $x = L$

Bulk Mean Temperature

Our analyses are greatly simplified by defining a single temperature that characterizes the thermal state of the fluid at any axial location, that is, the bulk mean temperature alluded to in the first item of our list.

Consider a fluid that is being heated as it flows through a tube. At any axial location, the fluid temperature is greatest at the wall and is a minimum at the centerline. The specific form of this temperature distribution is determined by details of the flow as described by the differential forms of mass, momentum, and energy conservation equations. For the time being, however, we are not concerned with these details but need only recognize that a temperature distribution $T(r, x)$ exists. To define the bulk mean temperature, imagine that we cut the tube at an arbitrary location x_O, and the fluid flows into a cup and mixes adiabatically. The temperature in the cup is the bulk mean temperature at x_O, that is, $T_m(x_O)$. Mathematically, the bulk mean temperature is the energy-averaged temperature of the fluid, as we see from the following formal definition:

$$\dot{m} c_{p,\text{av}} T_m = \int_{A_{x,\text{sec}}} c_p T(x, r) d\dot{m} = \int_0^R \rho V_x c_p T(x, r) 2\pi r \, dr$$

Should we assume constant thermophysical properties (ρ, c_p), this equation becomes:

$$T_m(x) = \frac{2\pi\rho \int_0^R V_x(x, r) V_x(x, r) T(x, r) r \, dr}{\dot{m}}$$

We see that this definition involves both the temperature and velocity distributions. Fortunately, we have no need to apply this formal definition. What is important for our purposes is the conceptual definition and the idea that a single temperature $T_m(x)$ represents the thermal state of the fluid at any arbitrary location x.

Uniform Heat Flux

Consider the control volume in Figure 15–3. The entering fluid is heated as it passes through the tube from $x = 0$ to $x = L$ by the application of a uniform heat flux $\dot{Q}_{\text{wall}}$ over

Figure 15–3

Constant-temperature differential along a tube.

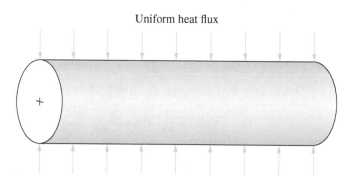

Uniform heat flux

Uniform wall temperature

Figure 15–4

Example of a taped throttling device.

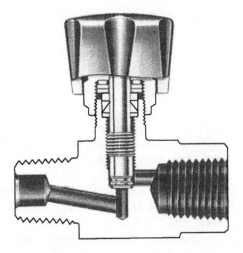

the circumferential area A_w ($= \pi DL$). This uniform heat flux boundary condition can be approximated in practice by using some form of heating. For example, an electrical heating tape can be spirally wound around the tube (Figure 15–4) and insulated on the other surface. Alternately, the tube itself can be heated by the passage of an electric current as a result of voltage difference between the ends.

The mathematical expression of energy conservation for this situation is given by the following equation:

$$\dot{Q}_{cv,net,in} - \dot{W}_{cv,net,out} = \dot{m}\left[(h_2 - h_1) + \frac{1}{2}\left(\alpha_2 V_{av,2}{}^2 - \alpha_1 V_{av,1}{}^2\right)\right] + g(z_2 - z_1)$$

where the enthalpies h_1 and h_2 are appropriately averaged over the inlet and exit stations. We simplify this relationship by noting that no work is performed by (or on) the control volume and that the outlet and inlet kinetic energy terms cancel out. Furthermore, we neglect the potential energy term since it is usually quite small compared to the thermal terms, which gives rise to the following expression:

$$\dot{Q}_{cv,net,in} \equiv \dot{Q}_{0-L} = \dot{m}(h_2 - h_1)$$

Using the preceding definition of the bulk mean temperature, we eliminate the mean enthalpies in the preceding equation to get:

$$\dot{Q}_{x-L} = \dot{m}c_p(T_{m,2} - T_{m,1})$$

This equation applies to the general problem of fluid heating or cooling in a tube. For the uniform heat flux condition, the net heat transfer from the surroundings to the control volume is simply the product of the uniform heat flux and the surface area over which it acts, i.e.,

$$\dot{Q}_{0-L} = \dot{Q}_w A_w = \dot{Q}_w \pi DL$$

Substituting this relationship and solving for T_m yield our final result:

$$T_{m,2} = T_{m,1} + \frac{\dot{Q}_w \pi DL}{\dot{m}c_p}$$

This equation can be transformed into a relationship expressing the axial variation of the bulk mean temperature $T_m(x)$ simply by replacing L with x, which gives:

$$T_m(x) = T_{m,1} + \frac{\dot{Q}_w \pi D}{\dot{m} c_p} x$$

As would be expected from the application of a uniform heat flux to the flow of a constant-property fluid, the bulk mean temperature increases linearly with distance.

Note that in the derivation of the preceding equation, we have implicitly neglected the effect of friction on the enthalpy change. This is justified for the same reason that led us to treat the friction-dominated problem as isothermal. The conversion of mechanical energy through friction results in only a small increase in the fluid temperature in most engineering applications. For example, head losses in a typical pipe-flow problem range from a few centimeters to, say, 10 meters. For a flow of water with a 10 m/s head loss, the corresponding temperature increase is 0.023 K. Such a small increase clearly justifies the assumption of isothermal flow. For nonisothermal flow, we can similarly justify neglecting elevation changes. To highlight this fact, a 100 m change in the elevation corresponds to a temperature change of 0.23 K in a flow of water.

Combining the preceding three equations, and knowing that the local heat flux is specified, we have in hand three of four relationships that we seek. Finding the wall-temperature distribution $T_w(x)$ is the only remaining task. We accomplish this by defining a local convection heat transfer coefficient $h_{\text{conv},x}$ in terms of the difference between the wall and the bulk mean temperature:

$$\dot{Q}_x = h_{\text{conv},x}[T_w(x) - T_m(x)]$$

We apply this definition to obtain the wall temperature distribution:

$$T_w(x) = T_m(x) + \frac{\dot{Q}_w}{h_{\text{conv},x}}$$

For a fully developed flow, the local heat transfer coefficient is constant, thus the wall temperature distribution differs from the bulk mean temperature distribution only by the constant $\dot{Q}_w / h_{\text{conv},x}$. The two distributions are thus parallel as illustrated in Figure 15–5.

Figure 15–5

Temperature distribution.

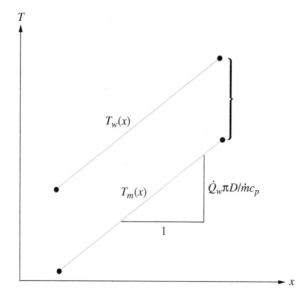

To determine a value for $h_{\text{conv},x}$ requires a detailed analysis of the flow field.

Constant Wall Temperature

The condition of constant wall temperature (Figure 15–3) is frequently approximated in practice when a fluid flowing over the outside surface of a tube or pipe undergoes a phase change. An example of this is the condensation of steam on the outside surface of a tube through which cold water flows.

In the following analyses, we derive useful relationships to calculate the total heat transfer rate, the bulk mean temperature distribution $T_m(x)$, and the local wall heat flux $\dot{Q}_{0,L}$.

To determine the total heat transfer rate, the basic analysis applied to the uniform heat flux situation also applies to the case of a constant tube wall temperature, giving rise to the following relationship:

$$\dot{Q}_{0-L} = \dot{m}c_p(T_{m,2} - T_{m,1})$$

Since the bulk mean inlet temperature $T_{m,1}$ is usually known, we have transformed the problem of finding $\dot{Q}_{0-L}$ into determining the bulk mean temperature at $x = L$ (i.e., $T_{m,2}$).

To determine the bulk mean temperature distribution requires that we consider a control volume of length Δx (Figure 15–6), rather than the original control volume. The enthalpies shown, again, are properly averaged values over the cross section. Conservation of energy is then expressed as follows:

$$\dot{Q}_{\text{cv}} = \dot{m}(h_{x+\Delta x}) - h_x$$

Figure 15–6

Heat transfer to a pipe by convection.

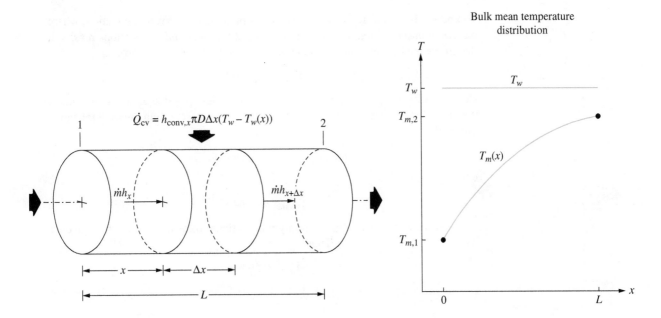

For a fluid with constant properties, we can express the mean enthalpy in terms of the bulk mean temperature as follows:

$$h(x) = c_p[T_m(x) - T_{\text{ref}}]$$

and the convection heat transfer is expressed as follows:

$$\dot{Q}_{\text{cv}} = h_{\text{conv},x} \pi D \Delta x[T_w - T_m(x)]$$

Substitution of these expressions yields:

$$h_{\text{conv},x} \pi D \Delta x[T_w - T_m(x)] = \dot{m} c_p[T_m(x + \Delta x) - T_m(x)]$$

Upon algebraic manipulation, we end up with the following expression:

$$\frac{h_{\text{conv},x} \pi D}{\dot{m} c_p}[T_w - T_m(x)] = \frac{dT_m(x)}{dx}$$

Integrating both sides, we get:

$$\ln \frac{T_m(x) - T_w}{T_m(0) - T_w} = \frac{-h_{\text{conv},x} \pi D}{\dot{m} c_p} x$$

or

$$T_m(x) = T_w + (T_m(0) - T_w) \exp\left[\frac{-h_{\text{conv},x} \pi D}{\dot{m} c_p} x\right]$$

Example 15–1

The pressure drop $(p_1 - p_2)$ associated with the flow of water through a 100-m-long horizontal run of 10-mm-inside-diameter pipe is 23.4 kPa. Determine the frictional force retarding the flow and the average shear stress in the fluid at the wall.

Solution

To find the frictional force, we apply the momentum conservation equation, where our assumption of identical velocity profiles at the inlet and exit stations results in the difference between the exit and inlet momentum flows being zero, i.e.,

$$\Sigma F_{x\text{-direction}} = 0$$

Here we identify pressure forces acting at the inlet (positive x-direction) and at the outlet (negative x-direction) and the frictional force retarding the flow (negative x-direction), thus:

$$p_1 A_{x-\text{sec}} - p_2 A_{x-\text{sec}} - F_{\text{friction}} = 0$$

which yields:

$$F_{\text{friction}} = 1.84 \text{ N}$$

The average wall shear stress is simply this frictional force divided by the area over which it acts, $A_w = (\pi DL)$. Thus:

$$\bar{\tau}_w = 0.59 \text{ bar}$$

Example 15–2

Water is heated as it flows through a 5-m-long tube. The tube is wrapped with an electrical heating tape, and a layer of insulation covers the tape. This arrangement results in a uniform heat flux of 2878 W/m^2 at the inside wall of the tube (of inside diameter of 20 mm). At an axial location $x = 3.75$ m, the bulk mean temperature of the water is 330 K, and the tube wall temperature is 350 K. Determine the local heat transfer coefficient at this location.

Solution

To find the local heat transfer coefficient $h_{conv,x}$, we apply its definition for internal flow as follows:

$$h_{conv,x} = \frac{\dot{Q}_x}{T_w(x) - T_m(x)} = 143.9 \text{ W/m}^2 \text{ K}$$

To determine the bulk mean temperature distribution requires that we consider a control volume of length Δx (Figure 15–6), rather than the original control volume in Figure 15–3. The enthalpies shown, again, are properly averaged values over the cross section. Conservation of energy yields:

$$\dot{Q}_{cv} = \dot{m}(h_{x+\Delta x} - h_x)$$

For a fluid with constant properties, we can express the mean enthalpy in terms of the bulk mean temperature as follows:

$$h(x) = c_p[T_m(x) - T_{ref}]$$

and the convection heat transfer is expressed as follows:

$$\dot{Q}_{cv} = h_{conv,x} \pi D \Delta x [T_w - T_m(x)]$$

By simple substitution, we get:

$$h_{conv,x} \pi D \Delta x [T_w - T_m(x)] = \dot{m} c_p [T(x + \Delta x) - T_m(x)]$$

Dividing by Δx, rearranging, and taking the limit, we get the following:

$$\frac{h_{conv,x} \pi D}{\dot{m} c_p}[T_w - T_m(x)] = \text{Limit} \frac{T_m(x + \Delta x) - T_m(x)}{\Delta x}$$

Recognizing the right-hand side as the definition of a derivative, we obtain the following first-order ordinary equation:

$$\frac{h_{conv,x} \pi D}{\dot{m} c_p}[T_w - T_m(x)] = \frac{dT_m(x)}{dx}$$

We solve this equation for the temperature distribution $T_m(x)$ by separating the variables and integrating from $x = 0$ to an arbitrary location x as follows:

$$\int_{T_m(0)}^{T_m(x)} \frac{dT_m(x)}{[T_w - T_m(x)]} = \frac{h_{conv,x} \pi D}{\dot{m} c_p} \int_0^x dx$$

Since the flow is fully developed, we note that the local convection heat-transfer coefficient is constant and can be removed from the integrand. Performing the indicated integrations and evaluating limits yield:

$$\ln \frac{T_m(x) - T_w}{T_m(0) - T_w} = \frac{-h_{conv,x} \pi D}{\dot{m} c_p} x$$

Removing the natural logarithm by exponentiation provides our final result:

$$T_m(x) = T_w + (T_m(0) - T_w) \exp \left[\frac{-h_{conv,x} \pi D}{\dot{m} c_p} x \right]$$

A sketch of this distribution is shown in Figure 15–6 for the case where:

$$T_w > T_m(0)$$

PROBLEMS

15–1   Consider a flow of air through a 2-mm-inside-diameter copper tube. At the location $x = 0$, the flow is fully developed, and the air has an average velocity of 12 m/s and a temperature of 300 K. The tube wall is maintained at a constant temperature of 400 K. The convection heat transfer coefficient and c_p of the air can be treated as constants with magnitudes of 532 W/m²·K and 1008 J/kg·K, respectively. The pressure is nominally 1 bar. Determine the length L of the tube segment, such that the overall heat transfer coefficient is 3.09 W. Also determine the local heat flux at $x = L$.

15–2   Small-bore glass capillary tubes can be used to construct inexpensive flowmeters. Consider the flow of air through a 1-m-long, 2-mm-inside-diameter tube. The air enters at 1 bar and 300 K. The pressure drop from entrance to exit is 585 Pa. Determine the mass flow rate of the air through the tube. Also determine the flow rate for a tube with 1-mm bore for the same conditions.

15–3   Using the conditions as in the previous problem, determine the head loss from its defining relationship and from its relationship to the mass flow rate. Also determine the friction factor.

15–4   Determine the pressure drop and the head loss associated with a 0.0269 kg/s flow of water through a smooth, 100-m-long, horizontal tube. The inside diameter of the tube is 2.0 cm. The temperature of the water is 300 K, and the flow is fully developed.

16

Fanno Flow Process for a Viscous Flow Field

Chapter Outline

16–1 Introduction **254**

16–2 Analysis for a General Fluid **254**

16–3 Limiting Point **256**

16–4 Momentum Equation **257**

16–5 Working Equations for an Ideal Gas **257**

16–6 Reference State and Fanno Tables **258**

16–7 Friction Choking **260**

16–8 Moody Diagram **262**

16–9 Losses in Constant-Area Annular Ducts **265**

Impeller

Figure 16-1

A control volume for
Fanno flow analysis.

16-1 Introduction

The fact should first be emphasized that area changes, friction, and heat transfer are some of the most important factors affecting the properties in a flow process. Up to this point, and with the exception of Chapter 15, we have only considered one of these factors; namely that of the cross-flow area variation.

In order to study the effects of friction, we will analyze the flow behavior in a constant-area duct (Figure 16–1), without any heat transfer. This corresponds to many practical flow situations which involve reasonably short ducts. We will first consider the flow of an arbitrary fluid and will discover that its behavior follows a definite pattern which is dependent upon whether the flow is in the subsonic or supersonic regime. Working equations will be developed for the case of an ideal gas, and the introduction of a reference point allows tables to be constructed. The tables will permit fast solutions to many problems which are collectively referred to as *Fanno flow problems*.

16-2 Analysis for a General Fluid

We first consider the general flow behavior of an arbitrary fluid. In order to isolate the effects of friction, we should make the following assumptions:

- Steady, one-dimensional flow
- An adiabatic flow process throughout
- Constant cross-flow area

Applying the continuity equation for this constant-area flow process, we have:

$$\rho V = G = \text{constant} = \text{the mass flux}$$

Now, applying the energy-conservation equation, we get:

$$h_t = h + \frac{G^2}{2\rho^2}$$

Now for any given flow process, both the total enthalpy (h_t) and the mass flux (G) are calculable. Therefore, the preceding equation establishes a unique relationship between the static magnitudes of enthalpy and density. Figure 16–2 shows a plot of this equation on the h-s frame of reference for different magnitudes of the mass flux, but all for the same value of total enthalpy. Each curve is called a Fanno line, and represents the flow at a particular velocity. Ducts of various sizes could pass the same mass-flow rate, but would have different velocities.

Once the fluid is known, we can also plot lines of constant entropy on the h-v diagram. It is much more instructive to plot these Fanno lines on the h-s diagram (commonly called the Mollier diagram). Such a diagram is shown in the Appendix for water. At this point,

Figure 16–2

Fanno line subsonic and supersonic flow branches.

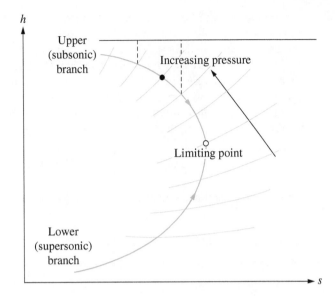

a significant fact becomes clear: Since we have assumed no heat transfer involvement, the only way entropy can be generated is through irreversibilities. Thus the flow can only progress toward increasing values of the entropy s.

Let us examine one Fanno line in greater detail. Figure 16–3 shows a given Fanno line together with typical pressure lines. All points on this line represent states with the same mass-flow rate per unit area and the same total enthalpy. Due to the irreversible nature of the frictional effects, the flow can only proceed to the right in this figure. Therefore, the Fanno line is divided into two distinct parts: an upper and lower branch, which are separated by a limiting point of maximum entropy.

In a constant-area duct, the frictional effects will show up as an internal generation of heat, with a corresponding decline in density. In order to pass the same mass-flow rate

Figure 16–3

Fanno lines for different volumetric flow rates.

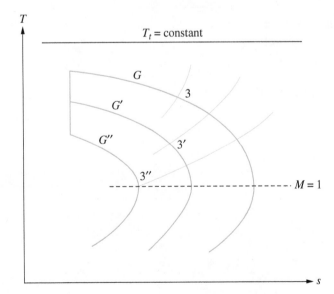

with a constant cross-flow area, continuity then forces the velocity to increase. The rise in kinetic energy must cause a decrease in enthalpy, since the total enthalpy is to remain constant. It is clear in this case that both the static and total pressures decline.

Now we move to the Fanno-line lower branch. Thermophysical properties across this branch do not vary in the manner predicted by intuition. Thus this must be a flow regime with which we are not familiar. Before we investigate the limiting point that separates these two flow structures, let us note that these two flow structures do have one thing in common. Recall the following form of the energy-conservation equation:

$$\frac{dp_t}{p_t} + T_t ds = 0$$

Therefore, any frictional effects must cause a decline in total pressure.

16-3 Limiting Point

Referring to Figure 16–4, and from the energy-conservation equation, we have:

$$h_t = h + \frac{V^2}{2}$$

Differentiating both sides, we get:

$$dh + VdV = 0$$

Also differentiating the continuity equation, we get:

$$\rho dV + Vdp = 0$$

Figure 16–4

The reference state for a Fanno process.

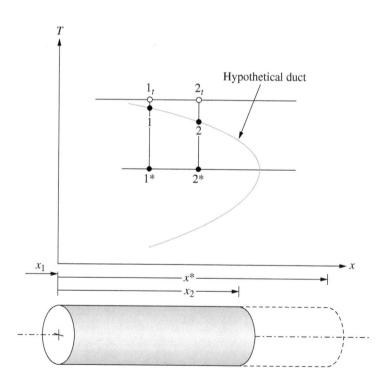

or:

$$dV = -V\frac{dp}{\rho}$$

$$dh_t = dh + V\,dV$$

Finally:

$$dh = \frac{V^2 d\rho}{\rho}$$

Now recall the property relationship:

$$T\,ds = \frac{V^2 d\rho}{\rho} = V^2\frac{d\rho}{\rho}$$

16–4   Momentum Equation

The preceding analysis utilized only the continuity and energy-conservation equations. We now proceed to apply the linear momentum equation to the control volume in Figure 16–1.

$$\Sigma F_x = (p_1 - p_2)A - F_f$$

where the last term on the right-hand side represents the friction force. It follows that:

$$p_1 + \rho_1 V_1{}^2 - \frac{F_f}{A} = p_2 + \rho_2 V_2{}^2$$

16–5   Working Equations for an Ideal Gas

Energy-Conservation Equation

For an ideal gas with a constant specific heat ratio:

$$T_t = T\left(1 + \frac{\gamma - 1}{2}M^2\right)$$

or:

$$\frac{T_2}{T_1} = \frac{1 + \frac{\gamma-1}{2}M_1{}^2}{1 + \frac{\gamma-1}{2}M_2{}^2}$$

Continuity Equation

$$\rho_1 V_1 = \rho_2 V_2$$

Introducing the equation of state, the definition of the sonic speed, and that of the Mach number, we get:

$$\frac{p_2}{p_1} = \frac{M_1}{M_2}\left(\frac{T_2}{T_1}\right)$$

Now we introduce the temperature ratio to obtain the following expression:

$$\frac{p_2}{p_1} = \frac{M_1}{M_2} \left[\frac{1 + \left(\frac{\gamma-1}{2}\right)M_1{}^2}{1 + \left(\frac{\gamma-1}{2}\right)M_2{}^2} \right]^{\frac{1}{2}}$$

Now the density relationship can be obtained as follows:

$$\frac{\rho_2}{\rho_1} = \frac{M_1}{M_2} \left[\frac{1 + \left(\frac{\gamma-1}{2}\right)M_2{}^2}{1 + \left(\frac{\gamma-1}{2}\right)M_1{}^2} \right]^{\frac{1}{2}}$$

Entropy Change

As indicated earlier in the thermodynamics part of this book:

$$\Delta s_{1-2} = c_p \ln \frac{T_2}{T_1} - R \ln p_2/p_1$$

or:

$$\frac{s_2 - s_1}{R} = \left(\frac{\gamma}{\gamma - 1}\right) - \ln p_2 p_1$$

Using the just-derived relationships of temperature and pressure ratios, as well as the entropy change expression, we get:

$$\frac{s_2 - s_1}{R} = \ln M_2 M_1 \left[\frac{1 + \frac{\gamma-1}{2}M_1{}^2}{1 + \frac{\gamma-1}{2}M_2{}^2} \right]^{\frac{\gamma+1}{2(\gamma-1)}}$$

Finally we can express the total-to-total pressure ratio as follows:

$$\frac{p_{t,2}}{p_{t,1}} = \frac{M_1}{M_2} \left[\frac{1 + \frac{\gamma-1}{2}M_2{}^2}{1 + \frac{\gamma-1}{2}M_1{}^2} \right]^{\frac{\gamma+1}{2(\gamma-1)}}$$

16–6 Reference State and Fanno Tables

The equations developed previously provide the means of computing the properties at one location in terms of those at another. The key to problem solving is predicting the Mach number at the new location. The solution of this equation for the unknown downstream Mach number presents a tough task as no explicit relationship is possible. Therefore, we turn to a technique similar to that used with isentropic flows.

We now introduce another reference state which is defined in the same manner as before "that thermodynamic state which would exist if the fluid reached a Mach number of unity by a particular process." In this case, we visualize that we continue the Fanno flow process until the velocity becomes equal to the sonic speed. Figure 16–5 shows a physical system, together with its *T-s* diagram for a subsonic Fanno flow. We know that if we continue along the Fanno line (remember that the production-of-entropy concept forces us toward the right), we will eventually reach the state of a unity Mach number. The dotted line in Figure 16–4 represents a hypothetical duct with a sufficient length to

Figure 16–5

Moody diagram for laminar and turbulent flows.

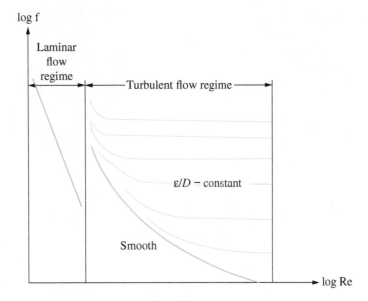

enable the flow stream to traverse the remaining portion of the upper branch and reach the limiting point. This is the reference state for the Fanno flow line.

One other fact is that should there be any entropy difference between two points (such as 1 and 2), then their isentropic * reference conditions are not the same, and we have always taken great care to label them separately as 1* and 2*. However, proceeding from either point 1 or 2 by Fanno flow will ultimately lead to the same place when the Mach number reaches unity.

We now rewrite the working equations in terms of the Fanno flow * reference condition. Consider first the following relationship:

$$\frac{T_2}{T_1} = \frac{1 + \frac{\gamma-1}{2}M_1{}^2}{1 + \frac{\gamma-1}{2}M_2{}^2}$$

Now let point 2 be an arbitrary point in the flow stream and let the Fanno * condition be point 1. With this, we have:

$$\frac{T}{T^*} = \frac{\frac{\gamma+1}{2}}{1 + \frac{\gamma-1}{2}M^2} = f(M, \gamma)$$

Similarly with the pressure ratio we have:

$$\frac{p}{p^*} = \frac{1}{M}\left[\frac{\frac{\gamma+1}{2}}{1 + \frac{\gamma-1}{2}M^2}\right]^{\frac{1}{2}} = f(M, \gamma)$$

Referring back to the momentum equation, we have:

$$\frac{f(x^* - x)}{D_e} = \frac{\gamma+1}{2\gamma}\ln A_1/A_2 + \frac{1}{\gamma}\left(\frac{1}{M^2} - 1\right) = f(M, \gamma)$$

Figure 16–6

Variation of the friction coefficient with surface roughness.

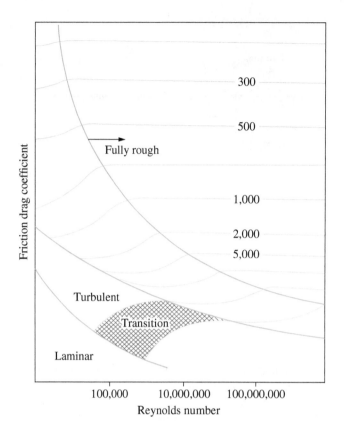

where:

$$A_1 = \frac{\gamma + 1}{2} M^2$$

$$A_2 = 1 + \frac{\gamma - 1}{2} M^2$$

The quantity $(x^* - x)$ (Figure 16–4) represents the length of the duct that has to be added to cause the flow to reach the Fanno * reference state. It can alternately be viewed as the maximum duct length which may be added without changing some flow conditions.

Figure 16–5 shows the so-called Moody diagram, which provides the friction factor for both laminar and turbulent flow regimes. Figure 16–6 shows the variation of the friction coefficient with the surface roughness.

16–7 **Friction Choking**

In studying subsonic-nozzle aerodynamics with constant inlet and exit conditions, we came to realize that as the back pressure was lowered, the flow rate through the nozzle rose. When the operating pressure ratio reached a certain magnitude, the section of minimum cross-flow area developed a Mach number of unity. The nozzle, then, was referred to as choked. Note that further reduction of the back pressure did not affect the mass-flow rate in any way.

Note that we have an isentropic flow at the duct's inlet, and then we move along a Fanno line. As the back pressure is decreased even more, the mass-flow rate and exit

Figure 16–7

Explanation of Fanno line on the basis of boundary layer.

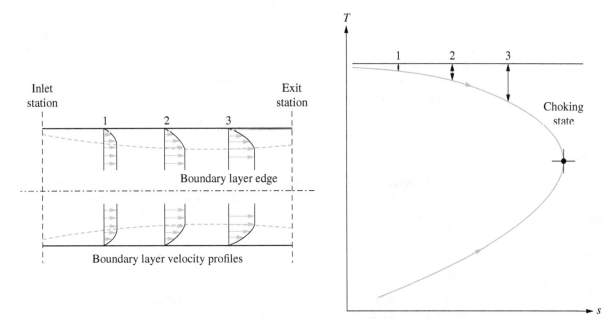

Mach number will both continue to increase while the system moves to a Fanno line of higher velocities. Eventually, when a certain pressure ratio is attained, the Mach number at the duct exit station will be unity. This is called "friction" choking (Figure 16–7), and any further reduction in the back pressure will not affect the flow conditions inside the system. What will occur as the flow leaves the duct and enters the region of reduced pressure? The answer is discussed next.

Let us consider the last case of choked flow with the exit pressure being equal to the receiver pressure. Now suppose that the receiver pressure is maintained at that value but more duct length is added to the system. What happens then? We know that we cannot move around the Fanno line, yet somehow we must reflect the added friction losses. This is done by moving to a new Fanno line at a decreased mass-flow rate. Note that the pressure equilibrium is still maintained at the exit station, but the system is no longer choked, although the mass-flow rate has decreased. If the back pressure is now reduced sufficiently, then a sonic velocity can exist at the exit station.

In summary, when a subsonic flow has become friction-choked and more duct is added to the system, the mass-flow rate must decrease. Just how much it drops and whether the exit velocity remains sonic both depend upon how much duct length is added as well as the receiver pressure imposed on the system.

The shock location is determined by the amount of added duct length. As more duct is added, the shock moves upstream and occurs at a higher Mach number. Eventually, the shock will move to that segment of the system which precedes the constant-area duct (most likely a converging-diverging nozzle was used to produce the supersonic flow). If a sufficient magnitude of length is added, the entire system will become subsonic, and then the mass-flow rate will decrease. Whether the exit velocity remains sonic will again depend on the back pressure.

16-8 **Moody Diagram**

This very well known diagram is shown in Figure 16–5. This chart provides the friction coefficient magnitude depending on whether the flow is that of a laminar or turbulent stream. Another factor on this graph is the surface roughness.

A connection between the growth of boundary layer (in fact the displacement thickness, as will be explained later) and the flow behavior on the *T-s* diagram (meaning the Fanno flow structure there) is shown in Figure 16–7. Examination of this figure reveals that the displacement effect, caused by the boundary layer growth, forms an initially constant-area duct into a converging nozzle, which explains why the velocity continues to rise along the duct.

Example 16–1

Figure 16–8 shows an adiabatic interstage constant cross-sectional-area duct that is, by definition, annular. The flow through the duct is swirling at an angle α of 28.4°, with a friction factor f of 0.02. At the duct inlet station (1), the following conditions apply:

- $p_1 = 12.1$ bars
- $M_1 = 0.37$

If the duct's "actual" length is 0.06 m, calculate the total pressure loss across the duct. Assume a specific heat ratio γ of 1.33.

Solution

We can calculate the exit static pressure using the Fanno line pressure-ratio expression to be 5.53 bars.

Using the fL_{eff} equation, we are confronted with a rather bulky equation to which the suitable procedure is a trial-and-error approach yielding, in the end, the exit Mach number:

$$M_2 \approx 0.57$$

In substituting the length of the duct, the swirling motion was accounted for, a process which further elongates the duct, that is, more than its "actual" length.

Now using the famous isentropic relationship giving the total pressure in terms of its static counterpart and the local Mach number, we get:

$$p_{t,1} = 13.24 \text{ bars}$$

$$p_{t,2} = 6.83 \text{ bars}$$

Comment

Note that the displacement thickness (δ^*) is a fraction of the boundary layer thickness, which is defined in a rather specific manner. This and the momentum thickness (θ) are two unique characteristics of the boundary layer.

Finally, we get the total pressure loss across the interstage duct as follows:

$$\Delta p_t = p_{t,1} - p_{t,2} = 6.41 \text{ bars}$$

Figure 16–8

Input variables for Example 16–1.

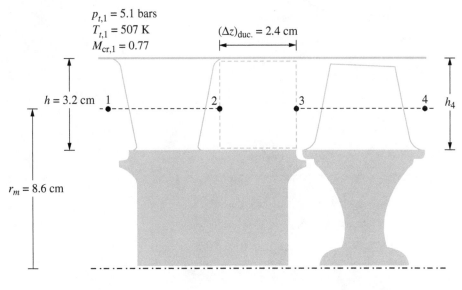

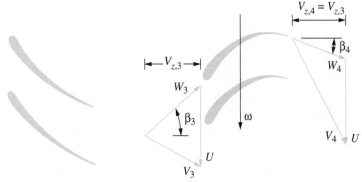

Example 16-2
Figure 16–9 shows the stator/rotor annular gap in an axial-flow turbine stage, the station designation, and the major dimensions. The gap has a constant cross-sectional area with an annulus height (Δr) of 2.2 cm, and a mean radius of 11.0 cm. Across the gap, the flow is swirling at an angle (α) of 78°, with a friction coefficient (f) of 0.05. The gap inlet magnitudes of total pressure and temperature are 7.6 bars and 782 K, respectively. Knowing that the gap inlet and exit Mach numbers are 0.7 and 0.72, respectively, calculate:

(a) The mass-flow rate

(b) The percentage of total pressure decline across the gap

Figure 16–9

Input variables for
Example 16–2.

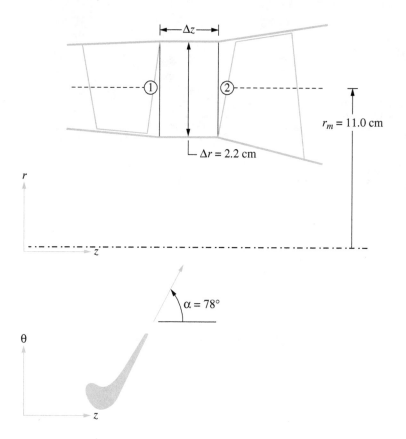

Solution

Part a:

We can calculate the variables $L_{C,1}$ and $L_{C,2}$ corresponding to the axial-gap inlet and exit Mach numbers as follows:

$$L_{C,1} = L_{M=0.7} = 0.223\frac{D_h}{f}$$

$$L_{C,2} = L_{M=0.72} = 0.184\frac{D_h}{f}$$

where the hydraulic diameter and friction factor are:

$$D_h = 2\Delta r = 4.4 \text{ cm}$$

$$f = 0.05 \text{ (given)}$$

This gives rise to the following:

$$L_{C,1} = 0.1967 \text{ m}$$

$$L_{C,2} = 0.1627 \text{ m}$$

Now the stator/rotor gap length Δz can be calculated as follows:

$$\frac{\Delta z}{\cos \alpha} = L_{C,1} - L_{C,2} = 0.034 \text{ m}$$

which yields:

$$\Delta z = 0.71 \text{ cm}$$

Part b:

Applying the continuity equation at the gap's inlet station, we get:

$$\dot{m} = 2.87 \text{ kg/s}$$

16-9 Losses in Constant-Area Annular Ducts

This section highlights the effects of boundary layer buildup over the endwalls in unbladed constant-area gaps within turbomachinery components. In order to quantify such effects, let us define a relevant boundary layer variable that is referred to as the *displacement thickness* δ^*. Referring to Figures 16–10 and 16–11, this variable is defined as follows:

$$\delta^* = \int_0^\delta \left(1 - \frac{\rho V}{\rho_{\text{edge}} V_{\text{edge}}} \right) dy$$

where:

δ^* is the local displacement thickness
δ is the boundary layer thickness

Figure 16–10

Graphical representation of the displacement thickness.

Graphical representation of the displacement thickness

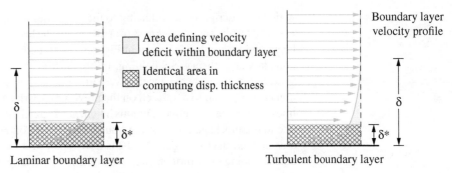

Laminar boundary layer   Turbulent boundary layer

Figure 16–11

Definition of the displacement thickness.

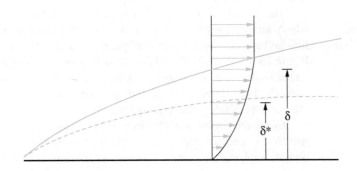

Figure 16–12

Interstage boundary layer growth.

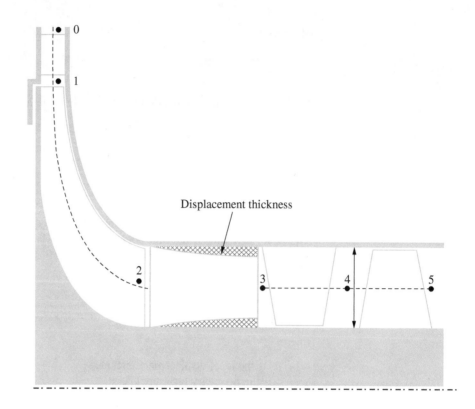

Displacement thickness

ρ_{edge} is the edge static density
V_{edge} is the local edge velocity

with the subscript (edge) referring to the properties at the boundary layer edge.

The displacement thickness (δ^*) represents the loss of mass flow across the boundary layer, and its buildup is shown in Figures 16–10 and 16–11. According to the classical boundary layer theory (Schlichting, 1979), the flow stream will sense the existence of the solid wall as if it were displaced toward the boundary layer edge through a distance that is numerically equal to δ^*. Based on the previous remarks, the distance between these two lines defines the "effective" annulus height at any axial location within the interstage duct due to the blockage effect of the boundary layer. In other words, the initially constant-area annular duct in Figure 16–12 is now replaced by an annular converging subsonic nozzle. Going even further, such phenomena as choking can theoretically prevail at the flow-passage exit station, depending on the operating conditions, friction factor, and interstage duct length.

The preceding discussion is intended to pave the way to the Fanno flow structure which is under consideration in this section. The process concerns a necessarily adiabatic flow and a constant-area duct configuration. Of these restrictions, the former is clear in Figure 16–13, in which the total-to-total flow process is represented by a horizontal line on the T-s diagram, as will be seen later in this section. In the following, we will proceed along the upper (subsonic) branch of the Fanno line discussing, in particular, the physical implications of key points on this branch.

Due to the viscosity-related irreversibility, the sequence of events on the T-s diagram will have to progress from left to right over the subsonic (actually also the supersonic) branch, implying (as is truly the case) a continuous entropy production throughout the

Figure 16–13

Effect of the boundary layer buildup in a rotor.

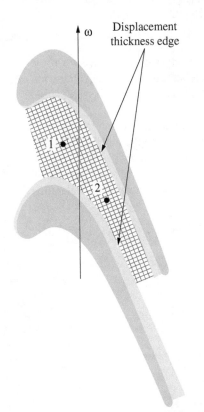

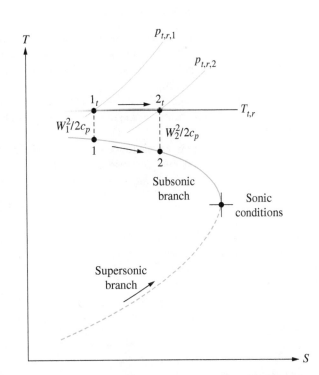

entire process. Construction of the subsonic Fanno-line branch reveals that the constant-area duct is now a subsonic-nozzle-like flow-accelerating passage, a feature that is consistent with the converging shape of the "effective" annular passage in Figure 16–13. At any point on this branch, the vertical distance away from the total temperature (horizontal) line is nothing but the local dynamic temperature ($\frac{V^2}{2c_p}$). The subsonic branch ends at the maximum-entropy point (identified as the choking point). According to the entropy-production version of the second law of thermodynamics, there is no way for the process path to enter the supersonic branch at this point, as this will imply entropy detraction, a violation of the second law of thermodynamics. In fact, the only way for the process to exist on the lower (supersonic) branch is to start with a supersonic flow, then proceed toward higher entropy magnitudes in what is now a supersonic diffuser. The different properties covering the Fanno line process have already been presented earlier in this section.

In the event of a swirling flow, the distance along the duct should be measured along the actual flow trajectory over the endwalls in particular. The trajectory elongation, in this case, can alternately be simulated in the form of an added surface roughness. This added roughness should be carefully calibrated in such a way as to yield a one-to-one equivalency of the axial distance (under such elevated roughness) and the distance along the actual (swirling) flow trajectory. A simpler means of achieving the same purpose is to proceed with the actual (not the axial) distance along the flow trajectory over the endwall of interest. Of course, each of the endwalls will have to be separately considered, since each endwall will be associated with its own flow-swirl angle.

The Fanno flow concept can be, by reference to Figure 16–13, applied to the blade-to-blade passage (already a subsonic nozzle) in an axial-flow turbine rotor (where relative thermophysical properties apply). Here, too, the displacement thickness growth will have the effect of increasing the inlet/exit cross-flow area ratio.

Example 16–3

Figure 16–14 shows the blade-to-blade passage in an axial-flow compressor rotor. Also shown is the mid-channel distribution of the relative-velocity tangential component at the mean radius. The rotor mean radius (r_m) is 8.0 cm, and the shaft speed (N) is 39,170 r/min.

By calculating the net magnitude of the radial acceleration, as a rough measure of the relative streamline radial shift, and sketch the streamline at the midspan location.

Solution

Let us consider the four locations that are identified on the W_θ versus nondimensional axial distance. Because the mean radius is constant, the centripetal acceleration component will remain constant as well.

$$(a)_{cent} = r_m \omega^2 = 1.35 \times 10^6 \text{ m/s}^2$$

Figure 16–14

A compressor rotor.

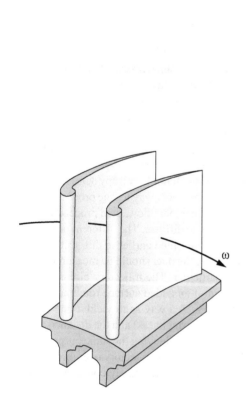

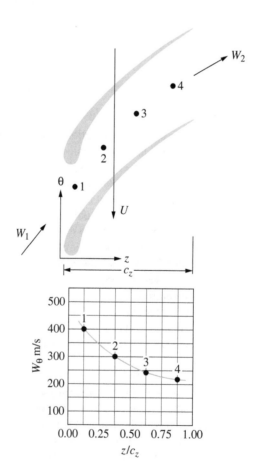

Figure 16–15

Radial shift of the mean
relative streamline.

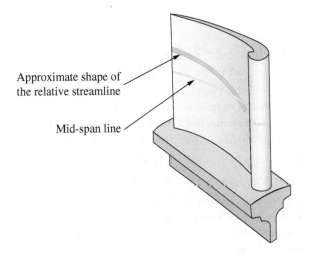

Approximate shape of
the relative streamline

Mid-span line

The Coriolis-acceleration radial component can be expressed as follows:

$$(a_r)_{\text{Coriolis}} = 2\omega W_\theta$$

Referring to Figure 16–14, note that $\vec{\omega} = -\omega e_z$, where e_z is the unit vector in the z-direction. The figure also shows that the relative-velocity tangential component (W_θ) is positive at all axial locations (by reference to the direction of the θ axis). As a result, the acceleration's radial component (above) will be positive (i.e., in the radially outward direction). Let us now compute this component at the four selected points in Figure 16–14:

$$[(a_r)_{\text{Coriolis}}]_1 = 2\omega W_{\theta 1} = 1.44 \times 10^6 \text{ m/s}^2$$

$$[(a_r)_{\text{Coriolis}}]_2 = 2\omega W_{\theta 2} = 1.03 \times 10^6 \text{ m/s}^2$$

$$[(a_r)_{\text{Coriolis}}]_3 = 2\omega W_{\theta 3} = 0.79 \times 10^6 \text{ m/s}^2$$

$$[(a_r)_{\text{Coriolis}}]_4 = 2\omega W_{\theta 4} = 0.66 \times 10^6 \text{ m/s}^2$$

The net magnitudes of the radial acceleration component at these four points can be obtained by simply adding the centripetal component, i.e.,

$$[(a_r)_{\text{net}}]_1 = 2.79 \times 10^6 \text{ m/s}^2$$

$$[(a_r)_{\text{net}}]_2 = 2.38 \times 10^6 \text{ m/s}^2$$

$$[(a_r)_{\text{net}}]_3 = 2.14 \times 10^6 \text{ m/s}^2$$

$$[(a_r)_{\text{net}}]_4 = 2.01 \times 10^6 \text{ m/s}^2$$

Taking these magnitudes to be indicative of the fluid-particle radial shift, the relative streamline should look like that in Figure 16–15. Due to the continuous streamwise decline in W_θ (Figure 16–15), the Coriolis. Note that the Coriolis-acceleration contribution and, therefore, the net radial acceleration component continue to decline as well. The relative streamline shape in Figure 16–15 reflects this fact in the form of a continually decreasing rate of radius gain along the streamline.

PROBLEMS

16-1   Given: $M_1 = 1.8$, p $= 2.72$ bars, and $M_2 = 1.2$. Determine p_2 and $\frac{f\Delta x}{D}$

16-2   Given: $M_2 = 0.94$, $T_1 = 400$ K, and $T_2 = 350$ K. Determine M_1 and the static pressure ratio.

16-3   A large chamber contains air at a temperature of 300 K and a pressure of 8 bars. The air enters a converging-diverging nozzle with an area ratio of 2.4. A constant-area duct is attached to the nozzle and a normal shock stands at the exit station. The receiver pressure is 3 bar. Assume the entire system to be adiabatic and neglect friction in the nozzle. Calculate the $\frac{f\Delta x}{D}$ for the duct.

16-4   Given: $M_1 = 1.80$, $p_1 = 2.7$ bars, and $M_2 = 1.20$. Find p_2 and $\frac{f\Delta x}{D}$.

16-5   Given: $M_2 = 0.94$, $T_1 = 400$ K, and $T_2 = 350$ K. Find M_1 and $\frac{p_2}{p_1}$.

16-6   A converging-diverging nozzle with an area ratio of 5.42 connects to a 1.13-m-long, constant-area, rectangular duct. The duct is a 3.1×1.6 cm in cross section and has a friction factor of $f = 0.02$. Find the minimum total pressure feeding the nozzle if the flow is supersonic throughout the entire duct and it exhausts to 1 bar.

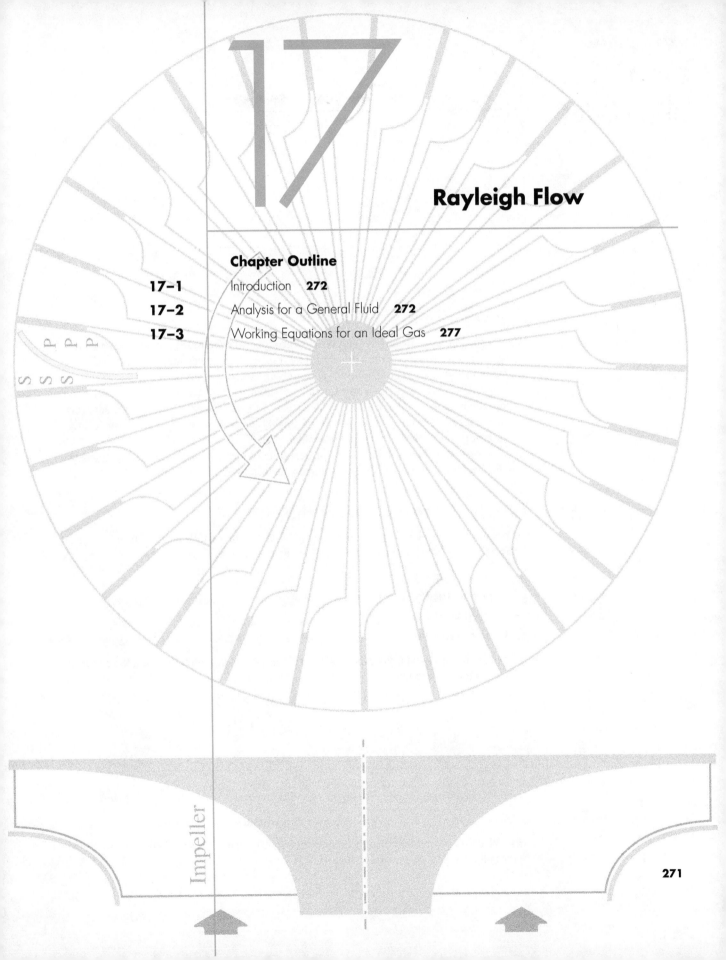

Rayleigh Flow

Chapter Outline

17–1   Introduction   **272**

17–2   Analysis for a General Fluid   **272**

17–3   Working Equations for an Ideal Gas   **277**

Impeller

Figure 17–1

Control volume for a
Rayleigh flow process.

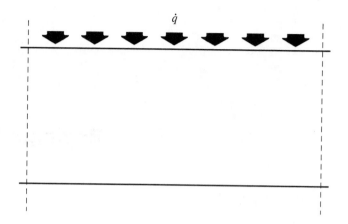

17–1 Introduction

In this section, we will consider the consequences of heat penetrating the system boundary (Figure 17–1). To isolate the effects of heat transfer from other major factors, we will assume the case of a flow field in a constant-area duct without the effects of friction. At first, this may seem to be an unrealistic situation, but actually it is a good first approximation to many real-life problems, as most heat exchangers have constant-area flow passages. It is also a simple and reasonably equivalent process for constant-area combustors, particularly those of the annular type.

In systems where the high rates of heat transfer occur, the entropy change caused by the heat transfer is much greater than that caused by friction.

17–2 Analysis for a General Fluid

We will first consider the general behavior of an arbitrary fluid. In order to isolate the effects of heat transfer, we make the following assumptions:

- Steady one-dimensional flow
- Negligible friction
- No shaft work
- Constant cross-flow area

We now proceed by applying the basic principles of mass, momentum, and the energy-conservation equations:

$$\rho V = G = \text{constant}$$

$$p_1 - p_2 = \rho V (V_2 - V_1) = G(V_2 - V_1)$$

$$h_{t,1} + q = h_{t,2}$$

$$p + GV = \text{constant}$$

where G is the magnitude of mass flux. An alternate form of the foregoing equations is:

$$p + G^2 v = \text{constant}$$

In both cases, we are led to equivalent results since both analyses deal with a cross-flow constant area, and assume negligible friction.

Figure 17–2

Rayleigh family of lines for different volumetric rates.

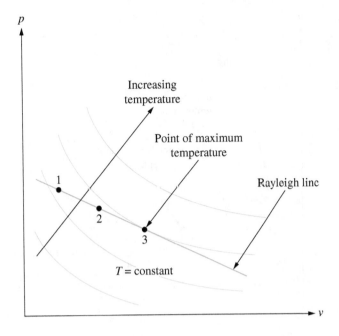

If we multiply the preceding equation by the constant area, we get:

$$pA + \dot{m}V = \text{constant}$$

The constant in this equation is called the "impulse" or the "thrust" function. Note that the thrust function remains constant for Rayleigh flow.

We would normally expect heating to increase the temperature and decrease the density. This appears to be in agreement with a process from state 1 to state 2, as marked in Figure 17–2. If we add more heat energy, we move further along the Rayleigh line, and the temperature increases even more. Soon point 3 is reached, where the temperature reaches its maximum.

Recall that the addition of heat causes the entropy to increase since:

$$ds = \frac{\delta q}{T}$$

Therefore, it appears that the real limiting condition involves entropy. We can continue to add heat until the flow reaches the state of maximum entropy.

It might be that this point of maximum entropy is reached before the point of maximum temperature, in which case we would never be able to reach state 3 in Figures 17–4 through 17–6.

For a constant static temperature line:

$$pv = RT = \text{constant}$$

For a constant-entropy line:

$$pv^{\gamma} = \text{constant}$$

or:

$$\frac{dp}{dv} = -\gamma \left(\frac{p}{v} \right)$$

Figure 17–3

Points of maximum
temperature and entropy.

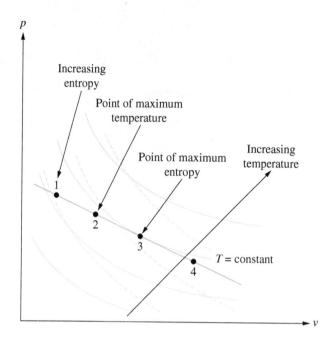

Comparing the two ordinary differential equations above, and noting that γ is always greater than unity, we see that not only can we reach the point of maximum temperature, but more heat can be added to take us beyond this point. If desired, we can move (by heating) all the way to the maximum entropy point by adding heat to the system and with its temperature continuing to decrease. Let us reflect further on the phenomenon that is occurring.

It was previously noted that the effects of heat addition are normally viewed as causing the fluid density to decrease. This requires the velocity to increase, since G is

Figure 17–4

Rayleigh flow on the h-s diagram.

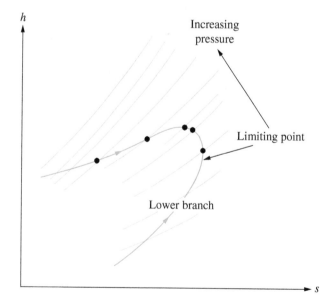

Figure 17–5

Branches of Rayleigh flow.

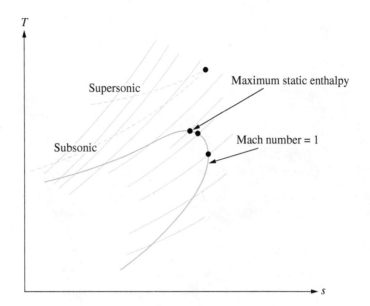

constant, by reference to the continuity equation. This velocity increase automatically boosts the kinetic energy. Some of the heat which is added to the system is converted into this increase in kinetic energy, with the heat energy in excess of this amount being available to increase the enthalpy.

Noting that the total kinetic energy is proportional to the kinetic energy per unit mass, we realize that higher velocities are reached by the addition of more heat, which is accompanied by an increase in kinetic energy. Eventually we reach the point where all of the added heat energy is required for the kinetic energy increase. At this point, there is no heat energy left over, and the system is at a point of maximum enthalpy

Figure 17–6

Limiting point for a Rayleigh flow process.

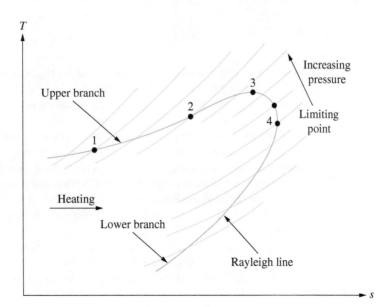

Figure 17–7

Input data for the Example 17–1.

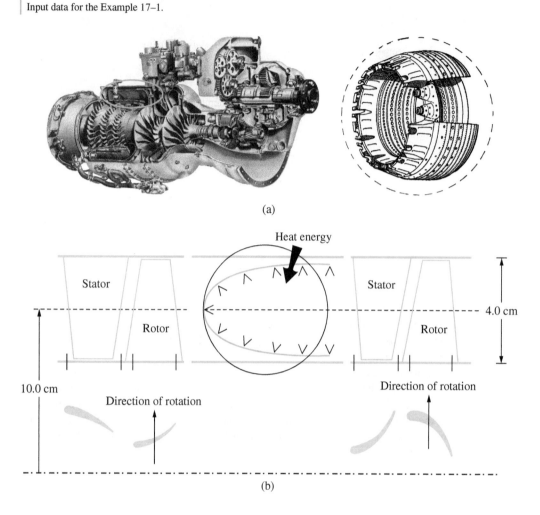

(a)

(b)

(also maximum temperature in the case of an ideal gas). Further addition of heat causes the kinetic energy to increase by an amount greater than the heat energy being added. Therefore, from this point on, the enthalpy must decrease to provide the proper energy balance.

Perhaps the foregoing discussion would be clearer if the Rayleigh lines were plotted on the *h-s* diagram. All points on this Rayleigh line represent states with some mass-flow rate per unit area and the same impulse (or thrust) function. For heat addition, the energy must increase, and the flow process moves to the right (second law of thermodynamics). Thus it appears that the Rayleigh line, like the Fanno line, is divided into two distinct (subsonic and supersonic) branches which are separated by a limiting point of maximum entropy.

As was the case for Fanno flow, note that the flow along the lower branch of the Rayleigh line appears to be the regime with which we are not familiar. The point of maximum entropy is some sort of a limiting point which separates these two flow structures.

17–3 Working Equations for an Ideal Gas

Noteworthy is the fact that we can progress along a Rayleigh line in *either* direction, depending upon whether the heat is being added to or removed from the system. We now proceed to develop relationships between properties at arbitrary sections. Recall that we want these working equations to be expressed in terms of the Mach number and the specific heat ratio. In order to obtain explicit relationships, we assume that the flow is that of an ideal gas.

Momentum Equation

$$p + GV = \text{constant}$$

or:

$$p + \rho V^2 = \text{constant}$$

Using the equation of state and the Mach number definition, we get:

$$\frac{p_2}{p_1} = \frac{1 + \gamma M_1{}^2}{1 + \gamma M_2{}^2}$$

Continuity Equation

$$\rho V = G = \text{constant}$$

Again, introducing the equation of state and the Mach number definition, we get:

$$\frac{pM}{\sqrt{T}} = \text{constant}$$

Written between two different states:

$$\frac{p_1 M_1}{\sqrt{T_1}} = \frac{p_2 M_2}{\sqrt{T_2}}$$

which can be solved for the temperature ratio as follows:

$$\frac{T_2}{T_1} = \frac{p_2{}^2 M_2{}^2}{p_1{}^2 M_1{}^2}$$

The introduction of pressure ratio results in the following equation for the static temperature ratio:

$$\frac{T_2}{T_1} = \left(\frac{1 + \gamma M_1{}^2}{1 + \gamma M_2{}^2}\right) \frac{M_2{}^2}{M_1{}^2}$$

Total Properties

The relationship between the total and static temperatures was earlier expressed in terms of the Mach number, and can be written as follows:

$$\frac{T_{t,2}}{T_{t,1}} = \left(\frac{1 + M_1{}^2}{1 + M_2{}^2}\right) \frac{M_2{}^2}{M_1{}^2} \left[\frac{1 + \frac{\gamma-1}{2} M_2{}^2}{1 + \frac{\gamma-1}{2} M_1{}^2}\right]$$

Similarly we can derive an expression for the total-to-total pressure ratio as follows:

$$\frac{p_{t,2}}{p_{t,1}} = \left(\frac{1 + \gamma M_1^2}{1 + \gamma M_2^2}\right)\left[\frac{1 + \frac{\gamma-1}{2}M_2^2}{1 + \frac{\gamma-1}{2}M_1^2}\right]$$

Figure 17–3 shows the points of maximum temperature and maximum entropy, while Figure 17–4 shows the limiting point on the *h-s* diagram. The point of maximum static enthalpy is shown in Figure 17–5, and the limiting point is emphasized in Figure 17–6.

Example 17–1

Figure 17–7 shows an annular constant cross-flow area combustor where the inlet conditions are as follows:

- Inlet total pressure $(p_{t,1}) = 12.0$ bars
- Inlet static temperature $(T_1) = 650$ K
- Inlet velocity $(V_1) = 150$ m/s

Heat is added across the combustor. At the flow location where the Mach number is 0.7, calculate the total pressure loss percentage. You may assume the specific heat ratio to be 1.4.

Solution

First we calculate the inlet Mach number as follows:

$$a_1 = \sqrt{\gamma R T_1} = 511.05 \text{ m/s}$$

$$M_1 = \frac{V_1}{a_1} = 0.294$$

Now using the two expressions for the total temperature ratio, we get:

$$T_{t,1} = T_1\left(1 + \frac{\gamma - 1}{2}M_1^2\right) = 661.2 \text{ K}$$

Now the total temperature at $M_1 = 0.7$ is 1382.0 K

$$p_{t,1} = 12.0 \text{ bars (given)}$$

Now we calculate $p_{t,2}$:

$$p_{t,2} = 10.07 \text{ bars}$$

Now we calculate the percentage of the total pressure loss:

$$\text{Total pressure loss percentage} = 16.08 \%$$

PROBLEMS

17–1

Shown in Figure 17–8 is a combustor that is situated between a centrifugal compressor stage and a turbine stage of the radial inflow turbine type. The combustor inlet conditions are as follows:

- Critical Mach number = 0.35
- Total pressure = 14.2 bars
- Total temperature = 763.0 K

Figure 17–8

Annular combustor in a turbine engine.

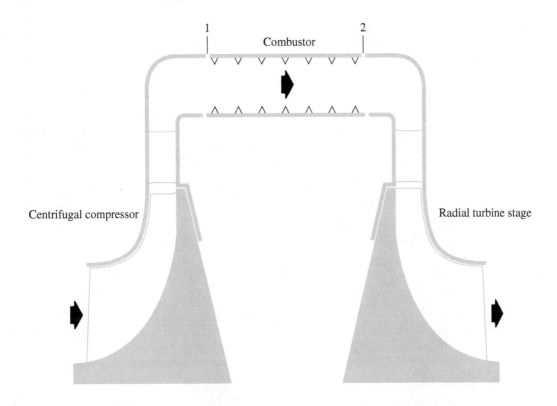

Figure 17–9

Geometrical configuration
of the example problem.

$p_1 = 12.0$ bars
$T_1 = 650$ K
$V_1 = 150$ m/s

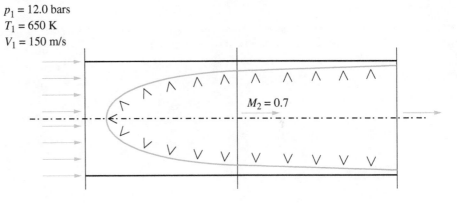

$M_2 = 0.7$

If the combustor length is sufficient for exit choking to materialize, calculate the amount of heat
added across the combustor.

17–2   Figure 17–9 shows a constant-area combustion chamber together with its inlet conditions. Calculate
the temperature at the location where the Mach number is 0.7.

III

Heat Transfer

As indicated earlier, there are three modes of heat transfer: conduction, convection, and radiation. In all cases the driving force is a temperature differential between the system and its surroundings or within the system itself. In this section, we discuss each of these modes separately. Under heat conduction, such real-life problems as contact resistance in a solid with multiple layers are discussed, with the subtopic of electrical analogies highlighted. Under heat convection, the distinction between free (or natural) and forced convection is clearly made, with practical applications. Also covered in detail is the general problem of heat exchangers, which is typically a real-life industry-related problem in terms of sizing and analysis. Aside from all factors governing radiation heat transfer, the topic of shape factor (which is basically how a heat-emitting surface "sees" another) receives particular emphasis.

18

Heat Conduction

Chapter Outline

18–1   Introduction **284**

18–2   Simple One-Dimensional Problems **285**

18–3   Equivalent Thermal Network **286**

18–4   Network Resistance for Heat-Convection Boundary Conditions **286**

18–5   Multilayer Plane Walls **287**

18–6   Thermal Contact Resistance **289**

18–7   Transient Heat Conduction **292**

18–1   Introduction

An isolated body is in thermal equilibrium if its temperature is identical at all points. If the temperature in a solid body is not uniform, heat will transfer by molecular activity from the high-temperature regions to those of lower temperatures. The process here is termed *heat conduction*, and is generally time-dependent and will continue until the point where the temperature is uniform throughout the entire body.

The conduction heat transfer process, either steady state or time-dependent, is governed by the first and second laws of thermodynamics. The rate form of the first law is written for a system that is composed of an incremental cube of the solid body. Fourier's law is used to represent the flux of heat energy crossing the boundaries of the cube (Figure 18–1). The differential energy equation for heat conduction is obtained by taking the limit of the expression as the cube's volume shrinks, theoretically, to zero. This expression and the boundary conditions represent the mathematical formulation of the model and will give the temperature distribution throughout the entire body at hand.

Nonuniform temperature fields are present in nearly all engineering applications. This internal temperature distribution and the flow of heat are important. The nonuniform temperature field may be created by sources of energy involving nuclear, chemical, or electrical resistance heat generation (or absorption). The temperature distribution in the system is governed, in part, by conduction.

The basic law governing the conduction of heat is due to Fourier. The temperature distribution in a given material is considered to be a function of the spatial location and, generally, time. Fourier postulated that the rate of conduction heat transfer per unit surface is proportional to the local temperature gradient. The Fourier law can be expressed as follows:

$$\dot{q} = -k\frac{\partial T}{\partial n}$$

where $\dot{q}$ is the rate of heat transfer per unit area (normally termed the heat flux), and $\vec{n}$ is the unit vector perpendicular to the surface. The thermal conductivity coefficient "k" is generally a function of temperature.

Let us now consider the larger three-dimensional picture in Figure 18–1. Assuming that there exists a heat source with strength $\dot{q}$ within the control volume shown in the

Figure 18–1

Three-dimensional heat conduction.

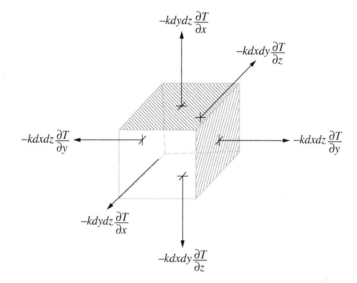

Figure 18–1, and applying Fourier's law across each surface area, and considering the time-independency case, in the limit we get:

$$k\left(\frac{\partial^2 T}{\partial x^2} + \frac{\partial T}{\partial y^2} + \frac{\partial^2 T}{\partial z^2}\right) + \dot{q} = 0$$

which can be rewritten as follows:

$$k\nabla^2 T + \dot{q} = 0$$

It is useful to re-express the preceding heat conduction equation in cylindrical coordinates as follows:

$$k\left(\frac{\partial^2 T}{\partial r^2} + \frac{1}{r}\frac{\partial T}{\partial r} + \frac{1}{r^2}\frac{\partial^2 T}{\partial \theta^2} + \frac{\partial^2 T}{\partial z^2}\right) + \dot{q} = 0$$

18–2 Simple One-Dimensional Problems

In the following, two examples are offered which are direct applications of Fourier's equation in both Cartesian and cylindrical frames of reference.

Example 18–1

The outside of the brick wall of a house is exposed to the outside temperature (T_o). The inside of the wall, however, is kept at a higher temperature T_i. Determine the temperature distribution across the wall.

Solution

Simplifying the foregoing Cartesian-coordinate Fourier equation, we have:

$$\frac{\partial^2 T}{\partial x^2} = 0$$

Integrating twice, we get:

$$T = ax + b$$

Utilizing the boundary conditions at $x = 0$ and $x = L$, we get:

$$T = \frac{T_o - T_i}{L}x + T_i$$

Example 18–2

Now the wall is cylindrical with the same inner and outer temperatures, T_i and T_o, respectively. It is required to derive an expression for the temperature in terms of radius this time.

Solution

Referring back to Fourier's heat conduction equation in a cylindrical frame of reference, we get:

$$\frac{d^2 T}{dr^2} + \frac{dT}{dr} = 0$$

Now define a new variable ($\bar{T}$), such that:

$$\bar{T} = \frac{dT}{dr}$$

Simple substitution in the preceding equation yields:

$$\frac{d\bar{T}}{dr} + \frac{\bar{T}}{r} = 0$$

Substituting the boundary conditions, we get:

$$T = T_i - \frac{r_i - r_o}{\ln \frac{r_o}{r_i}} \ln r$$

18–3 Equivalent Thermal Network

Shown in Figure 18–2 is an electrical network analogy that satisfies the boundary conditions. With the temperature representing the voltage, the only resistance here is representative of the conductivity, and can be expressed as follows:

$$R = \frac{L}{kA}$$

18–4 Network Resistance for Heat-Convection Boundary Conditions

When one of the walls of a solid body is exposed to a fluid in motion, it is said that this surface is exposed to heat convection with a convection-heat-transfer coefficient of h, where:

$$\dot{q} = hA(T - T_o)$$

where T_o is the fluid temperature. The thermal resistance in this situation is R, where:

$$R = \frac{1}{hA}$$

Figure 18–2

Electrical analogy.

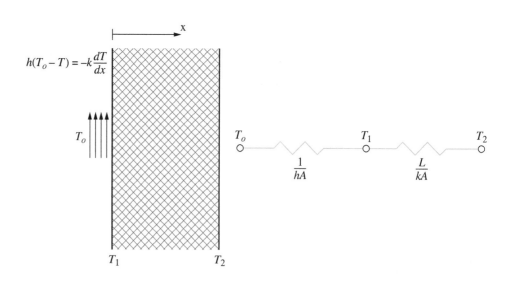

$$h(T_o - T) = -k\frac{dT}{dx}$$

18–5   Multilayer Plane Walls

In practice we often encounter plane walls which consist of several layers of different materials. The thermal resistance concept can still be used to determine the rate of steady heat transfer through such composite walls. This is done by simply noting that the conduction resistance of each wall is $\frac{L}{kA}$ connected in series, and using the electrical analogy. This is achieved by dividing the temperature difference between two surfaces at known temperatures by the total thermal resistance between them (Figure 18–3).

Consider a plane wall which consists of two layers (such as a brick wall with a layer of insulation). The rate of steady heat transfer through this two-layer composite wall can be expressed (Figure 18–3) as:

$$\dot{Q} = \frac{T_{o,1} - T_{o,2}}{R_{\text{total}}}$$

where R_{total} is the total thermal resistance, expressed as follows:

$$R_{\text{total}} = \frac{1}{h_1 A} + \frac{L_1}{k_1 A} + \frac{L_2}{k_2 A} + \frac{1}{h_2 A}$$

With h_1 and h_2 representing the heat-convection coefficients.

The subscripts 1 and 2 in the R_{wall} relations (above) indicate the first and second layers, respectively. Note, from the thermal resistance network, that the resistances are in series, and thus the total thermal resistance is simply the arithmetic sum of the individual thermal resistances in the path of heat flow.

The preceding result for the two-layer case is analogous to the single-layer case, except that an additional resistance is added for the additional layer. This result can be extended to plane walls that consist of three or more layers by adding an additional resistance for each additional layer.

Figure 18–3

Thermal resistance network.

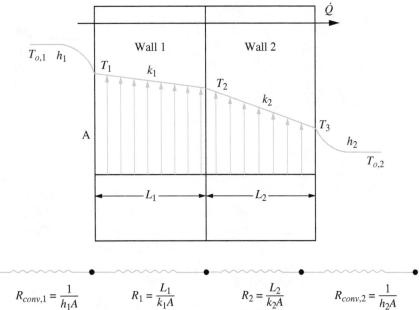

Once $\dot{Q}$ is known, an unknown surface temperature T_j at any surface j can be determined as follows:

$$\dot{Q} = \frac{T_i - T_j}{R_{\text{total},i-j}}$$

where T_i is a known temperature at location i and $R_{\text{total},i-j}$ is the total thermal resistance between the locations i and j. For example, when the fluid temperatures $T_{o,1}$ and $T_{o,2}$ for the two-layer case shown in Figure 18–3 are available and $\dot{Q}$ is calculated using the preceding equation, the interface temperature T_2 between the two walls can be determined from:

$$\dot{Q} = \frac{T_{o,1} - T_2}{R_{\text{conv},1} + R_{\text{conv},j}} = \frac{T_{o,1} - T_2}{\frac{1}{h_1 A} + \frac{L_1}{k_1 A}}$$

The temperature drop across a layer is easily determined from the preceding equation by multiplying $\dot{Q}$ by the thermal resistance of that layer:

- To find T_1: $\dot{Q} = \frac{T_{o,1} - T_1}{R_{\text{conv},1}}$
- To find T_2: $\dot{Q} = \frac{T_{o,1} - T_1}{R_{\text{conv},1} + R_1}$
- To find T_3: $\dot{Q} = \frac{T_3 - T_{o,2}}{R_{\text{conv},3}}$

The thermal resistance concept is widely used in practice because it is intuitively easy to understand and it has proven to be a powerful tool in the solution of a wide range of

Figure 18–4

Physical contact resistance.

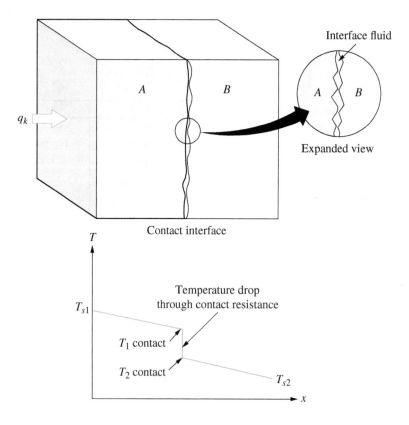

heat transfer problems. But its use is limited to systems through which the rate of heat transfer $\dot{Q}$ remains constant.

Thermal Contact Resistance

In the analysis of heat conduction through multilayer solids, we have assumed "perfect contact" at the interface of any two layers, and thus no abrupt temperature drop at the interface. This would be the case when the surfaces are perfectly smooth and are in perfect contact at each point, obviously a practically impossible situation. In reality, however, even surfaces that would appear smooth to the naked eye appear rather rough (Figure 18–4) when microscopically examined, with numerous peaks and valleys. Table 18–1 and Figure 18–5 quantify the thermal contact resistance phenomenon.

When two such surfaces are pressed against each other, the peaks will form a good material contact, but the valleys will form voids filled with air. Thus an interface offers some resistance to heat transfer, and this resistance is referred to as the *thermal contact resistance* R_c (Table 18–1). The value of R_c is determined experimentally, and there is a considerable scatter of data because of the difficulty in characterizing the surfaces. As expected, the contact resistance is observed to decrease with decreasing surface roughness and increasing interface pressure. Noting that $\dot{Q} = \frac{\Delta T}{R_c}$ for any two layers in contact with one another, the thermal contact resistance can be determined by measuring the temperature drop ΔT at the interface and multiplying it by the steady heat transfer rate $\dot{Q}$. Most experimentally determined values of the thermal contact resistance fall between 0.0001 and 0.001 m^2 (K·W).

The thermal contact resistance can be minimized by applying a thermally conducting liquid called a thermal grease, such as silicon oil, on the surfaces before they are pressed against each other.

Table 18–1

Thermal Contact Resistance

Approximate Range of Thermal Contact Resistance for Metallic Interfaces Under Vacuum Conditions

Interface Material	Resistance, R_i (m^2 K/W $\times 10^4$)	
	Contact Pressure 100 kN/m^2	Contact Pressure 10,000 kN/m^2
Stainless steel	6–25	0.7–4.0
Copper	1–10	0.1–0.5
Magnesium	1.5–3.5	0.2–0.4
Aluminum	1.5–5.0	0.2–0.4

Thermal Contact Resistance for Aluminum-Aluminum Interface with Different Interfacial Fluids

Interfacial Fluid	Resistance, R_i (m^2 K/W)
Air	2.75×10^{-4}
Helium	1.05×10^{-4}
Hydrogen	0.720×10^{-4}
Silicone oil	0.525×10^{-4}
Glycerin	0.265×10^{-4}

Figure 18–5

Contact resistance between dissimilar bare and metal surfaces.

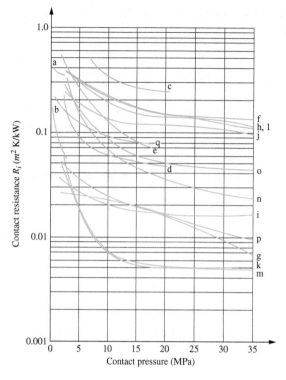

Curve	Material	Finish	Roughness rms (µm)	Temp. (°C)	Condition	Scatter of data
a	416 stainless 7075(75S)T6 Al	Ground	0.76–1.65	93	Heat flow from stainless to aluminum	± 26%
b	7075(75S)T6 Al to stainless	Ground	1.65–0.76	93–204	Heat flow from aluminum to stainless	± 30%
c	Stainless aluminum		19.94–29.97	20	Clean	
d	Stainless aluminum		1.02–2.03	20	Clean	
e	Bessemer steel foundry brass	Ground	3.00–3.00	20	Clean	
f	Steel Ct-30	Milled	7.24–5.13	20	Clean	
g	Steel Ct-30	Ground	1.98–1.52	20	Clean	
h	Steel Ct-30 aluminum	Milled	7.24–4.47	20	Clean	
i	Steel Ct-30 aluminum	Ground	1.98–1.35	20	Clean	
j	Steel Ct-30 copper	Milled	7.24–4.42	20	Clean	
k	Steel Ct-30 copper	Ground	1.98–1.42	20	Clean	
l	Brass aluminum	Milled	5.13–4.47	20	Clean	
m	Brass aluminum	Ground	1.52–1.35	20	Clean	
n	Brass copper	Milled	5.13–4.42	20	Clean	
o	Aluminum copper	Milled	4.47–4.42	20	Clean	
p	Aluminum copper	Ground	1.35–1.42	20	Clean	
q	Uranium aluminum	Ground		20	Clean	

The interface resistance is primarily a function of the surface roughness, the pressure holding the two surfaces in contact, the interface fluid, and the interface temperature. At the interface, the mechanism of heat transfer is complex. Conduction takes place through the contact points of the solid, while heat is transferred by convection and radiation across the trapped interfacial fluid. Figure 18–4 shows such a mechanism, as well as the temperature drop over the interface.

When two surfaces are in perfect thermal contact, the interface resistance approaches zero, and there is no temperature drop across the interface. For imperfect thermal contact, a temperature difference will take place at the interface as shown in Figure 18–4.

Numerous measurements have been made of the contact resistance at the interface between dissimilar metallic surfaces, but no satisfactory correlations have yet been found. Each situation must be treated separately. The results of many different conditions and materials are summarized in Figure 18–5. In this figure, some experimental results for the contact resistance between dissimilar base metal surfaces at atmospheric pressure are plotted as a function of the contact pressure.

Example 18–3

An instrument used to study the ozone depletion near the poles is placed on a large 2-cm-thick duralumin plate. To simplify the analysis, the instrument can be thought of as a stainless steel plate 1 cm tall with a 10 cm × 10 cm square base, as shown in Figure 18–6. The interface roughness of the steel and the duralumin is between 20 and 30 μm. Four screws at the corners provide fastening that exerts an average pressure of approximately 7.0 MPa. The top and sides of the instrument are thermally insulated. An integrated circuit placed between the insulation and the upper surface of the stainless steel plate generates heat. If this heat is to be transferred to the lower surface of the duralumin, estimated to be at a temperature of 0°C, determine the maximum allowable dissipation rate from the circuit if its temperature is not to exceed 40°C.

Solution

Since the top and sides of the instrument are insulated, all the heat generated by the circuit must flow downward. The thermal circuit will have three resistances (Figure 18–6): the stainless steel, the contact, and the duralumin. The thermal resistance of the metal plates is calculated as follows:

$$R_k = \frac{L_{ss}}{Ak_{ss}} = 0.07 \text{ K/W}$$

As for the duralumin, we have:

$$R_k = \frac{L_{AL}}{Ak_{AL}} = 0.012 \text{ K/W}$$

The contact resistance, in this case, is obtained from Figure 18–5. For this pressure, the unit contact resistance given by line c in this figure is 0.5 m²(K·kW). Therefore:

$$R_i = 0.05 \text{ K/W}$$

The thermal circuit, with all the resistances' magnitudes, is shown in Figure 18–5. The total resistance is 0.132 K/W, and the maximum allowable rate of heat dissipation is therefore:

$$q_{max} = \frac{\Delta T}{R_{total}} = 303.0 \text{ W}$$

Figure 18–6

Sketch of the instrument
and thermal circuit.

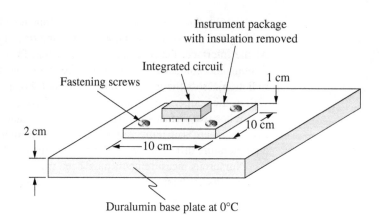

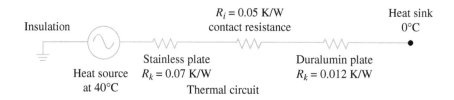

Example 18–4

Consider a 3-m-high, 5-m-wide and 0.3-m-thick wall whose thermal conductivity coefficient is $k = 0.9$ W/(m·°C). On a certain day, the temperatures of the inner and outer surfaces of the wall are measured to be 16°C and 2°C, respectively. Determine the rate of heat loss through the wall.

Solution

We assume a steady heat conduction across the wall. We also assume the heat transfer through the wall to be one-dimensional.

Noting that the heat transfer through the wall is by conduction and the surface of the wall is $A = 3$ m $\times$ 5 m $= 15$ m^2, the steady rate of heat transfer through the wall can be determined as follows:

$$\dot{Q} = kA\frac{T_1 - T_2}{L} = 630 \text{ W}$$

We could also determine the steady rate of heat transfer through the wall by making use of the thermal resistance concept:

$$\dot{Q} = \frac{\Delta T_{\text{wall}}}{R_{\text{wall}}} = 630 \text{ W}$$

18–7 Transient Heat Conduction

A heat conduction problem is transient whenever the temperature within the system being considered changes with time. There are many practical instances involving transient heat conduction. For example, many manufacturing processes require that the product be heated or cooled to convert it into a useful product.

Transient Conduction in a Semi-Infinite Solid

A semi-infinite solid is a large body with one surface. A good example here is the planet earth. If the temperature of the surface of the earth is changed, heat is conducted into the earth, and due to its infinite extent, the temperature is a function of distance from the earth surface x and time t, or expressed mathematically, $T = T(x, t)$. The conduction equation covering the lumped heat analysis can now be simplified as follows:

$$\frac{\partial^2 T}{\partial x^2} = \frac{1}{\alpha} \frac{\partial T}{\partial t}$$

where x is measured from the surface, and α is the thermal diffusivity.

Before solving this equation, we must specify two boundary conditions. The initial condition is:

$$T(x, 0) = T_o$$

That is, the entire semi-infinite solid is at a uniform temperature T_o at $t = 0$

One of the boundary conditions requires that the material at a large distance from the surface never changes in temperature or:

$$T(\text{infinity}, t) = T_o$$

Solutions are possible for several different choices for the second boundary condition. We will next consider two possible cases.

Case 1: Isothermal Boundary Condition

One possible boundary condition is relatively easy to achieve physically. This is to suddenly change the temperature of the surface ($x = 0$) to a value equal to T_s and maintain it at the constant value. The isothermal boundary condition can be mathematically specified as follows:

$$T(0, t) = T_s$$

The solution under such circumstances is:

$$\frac{T(x, t) - T_s}{T_o - T_s} = \text{er f}\left(\frac{x}{2\sqrt{\alpha t}}\right)$$

The symbol er f is the so-called Gauss error function, which frequently occurs in engineering applications, and is defined as follows:

$$\text{er f}\left(\frac{x}{2\sqrt{\alpha t}}\right) = \frac{2}{\sqrt{pi}} \int_0^{\frac{x}{\sqrt{\alpha t}}} e^{-\pi^2} d\eta$$

Variation of the temperature profile with time in the plane wall is illustrated in Figure 18–7. When the wall is first exposed to the surrounding medium at $T_o < T_1$ at $t = 0$, the entire wall is at its initial magnitude T_1. But the wall temperature at and near the surfaces starts to drop as a result of convection from the wall to the surrounding medium. This creates a temperature gradient in the wall and initiates heat conduction from the inner parts of the wall toward its outer surfaces. Note that the temperature at the center of the wall remains at T_1 until $t = t_2$, and the temperature profile within the wall remains symmetric at all times about the center plane. The temperature profile gets flatter as time passes as a result of heat transfer, and it eventually becomes uniform at $T = T_o$. That is, the wall reaches thermal equilibrium with its surrounding medium. At this point, the heat transfer stops since there is no longer a temperature difference. Similar discussions can be presented for the long cylinder or sphere.

Figure 18–7

Progression of the temperature distribution with time.

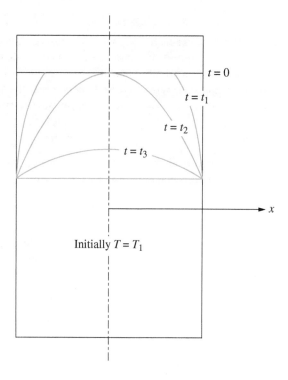

The formulation of the problem for the determination of the one-dimensional transient temperature distribution $T(x, t)$ in a wall results in a partial differential equation, which can be solved using advanced mathematical techniques. The solution, however, normally involves infinite series, which are inconvenient and time-consuming to evaluate. Therefore, there is a clear motivation to present the solution in tabular or graphical forms. However, the solution involves the parameters $x, L, t, k, \alpha, h, T_1$, and T_o, which are too many to make any graphical presentation of the results practical. In order to reduce the number of parameters, we nondimensionalize the problem by defining the following groupings:

- Dimensionless temperature: $\theta(x, t) = \frac{T(x,t) - T_o}{T_1 - T_o}$
- Dimensionless distance from the center: $X = \frac{x}{L}$
- Dimensionless heat transfer coefficient: $\text{Bi} = \frac{hL}{k} \ldots$ (Biot number)
- Dimensionless time: $\tau = \frac{\alpha t}{L^2} \ldots$ (Fourier number)

Nondimensionalization enables us to present the temperature in terms of three parameters only; namely X, Bi, and τ. This makes it practical to present the solution in a graphical form. The dimensionless quantities just defined for a plane wall can also be used for a cylinder or a sphere by replacing the space variable x with r, and the half-thickness L with the outer radius r_o. Note that the characteristic length in the definition of the Biot number is taken to be the half-thickness L for the plane wall, and the radius r_o for the long cylinder and the sphere.

The one-dimensional transient heat conduction described above can be solved exactly for any of the three geometries, but the solution involves infinite series, which are difficult to deal with. However, the terms in the solutions change rapidly with increasing time, and for $\tau > 0.2$, keeping the first term and neglecting all the remaining terms in the series results in an error that is below 2 percent. We are usually interested in the solution for times greater than $\tau > 0.2$, and thus it is very convenient to express the solution using the one-term approximation, which is given as follows:

$$\text{Plane wall: } \theta(x, t)_{\text{wall}} = \frac{T(x, t) - T_o}{T_1 - T_o} = A_1 e^{\frac{-\lambda}{\tau}} \cos\left(\frac{\lambda_1 x}{L}\right) \dots \tau > 0.2$$

$$\text{Cylinder: } \theta(x, t)_{\text{cyl}} = \frac{T(r, t) - T_o}{T_1 - T_o} = A_1 e^{\frac{-\lambda}{\tau}} J_o(\lambda_1 r / \lambda_o) \dots \tau > 0.2$$

$$\text{Sphere: } \theta(x, t)_{\text{sphere}} = \frac{T(r, t) - T_o}{T_1 - T_o} = A_1 e^{\frac{-\lambda}{\tau}} \frac{\sin(\lambda_1 r / r_o)}{\lambda_1 r / r_o} \dots \tau > 0.2$$

where the constants A_1 and λ_1 are functions of the Biot number only, and their values are listed against the Bi number for all three geometries. The function J_o is the zeroth-order Bessel function of the first kind, whose value can be determined from Table 18–2. Note that $\cos(0) = J_o(0) = 1$ and the limit of $\frac{\sin x}{x}$ is also 1. Therefore, the preceding relations simplify to the following at the center of the plane wall, cylinder, and sphere:

$$\text{Center of plane wall}(x = 0) : \theta_{0,\text{wall}} = A_1 e^{\frac{-\lambda}{\tau}}$$

$$\text{Center of cylinder}(r = 0) : \theta_{0,\text{cyl}} = A_1 e^{\frac{-\lambda}{\tau}}$$

$$\text{Center of sphere}(r = 0) : \theta_{0,\text{sphere}} = A_1 e^{\frac{-\lambda}{\tau}}$$

Once the Bi number is known, the preceding relations can be used to determine the temperature anywhere in the medium. Determination of the constants A_1 and λ_1 usually requires interpolation.

Example 18–5

A short brass cylinder with $k = 110 \text{ W/(m·°C)}$, $\alpha = 3.39 \times 10^{-5} \text{ m}^2/\text{s}$, with diameter $D = 10$ cm and height $H = 12$ cm is initially at a uniform temperature $T_1 = 120°C$ (Table 18–3). The cylinder is now placed in atmospheric air at 25 °C, where heat transfer takes place by convection, with a heat transfer coefficient of $h = 60 \text{ W/(m}^2\text{·°C)}$. Calculate the temperature at the center of the cylinder 15 minutes after the start of the cooling process.

Solution

This is a two-dimensional transient heat conduction problem, and thus the temperature will vary in both the axial and radial directions.

The dimensionless temperature at the center of the plane wall is:

$$\tau = \frac{\alpha t}{L^2} = 8.48$$

and the Biot number is:

$$\frac{1}{\text{Bi}} = \frac{k}{hL} = 30.6$$

Thus:

$$\theta_{\text{wall}}(0, t) = \frac{T(0, t) - T_o}{T_1 - T_o} = 0.8$$

Table 18–2

Comparison of the plane, cylindrical and spherical walls under transient boundary conditions.

Bi	Plane Slab λ_1	A_1	Cylinder λ_1	A_1	Sphere λ_1	A_1
0.01	0.0998	1.0017	0.1412	1.0025	0.1730	1.0030
0.02	0.1410	1.0033	0.1995	1.0050	0.2445	1.0060
0.04	0.1987	1.0066	0.2814	1.0099	0.3450	1.0120
0.06	0.2425	1.0098	0.3438	1.0148	0.4217	1.0179
0.08	0.2791	1.0130	0.3960	1.0197	0.4860	1.0239
0.1	0.3111	1.0161	0.4417	1.0246	0.5423	1.0298
0.2	0.4328	1.0311	0.6170	1.0483	0.7593	1.0592
0.3	0.5218	1.0450	0.7465	1.0712	0.9208	1.0880
0.4	0.5932	1.0580	0.8516	1.0931	1.0528	1.1164
0.5	0.6533	1.0701	0.9408	1.1143	1.1656	1.1441
0.6	0.7051	1.0814	1.0184	1.1345	1.2644	1.1713
0.7	0.7506	1.0918	1.0873	1.1539	1.3525	1.1978
0.8	0.7910	1.1016	1.1490	1.1724	1.4320	1.2236
0.9	0.8274	1.1107	1.2048	1.1902	1.5044	1.2488
1.0	0.8603	1.1191	1.2558	1.2071	1.5708	1.2732
2.0	1.0769	1.1785	1.5995	1.3384	2.0288	1.4793
3.0	1.1925	1.2102	1.7887	1.4191	2.2889	1.6227
4.0	1.2646	1.2287	1.9081	1.4698	2.4556	1.7202
5.0	1.3138	1.2403	1.9898	1.5029	2.5704	1.7870
6.0	1.3496	1.2479	2.0490	1.5253	2.6537	1.8338
7.0	1.3766	1.2532	2.0937	1.5411	2.7165	1.8673
8.0	1.3978	1.2570	2.1286	1.5526	2.7654	1.8920
9.0	1.4149	1.2598	2.1566	1.5611	2.8044	1.9106
10.0	1.4289	1.2620	2.1795	1.5677	2.8363	1.9249
20.0	1.4961	1.2699	2.2880	1.5919	2.9857	1.9781
30.0	1.5202	1.2717	2.3261	1.5973	3.0372	1.9898
40.0	1.5325	1.2723	2.3455	1.5993	3.0632	1.9942
50.0	1.5400	1.2727	2.3572	1.6002	3.0788	1.9962
100.0	1.5552	1.2731	2.3809	1.6015	3.1102	1.9990
∞	1.5708	1.2732	2.4048	1.6021	3.1416	2.0000

Moreover, at the center of the cylinder, we have:

$$\tau = 12.2$$

$$\frac{1}{Bi} = 36.7$$

Therefore:

$$\left[\frac{T(0, 0, t) - T_o}{T_1 - T_o} \right]_{short-cylinder} = \theta_{wall}(0, t) x \theta_{cylinder}(0, t) = 0.4$$

Table 18–2
(continued)

ξ	$J_0(\xi)$	$J_1(\xi)$
0.0	1.0000	0.0000
0.1	0.9975	0.0499
0.2	0.9900	0.0995
0.3	0.9776	0.1483
0.4	0.9604	0.1960
0.5	0.9385	0.2423
0.6	0.9120	0.2867
0.7	0.8812	0.3290
0.8	0.8463	0.3688
0.9	0.8075	0.4059
1.0	0.7652	0.4400
1.1	0.7196	0.4709
1.2	0.6711	0.4983
1.3	0.6201	0.5220
1.4	0.5669	0.5419
1.5	0.5118	0.5579
1.6	0.4554	0.5699
1.7	0.3980	0.5778
1.8	0.3400	0.5815
1.9	0.2818	0.5812
2.0	0.2239	0.5767
2.1	0.1666	0.5683
2.2	0.1104	0.5560
2.3	0.0555	0.5399
2.4	0.0025	0.5202

and:

$$T(0, 0, t) = T_o + 0.4(T_1 - T_o) = 63°C$$

Let us say that we wish to determine the transient temperature at point P in a cylinder of a finite length (Table 18–3). Point P is located by two coordinates (x, r) as shown in Table 18–3. The initial condition and boundary conditions are the same as those that apply to the one-dimensional charts. The cylinder is initially at a uniform temperature, T_0 at time $t = 0$. The entire surface, on the other hand, is subjected to a fluid with constant ambient temperature T_o, and the convection heat transfer coefficient between the cylinder surface area and the fluid is $\bar{h}_x$.

The radial temperature distribution for an infinitely long cylinder is given in Figure 18–8. For a cylinder with a finite length, the radial and axial temperature

Figure 18–8

Geometry for a cylinder
with a finite length.

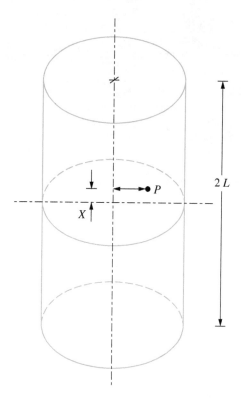

distributions are given by the product solution of an infinitely long cylinder and infinite plate as follows:

$$\frac{\theta_P(r, x)}{\theta_o} = C(r)P(x)$$

where $C(r)$ and $P(x)$ are the dimensionless temperatures for the infinite cylinder and infinite plate, respectively.

$$C(r) = \frac{\theta(r, t)}{\theta_o}$$

$$P(x) = \frac{\theta(x, t)}{\theta_o}$$

Case 2: Convection Boundary Condition

Instead of changing the surface temperature of the infinite solid, we could subject the surface to a fluid with an ambient temperature of T_f and an average convection heat transfer coefficient of h. The heat transferred into the solid must then be convected through the fluid and then conducted into the solid. The appropriate boundary condition for this type of problem can be written as follows:

$$h[T_f - T(0, t)] = -k\left(\frac{\partial T}{\partial s}\right)_{x=0}$$

The solution to this equation is subject to the same initial condition as in Case 1, and the two boundary conditions in the same case are given as:

$$\frac{T(x, t) - T_o}{T_f - T_o} = \text{er f}\psi - [\exp[B + \eta][1 - \text{er f}\psi + \sqrt{\eta}]]$$

where:

$$\psi = \sqrt{\frac{x^2}{4\alpha t}} = \text{Fo}^{\frac{-1}{2}}$$

$$\text{Fo} = \frac{\alpha t}{x^2}$$

$$\text{Bi} = \frac{hx}{k}$$

$$\eta = \frac{h\alpha t}{k^2} = (\text{Bi})^2 \text{Fo}$$

Therefore, the dimensionless temperature distribution in a semi-infinite solid with uniform initial temperature subjected to a fluid with temperature T_f at $t = 0$ is only a function of the Fourier and Biot numbers Fo and Bi, respectively, which will both be defined later in this section.

Linear and Homogeneous Problems

The terms *linear* and *homogeneous* should first be defined for better understanding of the types of equations which will be considered here.

A differential equation is linear if it contains no products of the dependent variable or its derivatives. That is, a term like T^2 is not permitted under this classification. A boundary condition is linear if it contains no products of the dependent variable or its derivatives.

A differential equation is homogeneous if, when it is satisfied by T, it is also satisfied by CT, where C is an arbitrary constant. A boundary condition (different from the inlet condition) is homogeneous if, when satisfied by T, it is also satisfied by CT, where C is a constant.

A Plane Wall Transient

To see what is involved in the separation-of-variables method, let us consider a rather simple problem. Let us consider a wall which is initially at a uniform temperature T_1 and is suddenly exposed on both surfaces to a fluid at a temperature T_o. The temperature-time history of the wall during the transient heat transfer process may be desired to see how long it takes the midspan temperature to reach a specified value.

For an infinitely long slab, the heat transfer is taken to be one-dimensional. The governing differential equation in this case is:

$$\frac{\partial^2 T}{\partial x^2} = \frac{1}{x}\frac{\partial T}{\partial t}$$

The boundary conditions are:

- $T = T_o$ at $x = 0$
- $T = T_o$ at $x = L$

These boundary conditions can be restated in the following form:

$$T(0, t) = T_o$$

$$T(L, t) = T_o$$

The initial condition is:

$$T(x, 0) = T_1$$

Example 18–6

A titanium spherical body [$c_p = 7.10$ kJ/(kg·K)] initially at 350 K is exposed to a heat source which conveys 200 W to the sphere for 15 seconds. If the sphere's mass is 4.5 kg, calculate the final temperature.

Solution

Treating the problem as that of "lumped" heat transfer, we have:

$$\dot{Q}\Delta t = mc_p\Delta T$$

which yields:

$$T_2 = 443.9 \text{ K}$$

Example 18–7

A solid steel sphere, AISI 1010, 1 cm in diameter, initially at 15°C is placed in a hot air stream $T_o = 60°C$. Calculate the sphere's temperature as a function of time after being placed in the hot air stream. The average convection heat transfer coefficient is 20 W/(m²·°C).

Solution

The thermophysical properties of the steel sphere are obtained from the tables as follows:

- $k = 63.9$ W/(m·°C)
- $\rho = 78.3$ kg/m³
- $c = 434$ J/(kg·°C)

The thermal diffusivity is:

$$\alpha = \frac{k}{\rho c} = 18.8 \times 10^{-8}$$

The characteristic length is:

$$L_c = \frac{V}{A} = 0.0017 \text{ m}$$

$$\text{Bi} = \frac{hL_c}{k} = 521.8 \times 10^{-6}$$

Now the Fourier number is:

$$\text{Fo} = \frac{\alpha t}{L_c^2} = 6.77t$$

Substitution of these relationships reveals the following:

$$T = \exp(-\text{BiFo})$$

$$T = 60 - 45\exp(-3.53 \times 10^{-3})t$$

Figure 18–8 shows this functional relationship.

PROBLEMS

18–1   A hot fluid at an average temperature of 200°C flows through a plastic pipe of 4.0 cm outer diameter and 3.0 cm inner diameter. The thermal conductivity of the plastic is 0.5 W/m·K, and the heat transfer coefficient at the inside is 300 W/(m^2·K). The pipe is located in a room at 30°C, and the heat transfer coefficient at the outer surface is 10 W/(m^2·K). Calculate the overall heat transfer coefficient and the heat loss per unit length of the pipe.

18–2   A long electrical heating element made of iron has a cross section of 10 cm × 1.0 cm. It is immersed in a heat transfer oil at 80°C. If heat is generated uniformly at a rate of 1,000,000 W/m^2 by an electric current, determine the heat transfer coefficient necessary to keep the temperature of the heater below 200°C. The thermal conductivity for iron at 200°C is 64 W/(m·K) by interpolation using the Appendix.

18–3   Compare the heat loss from an insulated and uninsulated copper pipe under the following conditions: The pipe [$k = 400$ W/(m·K)] has an internal diameter of 10 cm and an external diameter of 12.0 cm. Saturated steam flows inside the pipe at 110°C. The pipe is located in a space at 30°C, and the heat transfer coefficient on its outer surface is estimated to be 15 W/(m^2·K). The insulation available to reduce the heat losses is 5.0 cm thick, and its conductivity is 0.2 W/(m·K).

18–4   A spherical, thin-walled metallic container is used to store liquid nitrogen at 77 K. The container has a diameter of 0.5 m and is covered with an evacuated insulation system that is composed of silica powder [$k = 0.0017$ W/(m·K)]. The insulation is 25 mm thick, and its outer surface is exposed to ambient air at 300 K. The latent heat of evaporation h_{f-g} of liquid nitrogen is 2×10^5 J/kg. If the convection coefficient is 20 W/(m^2·K) over the outer surface, determine the rate of liquid boil-off of nitrogen per hour.

18–5   This problem concerns a graphite-moderated nuclear reactor. Heat is generated uniformly in uranium rods of 0.05 m diameter at the rate of 7.5×10^7 W/m^2. These rods are jacketed by an annulus in which water at an average temperature of 120°C is circulated. The water cools the rods, and the average heat convection coefficient is estimated to be 55,000 W/(m^2·K). If the thermal conductivity of uranium is 29.5 W/m k, determine the center temperature of the uranium fuel rods.

18–6   An experimental device that produces excess heat is passively cooled. The addition of pin fins to the casing of this device is being considered to augment the rate of cooling. Consider a copper pin fin 0.25 cm in diameter that protrudes from a wall at 95°C into ambient air at 25°C. The heat transfer is mainly by natural convection (discussed in detail next) with a coefficient that is equal to 10 W/(m^2·K). Calculate the heat loss, assuming that (a) the fin is infinitely long and (b) the fin is 2.5 cm long and the coefficient at the end of the fin is the same as that around the circumference. Finally, (c) how long would the fin have to be for the infinitely long solution to be correct within 5 percent.

19

Heat Convection

Chapter Outline

19–1   Introduction   **304**

19–2   Convection Heat Transfer Coefficient   **305**

19–3   Natural Convection   **309**

19–4   Lumped Parameter Analysis   **317**

19–5   Forced Convection Analysis   **320**

Introduction

A current of liquid or gas that absorbs heat at one place and then moves to another, where it mixes with a cooler portion of the fluid and rejects heat, is called a *convection current*. If the motion of the fluid is caused by a difference in density that is associated with a temperature difference, the phenomenon is called *natural* (or free) *convection*. If the fluid is made to move by the action of a pump, a fan, or a blower, we have a case of *forced convection*.

Before attempting to calculate a heat-transfer coefficient, we will examine the convection process in some detail. Consider a hot flat plate as it is cooled by a colder stream of air flowing over it. Shown in Figure 19–1 are the velocity and temperature distributions. The first point to note is that the velocity decreases in the direction toward the surface as a result of viscous forces. First we have the following governing relationship:

$$\dot{q}_C = -k\left(\frac{\partial T}{\partial y}\right)_{y=0} = h_C(T_s - T_o)$$

Although this viewpoint suggests that the process can be viewed as conduction, the temperature gradient at the surface, $\left(\frac{\partial T}{\partial y}\right)_{y=0}$ is determined by the rate at which the fluid farther from the wall can transport the energy into the mainstream. Thus the temperature gradient at the wall depends on the flow field, with higher velocities being able to produce larger temperature gradients and higher rates of heat transfer. At the same time, however, the thermal conductivity of the fluid plays a role. For example, the value of k_f for water is an order of magnitude higher than that of air; thus, the convection heat-transfer coefficient for water is larger than that of air.

Consider a fluid in contact with a flat wall whose temperature is higher than that of the main body of the fluid. Although the fluid may be in motion, there is a relatively thin film of stagnant fluid next to the wall, with a thickness that depends on the character of the motion of the main body of the fluid. The more turbulent the motion, the thinner the film will be. Heat is transferred from the wall to the fluid by a combination of conduction through the film and convection in the fluid. Now let us define the so-called heat convection coefficient (h) which includes the combined effect of conduction through the film and convection in the fluid:

$$\dot{Q} = hA\Delta T$$

Figure 19–1

Hydrodynamic vs.
thermal boundary layers.

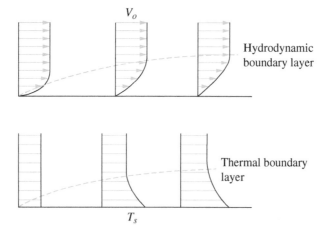

where $\dot{Q}$ is the rate of heat transfer by convection, A is the wall area, and ΔT is the temperature difference between the surface of the wall and the main body of the fluid. The fundamental problem of heat convection is to find the value of h that is appropriate to a particular piece of equipment.

Experiments show that the convection heat transfer coefficient depends on the following factors:

- Whether the wall is flat or curved
- Whether the wall is vertical or horizontal
- Whether the fluid in contact with the wall is a gas or a liquid
- The density, viscosity, specific heat ratio, and the fluid thermal conductivity
- Whether the fluid velocity is small enough to give rise to a laminar flow
- Whether evaporation or condensation is taking place

Since the physical properties of the fluid depend on the temperature and pressure, it is clear that the rigorous calculation of a convection heat transfer coefficient appropriate to a given wall and fluid is rather a tedious, practically nonending process. It was relatively recently that solutions to the problems that are good enough for practical purposes have been attained with the aid of dimensional analysis. Such analysis yields an expression for h containing the physical properties and velocity of the fluid as well as unknown constants and exponents. The constants and exponents are then obtained through experimentation.

19-2   Convection Heat Transfer Coefficient

Let us simply begin by defining h:

$$h = \frac{\dot{q}}{\Delta T}$$

The heat flux $\dot{q}$ is equal to the rate of heat transfer from the surface divided by the surface area. This is a vector that is perpendicular to the surface, and it is considered to be positive when the heat flows from the surface to the fluid. The temperature difference, ΔT, is the difference between the surface temperature T_w and the temperature of the fluid outside the boundary layer T_o.

The heat flux and surface temperature will depend on the location along the surface in an external flow; thus the local heat transfer coefficient is defined as follows:

$$h_x = \frac{\dot{q}_x}{(T_w - T_o)_x}$$

where x is the coordinate everywhere tangent to the heating surface. The local heat transfer coefficient can also be expressed in terms of the temperature gradient in the fluid at the fluid/surface interface using Fourier's law:

$$\dot{q}_x = -k\left(\frac{\partial T}{\partial y}\right)_{y=0}$$

to obtain:

$$h_x = \left[\frac{-k\left(\frac{\partial T}{\partial y}\right)_{y=0}}{(T_w - T_o)}\right]_x$$

$$\dot{Q} = hA(T_w - T_o)$$

The average heat transfer coefficient $\bar{h}$ is obtained by integrating the local coefficient over the complete surface length L:

$$\bar{h} = \frac{\int_0^L h_x dx}{L}$$

Over a flat plate, the local Nusselt number (Nu_x) is defined as follows:

$$\mathrm{Nu}_x = \frac{h_x x}{k}$$

where:

h_x is the local heat transfer coefficient.

x is the distance along the flat plate measured from the plate's leading edge.

k is the fluid thermal conductivity.

This local Nusselt number can always be integrated over the entire plate to find the overall Nusselt number.

Example 19–1

Over a flat plate, the local friction coefficient and Nusselt number at a location x from the leading edge for laminar flow are given by:

$$c_{f,x} = \frac{0.664^{\frac{1}{2}}}{\sqrt{\mathrm{Re}_x}}$$

$$\mathrm{Nu}_x = \frac{h_x x}{K} = 0.332 \, \mathrm{Re}_x^{\frac{1}{2}} Pr^{\frac{1}{2}} \text{ for } Pr \geq 0.6$$

where the Reynolds number $\mathrm{Re}_x = \frac{Ux}{\nu}$. Note that $c_{f,x}$ is proportional to $\sqrt{\frac{1}{\mathrm{Re}_x}}$, and thus $x^{\frac{-1}{2}}$.

Calculate the range of x_{cr} over which the flow remains laminar.

Solution

The average friction coefficient and the Nusselt number over the entire plate are determined by substituting the relationships above and performing the simple integration as follows:

$$c_f = \frac{1.328^{\frac{1}{2}}}{\mathrm{Re}_L}$$

$$\mathrm{Nu} = \frac{hL}{k} = 0.664 \, \mathrm{Re}_L^{\frac{1}{2}} Pr^{\frac{1}{3}}$$

The preceding relations give the average friction and heat transfer coefficient for the entire flat plate when the flow is laminar over the entire plate.

Taking the critical Reynolds number to be $5 \times 10^5 = \frac{Ux_{cr}}{\nu}$, the preceding relationship can be used for $x \leq x_{cr}$.

The boundary layer development on an isothermal surface for a fluid at several locations along the plate is shown in Figure 19–1. It is seen that as one moves downstream away from the plate's leading edge, the thickness of the laminar boundary layer increases while $\frac{\partial T}{\partial y}$ at $y = 0$ decreases. In the turbulent boundary layer region, the value of the fluid temperature gradient at the wall also decreases, as shown in Figure 19–1, as one moves downstream. But its magnitude, and thus the rate of heat transfer and the local

Figure 19–2

A fluid that is heated as it flows through a tube.

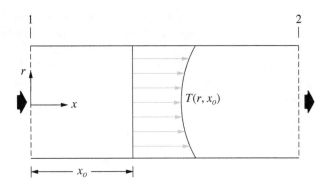

heat transfer coefficient, is considerably greater than that observed in the laminar flow region. The increase in the rate of heat transfer in the turbulent flow region is associated with the random fluctuations of the fluid particles which increase the mixing of the fluid, and thereby enhance the transfer of thermal energy between the surface and the fluid.

Figure 19–2 reveals the temperature distribution at a typical station in a high-surface-temperature, hot-walled cylinder. Note that the temperature attains its minimum magnitude at the cylinder centerline. An integral analysis of a laminar boundary layer flow indicates that the fluid boundary layer thickness and that of the thermal boundary layer are related by the following expression:

$$\frac{\delta}{\delta_T} = 1.026 \, \mathrm{Pr}^{\frac{1}{3}}$$

An important relationship between the convection heat transfer coefficient and the boundary layer thickness can be obtained by assuming that the temperature varies linearly across the thermal boundary layer. This assumption neglects the effect of the moving fluid on the temperature distribution, thus the thermal energy transfer across the thermal boundary layer is due entirely to conduction. Although this approximation is inaccurate, it does enable us to observe a significant effect. It follows that the local heat transfer coefficient can be approximated as follows:

$$h_x = \frac{\dot{q}_x}{(T_w - T_o)} = \frac{-k(T_o - T_w)}{\delta_T}(T_w - T_o)$$

or:

$$h_x \approx \frac{1}{\delta_T}$$

Introducing the distance from the plate's leading edge, x, as a characteristic length, we get:

$$\frac{h_x x}{k} = \frac{x}{\delta_T}$$

The quantity on the left-hand side of this equation is a dimensionless heat transfer coefficient and is called the Nusselt number. Being subscripted with x does nothing but indicate that it is a local quantity.

$$\mathrm{Nu}_x \equiv \frac{h_x x}{k}$$

The average Nusselt number is given as follows:

$$\mathrm{Nu} \equiv \frac{hL}{k}$$

where L is the plate length. Correlations to determine the convection heat transfer coefficient are usually expressed in terms of the Nusselt number. We now turn our attention to the turbulent-flow case.

The relationship between the velocity and thermal boundary layer thicknesses are such that they are identical to one another at a Prandtl number Pr of unity. This suggests that a similarity exists between the transfer of momentum and heat. A simple relationship between the average friction coefficient of a flat plate c_f and the average heat transfer coefficient is:

$$\frac{c_f}{2} = \frac{h}{\rho c_p V}$$

This is often referred to as the Reynolds analogy. The term on the right-hand side of this expression is dimensionless and is called the Stanton number:

$$St \equiv \frac{h}{\rho c_p V} = \frac{Nu}{Re_L} Pr$$

An ordinary fluid absorbs or releases a large amount of heat essentially at constant temperature during a phase-change process. The heat capacity rate of a fluid during a phase-change process must approach infinity since the temperature change virtually approaches zero. That is, $C = \dot{m} c_p$ tends to infinity when ΔT tends to infinity, so that the heat transfer rate $\dot{Q} = \dot{m} c_p \Delta T$ is a finite quantity. Therefore, in heat exchanger analyses, a condensing or boiling fluid is conveniently modeled as a fluid whose heat capacity rate is infinity.

Example 19–2

Water flows across a tube which is 2 cm in diameter, at a velocity of 1.0 m/s. The temperature of the water is 70°C. Vapor is condensed inside the tube, and the outer surface of the tube may be considered to be at a uniform temperature of 50°C. Determine the rate of heat transfer per unit length of the tube.

Solution

Using the water tables, we get:

$$\rho = 978 \text{ kg/m}^3$$

$$\mu = 0.404 \times 10^{-3} \text{ N} \cdot \text{s/m}^2$$

$$Pr = 2,57$$

$$k = 0.659 \text{ W/(m} \cdot \text{°C)}$$

The mean Nusselt number for the cylinder is obtained for $1 \leq Re_L \leq 10^5$:

$$\bar{Nu} = 0.3 + \sqrt{(Nu_{\text{laminar}})^2 + (Nu_{\text{turbulent}})^2}$$

The characteristic length $\frac{\pi d}{2} = 0.0314$ m. The Reynolds number is:

$$Re_L = \frac{\rho U L}{\mu} = 76.0 \times 10^3$$

The mean Nusselt number for laminar flow can be computed as follows:

$$(Nu)_{\text{laminar}} = 0.664\sqrt{Re_L} Pr^{\frac{1}{3}} = 250.8$$

For turbulent flow:

$$\bar{c}_f = 0.074(\text{Re})^{-0.2} = 7.818 \times 10^{-3}$$

and:

$$\bar{\text{Nu}}_{\text{turbulent}} = 450.2$$

The mean Nusselt number for the cylinder, therefore, is:

$$\bar{\text{Nu}} = 515.6$$

The mean convection heat transfer coefficient can be computed as follows:

$$\bar{h} = \frac{\text{Nu}k}{L} = 10,820 \ \text{W/(m}^2 \cdot {}^\circ\text{C})$$

Finally the rate of heat transfer per unit length of the tube is:

$$\dot{Q} = \bar{h}A(T_w - T_o) = -13.6 \ \text{W/m}$$

The Reynolds analogy can therefore be expressed as follows:

$$\frac{\bar{c}_f}{2} = \bar{S}t$$

The accuracy of this expression depends on the Prandtl number of the fluid. To increase the range of applicability, a modification can be made that yields a more reasonably accurate relationship in the Prandtl number range of 0.6 to 60.

$$\frac{c_f}{2} = \bar{S}t\text{Pr}^{\frac{2}{3}}$$

This is known as the Chilton-Colburn analogy, and which is valid for laminar flows over a flat plate and turbulent flows over surfaces of any shape.

19–3 Natural Convection

The movement of a fluid caused by the presence of a pressure gradient usually created by a fan or a pump and a body force that is referred to as *natural convection*. When the body force term is small compared to the force exerted on the fluid by the pressure gradient, an exchange of heat between the fluid and surface is classified as forced convection. If the heat is transferred from or to a fluid in which the body force term is much larger than that associated with the pressure gradient, the heat transfer (Figures 19–3 and 19–4) is termed free (or natural) convection. When the forces are of the same magnitude and heat is transferred, the process is identified as mixed or combined natural-forced convection.

For example, in Figure 19–3 we have a hot object on a flat plate. This body will continually cool down by transferring heat through "natural" convection to the environment, while creating the flow pattern shown in Figure 19–3. Here, the cold air adjacent to the hot body will heat up, with its density going down. As a result, that air will rise up (Figure 19–3). In another example (Figure 19–4), the opposite situation takes place, with the flow stream adjacent to the cold body directed downward.

Natural convection is just as effective in the heating of cold surfaces in a warmer environment as it is in the cooling of hot surfaces in a cooler environment.

In a gravitational field, there seems to be a net force that pushes a light fluid placed in a heavier fluid upward. The upward force exerted by a fluid on a body that is completely or partially immersed in it is called *buoyancy force*. The magnitude of this force is equal

Figure 19–3

Natural convection due to the presence of a hot body.

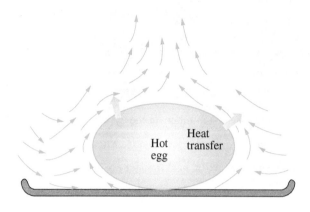

to the weight of the fluid displaced by the body. In other words:

$$F_{\text{buoyancy}} = \rho_{\text{fluid}} g V_{\text{body}}$$

where ρ_{fluid} is the average density of the fluid, g_c is the local gravitational acceleration, and V_{body} is the volume of the portion of the body which is immersed in the fluid (for bodies completely immersed in the fluid, it is the total volume of the body). In the absence of other forces, the net vertical force acting on the body is the difference between the weight of the body and the buoyancy force, i.e.,

$$F_{net} = W - F_{\text{buoyancy}} = \rho_{\text{body}} g_c V_{\text{body}} - \rho_{\text{fluid}} g_c V_{\text{body}} = (\rho_{\text{body}} - \rho_{\text{fluid}}) g_c V_{\text{body}}$$

Note that this force is proportional to the density difference between the fluid and the body immersed in it. Thus, a body immersed in a fluid will experience a "weight loss" at an amount that is equal to the weight of the fluid it displaces. This is known as Archimedes' principle.

Figure 19–4

Natural convection due to the presence of a cold body.

Figure 19–5

Flow behavior near a
heated plate.

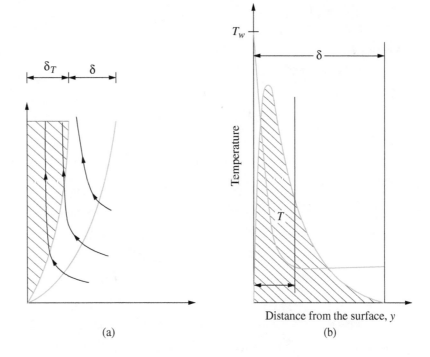

(a)
(b)

Natural Convection Over Surfaces

Natural convection heat transfer over a surface depends on the geometry of the surface as well as its orientation. It also depends on the variation of temperature on the surface and the thermophysical properties of the fluid involved.

Figure 19–5 reveals the streamlines and temperature distribution near a heated flat plate. As in forced convection, the thickness of the boundary layer increases in the flow direction. Unlike forced convection, however, the fluid velocity is zero at the outer edge of the velocity (or hydrodynamic) boundary layer as well as at the surface of the plate. This should be expected since the fluid beyond the boundary layer is stationary. The temperature of the fluid will be equal to the plate temperature at the surface, and it will gradually decrease to the surroundings' fluid temperature at a distance that is sufficiently far from the surface, as shown in Figure 19–5. In the case of cold surfaces, the shapes of the velocity and temperature profiles remain the same, but their direction is reversed.

The situation in Figure 19–6 is quite similar in free convection. The principal difference is that in forced convection the velocity approaches the free-stream magnitude imposed by an external force, whereas in free convection the velocity, at first, increases with increasing distance from the plate because the action of viscosity diminishes rather rapidly, while the density difference decreases more slowly. Eventually, however, the buoyancy force decreases as the fluid density approaches the value of the surrounding fluid. The temperature fields in free and forced convection have similar shapes, and in both cases, the heat transfer mechanism at the fluid/solid interface is conduction.

Figure 19–7 illustrates the mechanism behind natural convection. For instance, and by reference to Figure 19–6, cold air comes in contact with the hot body. As a result of heating, the density decreases and, with that, the fluid particles rise up.

Figure 19–6

Hydrodynamic and
thermal boundary layers.

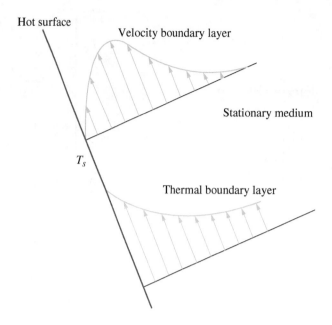

Figure 19–7

Temperature and velocity
boundary-layer profiles.

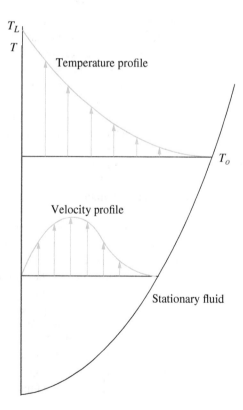

Vertical Flat Plate—Isothermal

A new dimensionless group is now introduced in natural convection heat transfer. This is called the Grashof number, and its local value is defined as follows:

$$\text{Gr}_x \equiv \frac{g_c \beta (T_w - T_o) x^3}{v^2}$$

or, for the entire surface of length L:

$$\text{Gr}_L \equiv \frac{g_c \beta (T_w - T_o) L^3}{v^2}$$

The spatial coordinate parallel to the plate in the direction of the flow stream is x, while y is the coordinate perpendicular to the plate. The coefficient of volumetric expansion is β, where:

$$\beta \equiv -\frac{1}{\rho} \frac{\partial \rho}{\partial T}_{\text{plate}}$$

For an ideal gas:

$$\beta = \frac{1}{T}$$

where T is the absolute film temperature $\frac{T_w + T_o}{2}$ in Kelvins.

$$\text{Ra}_L = \text{Gr}_L \text{Pr}$$

where Ra is the Rayleigh number. Note that the Rayleigh and Grashof numbers contain a characteristic length, which is identified by a subscript. For a vertical plate, the transition from laminar to turbulent flow occurs at $\text{Ra}_x = 10^9$. The following experimental correlation of the local and averaged Nusselt numbers is recommended by Churchill for a smooth vertical plate with a uniform surface temperature:

Laminar Flow

$$(\text{Nu})_x = 0.08 + 0.503[\text{Ra}_x \psi (\text{Pr})]^{\frac{1}{4}}$$

and:

$$(\text{Nu})_L = 0.08 + 0.67[\text{Ra}_L (\text{Pr})]^{\frac{1}{4}}$$

where:

$$\psi (\text{Pr}) = \left[1 + \left(\frac{0.492}{\text{Pr}} \right)^{\frac{9}{16}} \right]^{\frac{-16}{9}}$$

Turbulent Flow

$$\bar{\text{Nu}} = \int \text{Nu}_x = 0.15[\text{Ra}\psi (\text{Pr})]^{\frac{1}{3}}$$

The thermophysical properties of the fluid in the preceding equation are evaluated at the so-called film temperature $\frac{T_w + T_o}{2}$.

Several important observations should be stated at this point. First, since the Rayleigh number contains the temperature difference $(T_w - T_o)$, the convection heat transfer coefficient will be proportional to $(T_w - T_o)$. The expression for the heat flux is as follows:

$$\dot{q}_x = h_x (T_w - T_o)$$

which is proportional to $(T_w - T_o)^{\frac{5}{4}}$ for a laminar flow. In connection with a turbulent flow, the heat flux is proportional to $(T_w - T_o)^{\frac{4}{3}}$. The heat transfer is thus a nonlinear function of the local temperature difference. Secondly, the convection heat transfer coefficient in the turbulent flow region is independent of x, $\bar{h}$, and h_x

Horizontal Flat Plate

The rate of heat transfer from the top of a heated horizontal plate in a stagnant fluid is different from that at the bottom of the plate. The flow on the upper surface of the plate does not have a distinct boundary layer character. This is a continuous column of cold fluid moving downward toward the plate where the fluid is heated, and then reversing its direction to move upward. The fluid above the surface is therefore composed of displaced columns of hot and cold fluid moving in opposite directions. On the bottom surface, the cold fluid moves up to the surface, where it is heated, and then moves laterally to the edge of the plate before it continues its upward motion.

Hot Surface Up and Cold Surface Down

$$Nu_L = 0.54 \, Re_L^{\frac{1}{4}}$$

$$\text{for } 10^4 \leq Ra_L \leq 10^7$$

$$Nu_L = 0.15 \, Ra_L^{\frac{1}{3}}$$

$$\text{for } 10^7 \leq Ra_L \leq 10^{11}$$

Example 19–3

The surfaces of the side walls of a stove are at a temperature of 37.5°C when the temperature of the oven in the stove is set at 200°C. The stove is 0.75 m high, and each of its sides is 0.7 m wide. If the ambient temperature is 17.5°C, calculate the amount of heat lost through the stove sides.

Solution

At the average temperature of 27.5°C, the properties of air are:

$$\rho = 1.174 \text{ kg/m}^3$$

$$v = 15.8 \times 10^{-5} \text{ m}^2/\text{s}$$

$$c_p = 1.01 \text{ kJ/(kg} \cdot °\text{C)}$$

$$k = 26.2 \times 10^{-3} \text{ W/(m} \cdot °\text{C)}$$

$$Pr = 0.712$$

$$\beta = \frac{1}{300.7} = 0.0033 \text{ J/K}$$

The Rayleigh number for the flow is:

$$Ra_L = \frac{g_c \beta (T_w - T_o) L^3 Pr}{v^2} = 787 \times 10^6$$

The flow is laminar and the average Nusselt number can be calculated as follows:

$$Nu = 0.68 + 0.67(Ra \psi Pr)^{\frac{1}{4}} = 86.8$$

The magnitude of the average heat transfer coefficient can be calculated as follows:

$$h = \frac{\bar{Nu}k}{L} = 3.03 \text{ W/(m}^2 \cdot °\text{C)}$$

Finally, the total rate of heat lost by the four sides of the stove can be calculated as follows:

$$\dot{Q} = hA(T_w - T_o) = 127.3 \text{ W}$$

Example 19–4

A vertical 0.8-meter-high, 2-cm-wide, double-paned window consists of two sheets of glass separated by a 2-cm air gap at atmospheric pressure. If the glass surface temperatures across the air gap are measured to be 12°C and 2°C, determine the rate of heat transfer through the window.

Solution

We have a rectangular enclosure filled with air. The properties of air are:

$$T_{av} = \frac{T_1 + T_2}{2} = 7\,°C = 280\ K$$

$$k = 0.0246\ \text{W/(m} \cdot °\text{C)}$$

$$v = 1.40 \times 10^{-5}\ \text{m}^2/\text{s}$$

$$Pr = 0.717$$

$$\beta = \frac{1}{T_1} = 0.00357 K^{-1}$$

The characteristic length in this case is the distance between the two glasses, $\delta = 0.02$ m. Then the Rayleigh number becomes:

$$Ra = 1.024 \times 10^4$$

Then the natural convection Nusselt number, in this case, can be determined as follows:

$$Nu = 0.197\ Ra^{-1/9} \frac{H}{\delta^{-1/9}} = 1.32$$

then: $A = H \times L = 1.6\ \text{m}^2$, and:

$$\dot{Q} = k Nu A \frac{T_1 - T_2}{\beta} = 25.9\ \text{W}$$

Example 19–5

Air at an average temperature of 30°C flows over a fully rough flat plate that is 1 m long at a velocity of 100 m/s. Determine the average heat convection coefficient.

Solution

The properties of air at 30°C are:

$$\rho = 1.164\ \text{kg/m}^3$$

$$c_p = 1.006\ \text{kJ/(kg} \cdot °\text{C)}$$

$$Pr = 0.712$$

$$\mu = 16.0 \times 10^{-8}$$

$$k = 0.026\ \text{W/(m} \cdot °\text{C)}$$

$$Pr = 0.712$$

$$\beta = \frac{1}{300.7} = 0.0033\ \text{J/K}$$

The Rayleigh number for the flow is:

$$Ra_L = \frac{g_c \beta (T_w - T_o) L^3 Pr}{v^2} = 787 \times 10^6$$

The flow is laminar, and the average Nusselt number can be calculated as follows:

$$Nu = 0.68 + 0.67(Ra\psi Pr)^{\frac{1}{4}} = 86.8$$

The magnitude of the average heat transfer coefficient can be calculated as follows:

$$h = \frac{\bar{Nu}k}{L} = 3.03 \text{ W/(m}^2 \cdot {}^\circ\text{C)}$$

Finally the total rate of heat lost by the four sides of the stove can now be calculated:

$$\dot{Q} = hA(T_w - T_o) = 127.3 \text{ W}$$

Example 19–6

Engine oil at 60 °C flows over a 5-meter-long flat plate whose temperature is 20 °C at a velocity of 2.0 m/s. Determine the total drag force and the rate of heat transfer per unit width of the plate.

Solution

Let us assume that the critical magnitude of the Reynolds number is 5×10^5. The properties of the engine oil at the film temperature of:

$$T_f = \frac{T_w + T_o}{2} = 40°\text{C}$$

are:

$$\rho = 876 \text{ kg/m}^3$$

$$k = 0.144 \text{ W/(m} \cdot {}^\circ\text{C)}$$

$$Pr = 2870.0$$

$$\nu = 242 \times 10^{-6}$$

Noting that $L = 5$ m, the Reynolds number at the end of the plate becomes:

$$Re = \frac{UL}{\nu} = 41{,}300$$

which is less than the critical Reynolds number. Thus we have a laminar flow over the entire plate, and the average friction coefficient is determined as follows:

$$c_f = 1.328(Re_L)^{-0.5} = 0.0065$$

The drag force acting on the flat plate per unit width can be computed as follows:

$$F_D = c_f A \frac{\rho U^2}{2} = 57.2 \text{ N}$$

Similarly, the Nusselt number is determined using the laminar flow relations for a flat plate:

$$Nu = \frac{hL}{k} = 1{,}918$$

and:

$$h = \frac{k}{L} Nu = 55.2 \text{ W/(m}^2 \cdot {}^\circ\text{C)}$$

Finally:

$$\dot{Q} = hA(T_o - T_s) = 11{,}040 \text{ W}$$

The rate of heat transfer between two isothermal surfaces at T_1 and T_2 for a two-dimensional configuration is inversely proportional to the thermal resistance. The thermal resistance for two-dimensional steady-state heat conduction can be expressed in terms of the so-called conduction shape factor as follows:

$$R = \frac{1}{Sk}$$

A thermal network can be used with two-dimensional heat conduction to obtain approximate solutions for the rate of heat transfer. The accuracy of these calculations depends on the magnitude of temperature variation over the isothermal surfaces specified in the expression for the conduction shape factor. If the heat flow across the surface is one-dimensional and if the resistance in contact with the surfaces of the two-dimensional configurations is small compared to that of the two-dimensional body, the results obtained using a thermal network will usually be engineeringly acceptable.

Constant Surface Heat Flux

In the case where $\dot{q}_x$ is constant, the rate of heat transfer, for the case of a tube, can be expressed as follows:

$$\dot{Q} = \dot{q}A = \dot{m}c_p(T_e - T_i)$$

Then the mean fluid temperature at the tube exit becomes:

$$T_e = T_i + \frac{\dot{q}A}{\dot{m}c_p}$$

Note that the mean fluid temperature increases linearly in the flow direction in the case of a constant surface heat flux, since the surface area increases linearly in the flow direction (A is equal to the perimeter, which is constant, times the tube length).

The surface temperature in this case can be determined from the relationship:

$$\dot{q} = h(T_s - T_m)$$

Note that when h is constant, $T_s - T_m = $ constant, and thus the surface temperature will also increase linearly in the flow direction. Of course, this is true when the variation of the specific heat c_p with T is disregarded and c_p is assumed to remain constant.

19-4   Lumped Parameter Analysis

This approximation is valid when the internal resistance to heat transfer is much less than the resistance to heat transfer at the boundary. For a solid immersed in a convective environment, the Biot number Bi, a dimensionless parameter, expresses the ratio of internal to external thermal resistances:

$$\text{Bi} = \frac{\bar{h}_{conv}L_{char}}{k_s}$$

where $\bar{h}_{conv}$ is the average convection heat transfer coefficient at the system boundary, k_s is the system thermal conductivity, and L_{char} is a characteristic length. This characteristic length is defined as the ratio of the volume of the system to the surface area exposed to the convective environment, that is:

$$L_{char} = \frac{V_{system}}{A_{conv}}$$

For example, L_{char} for a sphere of radius R is:

$$L_{\text{char}} = \frac{(4/3)\pi R^3}{4\pi R^2} = R/3$$

The usual engineering criterion to invoke the lumped parameter approximation is:

$$\text{Bi is less than } 0.1$$

The smaller the Biot number is, the more uniform the system temperature must be during a transient process.

To introduce the temperature as an explicit variable in our conversation of energy expression, we recognize that:

$$\frac{dU}{dt} = m\frac{du}{dt}$$

and we apply the chain rule and the definition of the specific heat (included in Part 1 of this text) to yield:

$$\frac{du}{dt} \equiv \left(\frac{\partial u}{\partial T}\right)\frac{dT}{dt} = c_v\frac{dT}{dt}$$

With the assumption of incompressibility, $c_v = c_p = c$, our final result is:

$$\dot{Q}_{\text{in,net}} + \dot{W}_{\text{elec}} = m_{\text{system}}c\frac{dT}{dt}$$

We emphasize that the preceding equation is only applicable to a macroscopic system if the temperature is uniform, or sufficiently so for an engineering approximation. As a practical note, $\dot{W}_{\text{elec}}$ can be related to the electrical current i and the system electrical resistance $R_{\text{electrical}}$ as follows:

$$\dot{W}_{\text{electrical}} = i^2 R_{\text{electrical}}$$

This equation expresses the concept of Joule's heating, named in the honor of its discoverer James Prescott Joule. A body initially at a uniform temperature T_o suddenly experiences a change in its thermal environment. The rate at which this change is sensed in the body's interior will depend upon the resistance to heat transfer offered at its boundaries and the resistance to heat transfer internally offered within the material. If the thermal resistance offered at the boundaries is much greater than the internal thermal resistance, the temperature distribution within the body will be nearly uniform. The limiting case occurs when the thermal conductivity of the body is infinite. The temperature of the body is uniform, and the time-dependent temperature can be determined by using the lumped parameter analysis.

The first law of thermodynamics statement for the irregularly shaped body is:

$$\frac{dU}{dt} = \dot{Q}$$

The body is initially at a uniform temperature T_o. When t is greater than zero, the temperature of the surrounding fluid is changed to that of the fluid. If the internal thermal resistance is neglected, the temperature of the body is uniform, and the expression of the first law becomes:

$$\rho c V\frac{dT}{dt} = hA(T_f - T)$$

where V is the volume and A is the body surface area.

The initial condition is:

$$T = T_o$$

The temperature distribution within the body is obtained by integrating the preceding equation:

$$dT(T - T_f) = \frac{hA}{\rho c V} dt$$

Integrating both sides

$$\ln (T - T_f) = \frac{hA}{\rho c V} t + A$$

where A is the integration constant. The initial condition is used to determine that constant:

$$A = \ln (T_o - T_f)$$

The final expression for the temperature of the body is:

$$\frac{T - T_f}{T_o - T_f} = \exp \left(\frac{hAt}{\rho c V} \right)$$

This expression can be transformed into its dimensionless counterpart by introducing several dimensionless groups:

$$T = \frac{T - T_f}{T_o - T_f}$$

$$L_C = \frac{V}{A}$$

$$\alpha = \frac{k}{\rho c_p} = \left(\frac{k}{\rho c} \right)_{\text{solid}}$$

The exponent in the previous equation can be expressed in terms of two nondimensional groups frequently used in heat transfer. These are the Biot and Fourier numbers:

$$\text{Bi} = \frac{hL_c}{k}$$

$$\text{Fo} = \frac{\alpha t}{L_c^2}$$

Now the nondimensionalized form of the equation becomes:

$$T = \exp (-\text{BiFo})$$

The Biot number can be used to determine if significant errors are introduced in the calculation of the transient response of a body using the lumped parameter analysis.

The rate of heat transfer at any instant in time may be found as follows:

$$Q = \rho c V (T - T_o) = \rho c V (T_f - T_o) \left[1 - \exp \left(\frac{-hAt}{\rho c V} \right) \right] = Q_o \left[1 - \exp \left(\frac{-hAt}{\rho c V} \right) \right]$$

where $Q_o = \rho c V (T_f - T_o)$. This represents the maximum amount of energy that can be gained by the body.

If there is internal heat generation within the body that is activated at $t = 0$, the differential energy equation becomes:

$$\rho c V \frac{dT}{dt} = hA(T_f - T) + q''' V$$

The initial conditions remain the same.

Example 19–7

A solid steel sphere, AISI 1010, 1 cm in diameter and initially at 15°C is placed in a hot air stream, $T_o = 60°C$. Estimate the temperature of this sphere as a function of time after it is placed in the hot air stream. The average convection heat transfer coefficient is 20 W/(m²·°C).

Solution

The thermophysical properties of the steel are obtained from the tables of the Appendix as follows:

$$k = 63.9 \text{ W/(m} \cdot °C)$$

$$\rho = 7832 \text{ kg/}m^3$$

$$c = 434 \text{ J/(kg} \cdot °C)$$

The thermal diffusivity is:

$$\alpha = \frac{k}{\rho c} = 18.8 \times 10^{-6} \text{ m}^2/s$$

The characteristic length is:

$$L_c = \frac{V}{A} = \frac{(\pi/6)d^3}{\pi d^2} = 1.667 \times 10^{-3} \text{ m}$$

The Biot number can be calculated as follows:

$$\text{Bi} = \frac{hLc}{k} = 521.8 \times 10^{-6}$$

and it indicates that the lumped parameter analysis can be used without introducing an appreciable error in the calculations. The Fourier number can be obtained as follows:

$$\text{Fo} = \frac{\alpha t}{L_c^2} = 6.765t$$

Through simple substitution, we get:

$$T = \exp(-\text{Bi}, \text{Fo})$$

$$T = T_o + (T_{\text{sphere}} - T_o)\exp[-521.8 \times 10^{-6}(6.765t)]$$

$$= 60 - 45\exp(-3.53 \times 10^{-3} t)$$

This relationship can easily be graphically represented.

19–5 Forced Convection Analysis

Experiments on forced convection through circular pipes, combined with dimensional analysis, have shown that for a turbulent flow, the following equation holds:

$$\frac{hD}{k} = 0.023\left(\frac{\rho VD}{\mu}\right)^{0.2}\left(\frac{c_p\mu}{k}\right)^{0.8}$$

This equation is written in terms of dimensionless ratios. The left-hand side is called the Nusselt number as represented earlier. On the right-hand side, the first term will be recognized as the Reynolds number. The second term is the Prandtl number.

Experiments on many gases and vapors show that, over moderate temperature and pressure ranges, the ratio of viscosity to thermal conductivity remains approximately constant. The temperature variation of the thermal conductivity of helium is very similar to that of the viscosity. The specific heat at constant pressure c_p of helium is quite constant, so the Prandtl number calculated is approximately constant. It is customary to take an acceptable average for all gas and vapor values:

$$\text{Nu}_{\text{average}} = \frac{c_p \mu}{k} = 0.78$$

Introducing this value and solving for h, we get:

$$h = 0.027 c_p (\rho V)^{0.8} \left(\frac{\mu}{D}\right)^{0.2}$$

As much as we will consider forced convection only in pipes, it is convenient at this point to define a convection coefficient per unit length h_L instead of the usual coefficient per unit area h. For a pipe of length L and surface area A, the fundamental equation of convection can be expressed as follows:

$$\dot{Q} = h A \Delta T$$

and may be written as follows:

$$\dot{Q} = h_L L \Delta T$$

Therefore:

$$h A = h_L L$$

and since $A = \pi D L$,

$$h = \frac{h_L}{\pi D}$$

Another change is also worthwhile. The mass flow rate is given by:

$$\dot{m} = \rho \frac{\pi D^2}{4} V$$

Making these substitutions, we finally get:

$$h_L = 0.10 c_p \mu^{0.2} \left(\frac{\dot{m}}{D}\right)^{0.8}$$

PROBLEMS

19-1   When a thermocouple is moved from one medium to another at a different temperature, the thermocouple must be given sufficient time to come to thermal equilibrium with the new conditions before a reading is taken. Consider a 0.1-cm-diameter copper thermocouple wire originally at 150°C. Determine the temperature response when the wire is suddenly immersed in: (a) water at 40°C [$\bar{h}_c = 80$ W/(m²·K)], and (b) air at 40°C [$\bar{h}_c = 10$ W/(m²·K)].

19-2   Air at a pressure of 2.0 bar and a temperature of 490 K is flowing through a 2-cm-inner-diameter tube at a velocity of 10 m/s. Calculate the heat transfer coefficient if the tube temperature is 510 K. Also estimate the rate of heat transfer per unit length if a uniform heat flux is maintained.

19-3   Water at an inlet temperature of 333 K flows at a velocity of 0.2 m/s through a 0.3-m-long capillary tube having 0.00254 m inner diameter. Determine the overall heat transfer coefficient.

19-4   A liquid metal flows at a mass flow rate of 3.0 kg/s through a constant heat flux 5-cm-inner-diameter tube in a nuclear reactor. The fluid at 473 K is to be heated with the tube wall 30 K above the fluid temperature. Determine the length of the tube required for a one-degree-Kelvin rise in the bulk fluid temperature, using the following properties:

$$\rho = 7.7 \times 10^3 \text{ kg/m}^3$$

$$\nu = 8.0 \times 10^{-8} \text{ m}^2/\text{s}$$

$$c_p = 130 \text{ J/(kg} \cdot \text{K)}$$

$$k = 12 \text{ W/(m} \cdot \text{K)}$$

$$\text{Pr} = 0.011$$

For the flow over a slightly curved surface, the local shear stress is given by the following relationship:

$$\tau_s(x) = 0.3 \left(\frac{\rho \mu}{x} \right)^{0.5} U_o^{1.5}$$

Using this dimensional equation, derive nondimensional relations for the local and average friction coefficients.

20

Heat Exchangers

Chapter Outline

20–1	Introduction	**324**
20–2	Log Mean Temperature Difference	**325**
20–3	Overall Heat Transfer Coefficient	**329**
20–4	Fouling Factor	**331**
20–5	Analysis of Heat Exchangers	**334**
20–6	The Similitude Principle	**337**
20–7	Dynamic Similarity	**337**
20–8	Dimensionless Parameters	**338**

Introduction

Heat exchangers are devices that facilitate the exchange of heat between two fluids that are at different temperatures while keeping them from mixing with each other. Heat exchangers are commonly used in a wide range of applications, from heating and air-conditioning systems in a household, to chemical processing and power production in large plants. Heat exchangers differ from mixing chambers in that they do not allow the two fluids involved to mix. In a car radiator, for instance, heat is transferred from the hot water flowing through the radiator tubes to the air flowing through the closely spaced thin plates.

Heat transfer in a heat exchanger usually involves convection in each fluid and conduction through the wall separating the two fluids. In the analysis of heat exchangers, it is convenient to work with an overall heat transfer coefficient U that accounts for the contributions of all of these effects on heat transfer. The rate of heat transfer between the two fluids at a location in a heat exchanger depends on the magnitude of the temperature difference at that location, which varies along the heat exchanger. In the analysis of heat exchangers, it is usually convenient to work with the logarithmic mean temperature difference LMTD, which is an equivalent mean temperature difference between the two fluids for the entire heat exchanger.

Heat exchangers are manufactured in a variety of types, and thus we start this chapter with the classification of heat exchangers. We then discuss the determination of the overall heat transfer coefficient for heat exchangers, and the LMTD for some configurations. Next we introduce the correction factor F to account for the deviation of the LMTD in complex configurations.

The simplest type of heat exchanger consists of two concentric pipes of different diameters, as shown in Figure 20–1, and is called the double-pipe heat exchanger. One fluid in a double-pipe heat exchanger flows through the smaller pipe while the other

Figure 20–1

Two types of heat exchangers

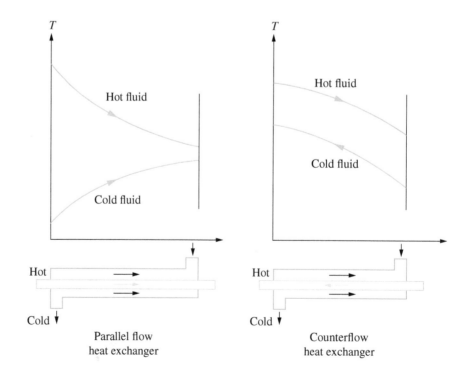

Parallel flow
heat exchanger

Counterflow
heat exchanger

Figure 20–2

Temperature variation in a heat exchanger when one of the fluids condenses or boils.

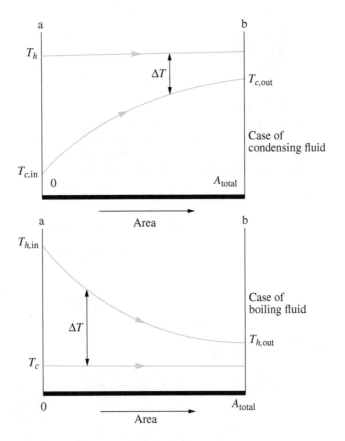

fluid flows through the annular space between the two pipes. Two types of flow arrangements are possible in this type of heat exchanger: in parallel flow, both the hot and cold fluids enter the heat exchanger at the same end and move in the same direction. In the counterflow type, the hot and cold fluids enter the heat exchanger at opposite ends.

The temperature profiles across another simple heat exchanger are shown in Figure 20–2. The upper half of this figure represents the case of condensing fluid stream (i.e., a fluid within the liquid plus vapor $L + V$ dome). The lower half of Figure 20–2 represents the case where the fluid is boiling (i.e., within the $L + V$ dome as well). In both cases the fluid temperature will remain constant during the heat exchange process.

Figure 20–3 shows a schematic of a shell-and-tube heat exchanger. Such an arrangement can be operated either in the counterflow or in the parallel flow arrangement, with either the hot or cold fluid passing through the annular space and the other fluid passing through the inside of the inner pipe.

20–2 Log Mean Temperature Difference

The temperatures of fluids in a heat exchanger are not constant but vary from one point to another as heat flows from the hotter to the colder fluid. Even for a constant thermal resistance, the rate of heat flow will, therefore, vary along the path of the exchanger because its value depends on the temperature difference between the hot and the cold fluid in a given section. Figure 20–4 illustrate the changes in temperature that may occur in both fluids in a shell-and-tube exchanger.

Figure 20–3

A simple shell-and-tube heat exchanger.

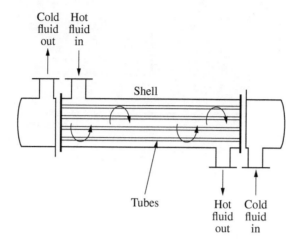

Figure 20–4

Parallel and counter flow heat exchanger's temperature variation.

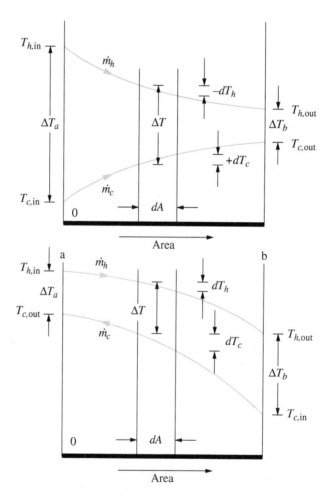

In summary, Figure 20–2 illustrates the case in which a vapor is condensing at a constant temperature while the other fluid is being heated. Shown in the same figure is the case where a liquid is evaporating at a constant temperature while heat is flowing from a warmer fluid whose temperature decreases as it passes through the heat exchanger. For both of these cases, the direction of flow of either fluid is immaterial, and the constant-temperature medium may also be at rest. Figure 20–1 represents conditions in a parallel-flow exchanger and a counterflow exchanger. No change in phase occurs in the latter two cases. Inspection of Figure 20–4 shows that no matter how long the exchanger is, the final temperature of the colder fluid can never reach the exit temperature of the hotter fluid in parallel flow. For counterflow, on the other hand, the final temperature of the cooler fluid may exceed the outlet temperature of the hotter fluid, since a favorable temperature gradient exists all along the heat exchanger. An additional advantage of the counterflow arrangement is that for a given rate of heat flow, less surface area is required than in parallel flow. In fact, the counterflow arrangement is the most effective of all heat exchanger arrangements.

To determine the rate of heat transfer in any of the above-mentioned cases, the general governing equation is:

$$\Delta q = U \, dA \, \Delta T$$

This equation must be integrated over the heat transfer area A along the length of the exchanger. If the overall heat transfer coefficient U is constant, the changes in kinetic energy are neglected, and if the shell of the exchanger is perfectly insulated, then the preceding equation can be easily integrated analytically for either parallel or counterflow arrangements. An energy balance over a differential area (dA) yields:

$$\Delta q = -\dot{m}_h c_{p,h} \, dT_h = \dot{m}_c c_{p,c} \, dT_c = U \, dA(T_h - T_c)$$

where $\dot{m}$ is the mass flow rate in kg/s, c_p is the specific heat at constant pressure in J(kg.K), and T is the average bulk temperature of the fluid in degrees Kelvin. The subscripts h and c refer to the hot and cold fluids, respectively. If the specific heats of the fluids do not vary with temperature (e.g., the case of a modest temperature change), we can write a heat balance from the inlet station to an arbitrary cross section in the exchanger.

$$-c_h(T_h - T_{h,\text{in}}) = c_c(T_c - T_{c,\text{in}})$$

where $c_h \equiv \dot{m}_h c_{p,h}$, the heat capacity rate of the hotter fluid in W/K, while $c_c \equiv \dot{m}_c c_{p,c}$, the heat capacity rate of the colder fluid, also in W/K. Solving this equation for T_h yields:

$$T_h = T_{h,\text{in}} - \frac{c_c}{c_h}(T_c - T_{c,\text{in}})$$

from which we obtain:

$$T_h - T_c = -\left(1 + \frac{c_c}{c_h}\right)T_c + \frac{c_c}{c_h}T_{c,\text{in}} + T_{h,\text{in}}$$

Through simple substitution, and upon some rearrangement, we get:

$$\frac{dT_c}{-[1 + (c_c/c_h)]T_c + (c_c/c_h)T_{c,\text{in}} + T_{h,\text{in}}} = \frac{U \, dA}{c_c}$$

Integrating this equation over the entire length of the heat exchanger (meaning from $A = 0$ to $A = A_{\text{total}}$) yields:

$$\ln\left[\frac{(1 + c_c/c_h)(T_{c,\text{in}} - T_{c,\text{out}}) + T_{h,\text{in}} - T_{c,\text{in}}}{T_{h,\text{in}} - T_{c,\text{in}}}\right] = -\left(\frac{1}{c_c} + \frac{1}{c_h}\right)UA$$

Also, from the heat balance equation, we have:

$$\frac{c_c}{c_h} = \frac{T_{h,\text{out}} - T_{h,\text{in}}}{T_{c,\text{out}} - T_{c,\text{in}}}$$

which can be used to eliminate the heat capacity rates. After some rearrangement, we get:

$$\ln\left(\frac{T_{h,\text{out}} - T_{c,\text{out}}}{T_{h,\text{in}} - T_{c,\text{in}}}\right) = [(T_{h,\text{out}} - T_{c,\text{out}}) - (T_{h,\text{in}} - T_{c,\text{in}})]\frac{UA}{q}$$

since:

$$q = c_c(T_{c,\text{out}} - T_{c,\text{in}}) = c_h(T_{h,\text{in}} - T_{h,\text{out}})$$

Letting $T_h - T_c = \Delta T_c$, we get the following relationship:

$$q = UA\frac{\Delta T_a - \Delta T_b}{\ln(\Delta T_c/\Delta T_h)}$$

where the subscripts a and b refer to the respective ends of the heat exchanger and ΔT_a is the temperature difference between the hot and cold fluid streams at the inlet, while ΔT_b is the temperature difference at the outlet end. In practice, it is convenient to use an average effective temperature difference $\Delta \bar{T}$ for the entire heat exchanger, defined by:

$$q = UA\Delta \bar{T}$$

Now we are in a position to derive the following relationship:

$$\Delta \bar{T} = \frac{\Delta T_a - \Delta T_b}{\ln(\Delta T_a/\Delta T_b)}$$

The average temperature difference $\Delta \bar{T}$ is referred to as the *logarithmic mean temperature difference* (LMTD). The term LMTD also applies when the temperature of one of the fluids is constant. When $\dot{m}_h c_{p,h} = \dot{m}_c c_{p}c$, the temperature difference is constant in the counterflow configuration, and $\Delta \bar{T} = \Delta T_a = \Delta T_b$. If the temperature difference ΔT_a is not more than 50 percent greater than ΔT_b, the arithmetic mean temperature difference will be within 1 percent of the LMTD, and this may be used to simplify the calculations.

Use of the logarithmic mean temperature is only an approximation in practice because U is neither uniform nor constant. In design work, however, the overall heat transfer coefficient is usually evaluated at a mean section halfway between the ends and treated as constant.

For more complex heat exchangers such as the shell-and-tube arrangements with several shell or tube passes and with crossflow exchangers having mixed and unmixed flow, the mathematical derivation of an expression for the mean temperature difference becomes quite complex. The usual procedure is to modify the simple LMTD expression using correction factors. The ordinate of each is the correction factor F. To obtain the true mean temperature for any of these two arrangements, the LMTD calculated for counterflow must be multiplied by the appropriate correction factor, i.e.,

$$\Delta T_{\text{mean}} = (F)(\text{LMTD})$$

The correction factor F is usually tabulated as a function of the shell-and-tube configuration. In applying the correction factors, it is immaterial whether the warmer fluid flows through the shell or the tubes. If the temperature of either of the fluids remains constant, the direction of the flow is also immaterial.

20–3   **Overall Heat Transfer Coefficient**

A heat exchanger typically involves two flowing fluids separated by a solid wall. Heat is first transferred from the hot fluid to the wall by convection, through the wall by conduction, and from the wall to the cold fluid again by convection. Any radiation effects are usually included in the convection heat transfer coefficients.

The thermal resistance network associated with this heat transfer process involves two convection resistances and one conduction resistance (Figure 20–5). Here the subscripts i and o represent the inner and outer surfaces of the inner tube. For a double-pipe heat exchanger, we have $A_1 = \pi D_i L$ and $A_o = \pi D_o L$, and the thermal resistance R_{wall} of the tube wall in this case is:

$$R_{\text{wall}} = \frac{\ln(D_o/D_i)}{2\pi k L}$$

Figure 20–5

Details of a parallel-flow heat exchanger.

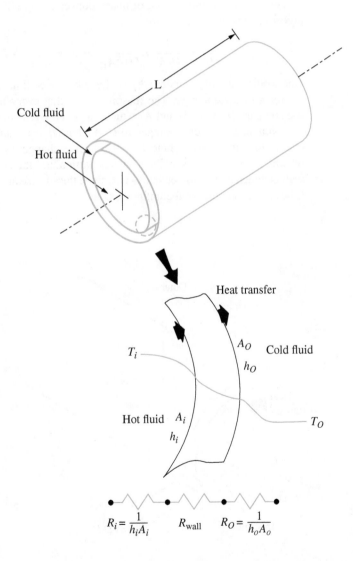

where k is the thermal conductivity of the wall material and L is the length of the tube. Then the total thermal resistance becomes:

$$R = R_{total} = R_i + R_{wall} + R_o = \frac{1}{h_i A_i} + \frac{\ln(D_o/D_i)}{2\pi k L} + \frac{1}{h_o A_o}$$

where A_i is the area of the inner surface of the wall that separates the two fluids, and A_o is the area of the outer surface of the wall. In other words, A_i and A_o are surface areas of separating wall wetted by the inner and outer fluids, respectively (Figure 20–6). When one fluid flows inside a circular tube and the other outside it, we have: $A_i = \pi D_i L$, and $A_o = \pi D_o L$.

In the analysis of heat exchangers, it is convenient to continue all the thermal resistances in the path of heat flow (from the hot fluid to the cold one) into a single resistance as shown in Figure 20–5 and to express the rate of heat transfer between the two fluids as:

$$\dot{Q} = \frac{\Delta T}{R} = UA\Delta T = U_i A_i \Delta T = U_o A_o \Delta T$$

where U is the overall heat transfer coefficient, whose unit is $W/(m^2 \cdot °C)$. This, in turn, is identical to the unit of the ordinary convection coefficient h. Cancelling ΔT, the preceding equation reduces to:

$$\frac{1}{UA} = \frac{1}{U_i A_i} = \frac{1}{U_o A_o} = R = \frac{1}{h_i A_i} + R_{wall} + \frac{1}{h_o A_o}$$

One would naturally wonder why we have two overall heat transfer coefficients U_i and U_o for a heat exchanger. The reason here is that every heat exchanger has two heat transfer surface areas A_i and A_o which, in general, are not equal to one another.

Note that $U_i A_i$ can be equal to $U_o A_o$, but $U_i \neq U_o$ unless $A_i = A_o$. Therefore, the overall heat transfer coefficient U of a heat exchanger is meaningless unless the area on which it is based is specified. This is particularly the case when one side of the tube wall is finned and the other side is not, since the surface area of the finned side is several times more than that of the unfinned side.

Figure 20–6

Inner and outer fluid heat exchange.

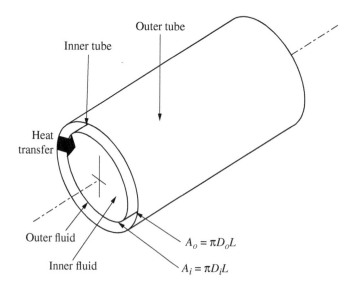

When the wall thickness of the tube is small and the thermal conductivity of the tube material is high, as is usually the case, the thermal resistance of the tube is negligible ($R_{\text{wall}} \approx 0$), and the inner and outer surfaces of the tube are almost identical (i.e., $A_i \approx A_o \approx A$). Then the preceding equation for the overall heat transfer coefficient simplifies to:

$$\frac{1}{U} \approx \frac{1}{h_i} + \frac{1}{h_o}$$

where $U \approx U_i \approx U_o$. The individual convection heat transfer coefficients inside and outside the tube, h_i and h_o, are determined using the convection relations discussed earlier in this section.

The overall heat transfer coefficient U is dominated by the smaller convection coefficient, since the inverse of a large number is small. When one of the convection coefficients is much smaller than the other (say h_i is much smaller than h_o), we have $\frac{1}{h_i}$ being much greater than $\frac{1}{h_o}$ and thus $U \approx h_i$. Therefore, the smaller heat transfer coefficient creates a "bottleneck" on the path of the heat flow and seriously impedes heat transfer. The situation arises frequently when one of the fluids is a gas and the other is a liquid. In such cases, fins are commonly used on the gas side to enhance the product UA, and thus the heat transfers on that side.

When the tube is finned on one side to enhance heat transfer, the total heat transfer surface area on the finned side becomes:

$$A = A_{\text{total}} = A_{\text{finned}} + A_{\text{unfinned}}$$

where A_{finned} is the surface area of the fins and A_{unfinned} is the area of the unfinned portion of the tube surface. For short fins of high thermal conductivity, we can use this total area in the convection resistance relation $R_{\text{conv}} = \frac{1}{hA}$ since the fins, in this case, will be very nearly isothermal. Otherwise, we should determine the effective surface area from:

$$A = A_{\text{unfinned}} + \eta_{\text{fin}} A_{\text{fin}}$$

where η_{fin} is the fin efficiency. This way, the temperature drop along the fins is accounted for. Note that $\eta_{\text{fin}} = 1$ for isothermal fins.

Worth noting here is that the Reynolds number interface between laminar and turbulent flows is 4000.

20-4   Fouling Factor

The performance of heat exchangers usually deteriorates with time as a result of the accumulation of deposits on heat transfer surfaces. The layer of deposits represents additional resistance to heat transfer, and it causes the rate of heat transfer in a heat exchanger to decrease. The net effect of these accumulations on heat transfer is represented by the so-called fouling factor R_f, which is a measure of thermal resistance introduced by fouling.

The most common type of fouling is the precipitation of solid deposits in a fluid on the heat transfer surface. One can observe this type of fouling even inside the house. If you check the inner surface of your teapot after prolonged use, you will probably notice a layer of calcium-based deposits on the surfaces at which boiling occurs. This is especially the case in areas where the water is hard. The scales of such deposits come off by scratching, and the surfaces can be cleaned of such deposits by chemical treatment. Now imagine those mineral deposits forming on the inner surfaces of fine tubes in a heat exchanger, and the detrimental effect it may have on the flow passage area on heat transfer. To avoid this potential problem, water in power and process plants is highly treated, and its solid contents are removed before they are allowed to circulate through the system. The solid ash particles in the gases accumulating on the surfaces of air preheaters create similar problems.

Another form of fouling, which is common in the chemical process industry, is corrosion and other chemical fouling. In this case, the surfaces are fouled by the accumulation of the products of chemical reactions on the surfaces. This form of fouling can be avoided by coating metal pipes with glass or using plastic pipes instead of metal ones. This type of fouling is called chemical reaction fouling, and it can be prevented by chemical treatment.

The overall heat transfer coefficient relation given earlier in this chapter is valid for clean surfaces and needs to be modified to account for the effects of fouling on both the inner and outer surfaces of the tube. For an unfinned shell-and-tube heat exchanger, this relationship can be expressed as follows:

$$\frac{1}{U} = \frac{1}{U_i A_i} = \frac{1}{U_o A_o} = R = \frac{1}{h_i A_i} + \frac{R_{f,i}}{A_i} + \frac{\ln(D_o/D_i)}{2\pi k L} + \frac{R_{f,o}}{A_o} + \frac{1}{h_o A_o}$$

where $A_i = \pi D_i L$, and $A_o = \pi D_o L$ are the areas of the inner and outer surfaces, and $R_{f,i}$ and $R_{f,o}$ are the fouling factors at those surfaces.

Example 20–1

Hot oil is to be cooled in a double-tube counterflow heat exchanger. The copper inner tubes have a diameter of 2 cm and negligible thickness. The inner diameter of the outer tube is 3 cm. Water flows through the tube at a rate of 0.5 kg/s, and oil through the shell at a rate of 0.8 kg/s. Taking the average temperatures of the water and the oil to be 45°C and 80°C, respectively, determine the overall heat transfer coefficient of this heat exchanger.

Solution   *Assumptions*

- The properties of water at 45°C are:

$$\rho = 990 \text{ kg/m}^3$$

$$k = 0.637 \text{ W/(m} \cdot °\text{C)}$$

$$\text{Pr} = 3.91$$

$$\nu = \frac{\mu}{\rho} = 0.602 \times 10^{-6} \text{ m}^2\text{/s}$$

- The properties of oil at 80°C are:

$$\rho = 852 \text{ kg/m}^3$$

$$k = 0.138 \text{ W/(m} \cdot °\text{C)}$$

$$\text{Pr} = 490$$

$$\nu = 37.5 \times 10^{-6} \text{ m}^2\text{/s}$$

- Both the oil and water flows are fully developed.
- Properties of the oil and water are constant.

Analysis

The overall heat transfer coefficient U can now be determined:

$$\frac{1}{U} \approx \frac{1}{h_i} + \frac{1}{h_o}$$

where h_i and h_o are the convection heat transfer coefficients inside and outside the tube, respectively, which are to be determined using the forced convection relations.

Now we repeat the analysis for oil. The hydraulic diameter for a circular tube is the tube diameter itself, meaning 0.02 m. The average velocity of water in the tube and the Reynolds number are:

$$V = \frac{\dot{m}}{\rho A} = 1.61 \text{ m/s}$$

and

$$\text{Re} = \frac{V D_h}{\nu} = 53{,}490$$

which is greater than 4000. Therefore, the flow of water is turbulent. Assuming the flow to be fully developed, the Nusselt number can be determined as follows:

$$\text{Nu} = \frac{h D_h}{k} = 0.23 \text{Re}^{0.8} \text{Pr}^{0.4} = 240.6$$

Then:

$$h = \frac{k}{D_h} \text{Nu} = 7663 \text{ W/(m}^2 \cdot {}^\circ\text{C)}$$

Now we repeat the preceding analysis for oil:

$$V = \frac{\dot{m}}{\rho A} = 2.39 \text{ m/s}$$

and

$$\text{Re} = 637$$

which is less than 4000. Therefore, the flow of oil is laminar. Assuming fully developed flow, the Nusselt number on the tube side of the annular space, Nu, corresponding to $\frac{D_i}{D_o} = 0.667$, can be determined from the tables by interpolation to be 5.45, and:

$$h_o = \frac{k}{D_h} Nu = 75.2 \text{ W/(m}^2 \cdot {}^\circ\text{C)}$$

Then the overall heat transfer coefficient for this heat exchanger becomes:

$$U = \frac{1}{\frac{1}{h_i} + \frac{1}{h_o}} = 74.5 \text{ W/(m}^2 \cdot {}^\circ\text{C)}$$

Example 20–2   A double-pipe (shell-and-tube) heat exchanger is constructed of stainless steel $[k = 15.1 \text{ W/(m} \cdot {}^\circ\text{C)}]$ with an inner tube of inner diameter $D_i = 1.5$ cm, an outer diameter $D_o = 1.9$ cm, and an outlet shell of inner diameter 3.2 cm. The convection heat transfer coefficient is given as 1200 W/m^2, $h_i = 800$ W/(m$^2 \cdot {}^\circ$C) on the inner surface of the tube, and $h_o = 1200$ W/(m²·°C) on the outer surface. For a fouling factor of $R_{f,i} = 0.0004$ (m²·°C)/W on the tube side and $R_{f,o} = 0.0001$ (m²·°C)/W on the outer side, determine:

a) The thermal resistance of the heat exchanger per unit length.

b) The overall heat transfer coefficients U_i and U_o based on the inner and outer surface areas of the tube.

Solution　The heat transfer coefficients and the fouling factors on both the tube and shell sides of the heat exchanger are given. The thermal resistance and the overall heat transfer coefficients based on the inner and outer areas are to be determined.

a) The thermal resistance for an unfinned shell-and-tube heat exchanger with fouling on both heat transfer surfaces is computed as follows:

$$R = \frac{1}{UA} = \frac{1}{U_i A_i} = \frac{1}{U_o A_o} = \frac{1}{h_i A_i} + \frac{R_{f,i}}{A_i} + \frac{\ln(D_o/D_i)}{2\pi k L} + \frac{R_{f,o}}{A_o} + \frac{1}{h_o A_o}$$

where:

$$A_i = \pi D_i L = 0.0471 \text{ m}^2$$

$$A_o = \pi D_o L = 0.0597 \text{ m}^2$$

Substituting, the total thermal resistance is calculated to be 0.0532 °C/W.

b) Knowing the total thermal resistance and the heat transfer surface areas, the overall heat transfer coefficient based on the inner and outer surfaces of the tube are determined as follows:

$$U_i = \frac{1}{R A_i} = 399.1 \text{ W/(m}^2 \cdot °\text{C)}$$

and

$$U_o = \frac{1}{R A_o} = 314.9 \text{ W/(m}^2 \cdot °\text{C)}$$

20–5　Analysis of Heat Exchangers

In the following, we will discuss the two methods used in the analysis of a heat exchanger. Of these, the log mean temperature difference (or LMTD) method is best suited for the first task and the effectiveness method for the second. First, however, we present some general considerations.

Heat exchangers usually operate for long periods of time with no change in their operating conditions. Therefore, they can be modeled as steady-flow devices. As such, the mass flow rate of each fluid remains constant, and the fluid properties such as temperature and velocity at any inlet or outlet stations remain fixed. Also, the fluid streams experience little or no change in their velocities and elevations, and thus the kinetic and potential energy changes are negligible. The specific heat of a fluid, in general, changes with temperature. But in a specified temperature range, it can be treated as a constant at some average magnitude with little loss in accuracy. Axial heat conduction along the tube is usually insignificant and can be ignored. Finally, the outer surface of a heat exchanger is assumed to be perfectly insulated, so there is no heat loss to the surrounding medium, and any heat transfer occurs between the two fluids only.

The idealizations just stated are closely approximated in practice, and they greatly simplify the analysis of a heat exchanger with little sacrifice in accuracy. Therefore, they are commonly used. Under these assumptions, the first law of thermodynamics requires that the rate of heat transfer from the hot fluid be equal to the rate of heat transfer to the cold one. That is,

$$\dot{Q} = \dot{m}_c c_{p,c} (T_{\text{cold,out}} - T_{\text{cold,in}})$$

and

$$\dot{Q} = \dot{m}_h c_{p,h} (T_{\text{hot,in}} - T_{\text{hot,out}})$$

where:

$$\dot{m}_c, \dot{m}_h \text{ are the mass flow rates.}$$

$$c_{p,c}, c_{p,h} \text{ are the specific heats.}$$

Note that the heat transfer rate $\dot{Q}$ is a positive quantity, and its direction is understood to be from the hot fluid to the cold one.

In heat exchanger analysis, it is often convenient to combine the product of the mass flow rate and specific heat of a fluid into a single quantity. This quantity is called the *heat capacity rate*, and it is defined as follows:

$$c = \dot{m} c_p$$

The heat capacity rate of a fluid stream represents the rate of heat transfer needed to change the temperature of the fluid stream by 1°C as it flows through the heat exchanger. Note that in a heat exchanger, the fluid with a large heat capacity rate will experience a small temperature change. Therefore, doubling the mass flow rate of a fluid while leaving everything else unchanged will halve the temperature change of that fluid.

With the preceding definition of the heat capacity rate, we end up with the following two expressions:

$$\dot{Q} = c_c (T_{\text{cold,out}} - T_{\text{cold,in}})$$

and

$$\dot{Q} = c_h (T_{\text{hot,in}} - T_{\text{hot,out}})$$

Figure 20–7 shows the two cases of interest. First, the upper part shows a long cylinder whose surface is exposed to a uniform heat flux. The lower part in this figure shows the cylinder which is now having a surface with the condition of a fixed temperature T_w.

A rather interesting situation, in discussing heat exchangers, is shown in Figure 20–2. In both cases, one fluid remains at a constant temperature (e.g., the water condensation case inside the liquid + vapor dome discussed in Part I), while the water substance in this figure is, again, at a constant temperature as it evaporates.

In other words, the heat transfer rate in a heat exchanger is equal to the heat capacity rate of either fluid multiplied by the temperature change of the fluid. Note that the only time the temperature rise of a cold fluid is equal to the temperature drop of the hot fluid is when the heat capacity rates of the two fluids are equal to each other.

Two special types of heat exchangers commonly used in practice are boilers and condensers (Figure 20–2). One of the fluids in a condenser or a boiler undergoes a phase-change process, and the rate of heat transfer can be expressed as follows:

$$\dot{Q} = \dot{m} h_{f-g}$$

Figure 20–7

Uniform heat flux vs. a constant surface temperature.

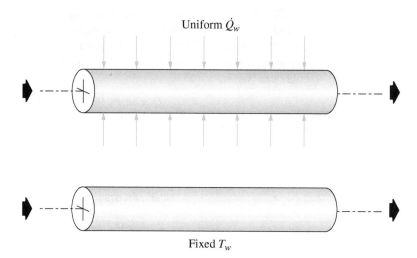

where $\dot{m}$ is the rate of evaporation or condensation of the fluid and h_{f-g} is the enthalpy of evaporation of the fluid at the specified temperature and pressure.

In a condensation as well as a boiling process, the pure substance temperature remains constant within the liquid + vapor dome. In the case of $\dot{q}_s$ = constant, the rate of heat transfer can also be expressed as follows:

$$\dot{Q} = \dot{q}_s A = \dot{m} c_p (T_e - T_i)$$

Note that the mean fluid temperature increases linearly in the flow direction in the case of constant surface heat flux (Figure 20–7), since the surface area increases linearly in the flow direction (A is equal to the perimeter, which is constant, times the tube length).

The surface temperature in this case can be determined from $\dot{q} = h(T_s - T_m)$. Note that when h is constant, $T_s - T_m$ = constant, and thus the surface temperature will also increase linearly in the flow direction. Of course, this is true when the variation of the specific heat c_p with T is disregarded and c_p is assumed to remain constant.

An ordinary fluid absorbs or releases a large amount of heat essentially at constant temperature during a phase-change process, and its heat capacity rate must approach infinity since the temperature change is practically zero. Therefore, in heat exchanger analysis, a condensing or boiling fluid is conveniently modeled as a fluid whose heat capacity rate is infinity.

The rate of heat transfer in a heat exchanger can also be expressed in an analogous to Newton's law of cooling as follows:

$$\dot{Q} = UA\Delta T_m$$

where U is the overall heat transfer coefficient, A is the heat transfer area, and ΔT_m is an appropriate average temperature difference between the two fluids. Here the surface area A can be precisely determined using the dimensions of the heat exchanger. However, the overall heat transfer coefficient U and the temperature difference ΔT between the hot and cold fluids, in general, are not constant, and they vary along the heat exchanger.

The average value of the overall heat transfer coefficient can be determined as described in the preceding section by using the average convection coefficients for each fluid. It turns out that the appropriate form of the mean temperature difference between the two fluids is logarithmic in nature.

20-6   The Similitude Principle

The engineering design of sophisticated devices is a complicated process, often involving many iterations to ensure that the final device performs as desired, is safe in its operation, and can be manufactured economically. For many devices, particularly those of large scale, construction and testing of full-size prototypes as part of the design process is not practical. In such cases, small-scale, geometrically similar models are used to provide the information needed to improve and evaluate the design. For instance, wind tunnel testing of models is routine in aircraft design. In fact, wind tunnel testing has always been an integral part of the aircraft design process. Models are used to understand wind loads on buildings. Every field of engineering offers many examples.

How can such model tests be planned and executed so that the results are somehow useful to full-scale devices? Indeed, how can model results be applied to a full-size prototype? The answers to both questions are contained in the following statement:

> **If all appropriate dimensionless variables have been defined, similitude occurs between geometrically similar models and prototypes when the same values of these dimensionless variables are attained for both.**

This statement embodies the concepts of geometric and dynamic similarity. Geometric similarity implies that the model is a scaled version of the prototype, that is, any linear dimension of the model is a constant fraction of the corresponding linear dimension of the prototype. For example, the use of quarter scale models is common in the aerodynamic design of automobiles. You are likely familiar with toy plastic models of aircraft, automobiles, and ships. In these toy models, most of the dimensions are scaled from the full-size devices. However, many small features, such as rivet details, may be grossly out of scale. The failure to achieve total geometric similarity in such details can be quite important in an engineering design.

20-7   Dynamic Similarity

This obviously relates to motion. In thermal-fluid applications, this motion is usually with a fluid moving around or through a device. In some applications, achieving complete dynamic similarity is not possible. Regardless, the idea that the various dimensionless variables that contain a velocity (or a time) must be equal is a statement of dynamic similarity. Even though we have yet to formally define any of these dimensionless variables, this statement of similarity, or scaling, can be best understood by applying it to a concrete example.

In this case, three dimensionless parameters are important:

- The lift coefficient, $c_L \equiv \frac{F_L/L^2}{\rho V^2/2}$
- The Reynolds number, $\mathrm{Re} \equiv \frac{\rho V L}{\mu}$, and
- The angle of attack, α

In these parameters, F_L is the lift force, L is a characteristic length, V is the velocity of the air stream relative to the aircraft, and μ and ρ are the dynamic viscosity coefficient and density, respectively. Data from wind tunnel tests of a particular scale model are collected and plotted to show the lift coefficient as a function of the angle of attack, using the Reynolds number as a parameter. We can use these data points to obtain the lift force for the full-scale prototype by equating the dimensionless parameters of the model and prototype, that is, we set:

$$\frac{F_{L,m}}{L_m^2}\rho_m V_m^2/2 = \frac{F_{L,p}/L_p^2}{\rho_P V_P^2/2}$$

when

$$\frac{\rho_m V_m L_m}{\mu_m} = \rho_P V_P L_P \mu_P$$

and,

$$\alpha_m = \alpha_P$$

Rearranging terms, we explicitly express the dimensional lift force on the prototype as follows:

$$F_{L,P} = F_{L,m}\frac{L_p^2 \rho_P V_p^2}{L_M^2 \rho_m N_m^2}$$

From this equation, we see that the lift force on the prototype scales with the square of the geometric scale, that is, $(L_P/L_m)^2$. Thus, if the model is 1/20 the size of the prototype, the lift force measured in the wind tunnel should be multiplied by $400 V_P^2/V_m^2$ to obtain the corresponding lift force on the full-scale aircraft wing, provided that the air densities are the same and that the Reynolds number and angle of attack are also the same. To achieve identical Reynolds numbers, however, is very difficult in aircraft testing since both V and L are quite large for the prototype compared to practically achievable magnitudes in the wind tunnel.

From the preceding discussion, the need for and use of dimensionless parameters should be quite clear. We now focus on these.

20–8   Dimensionless Parameters

Origins

The question we seek to answer in this section is: "How does one determine dimensionless parameters which are important in any particular problem?" We will discuss two methods to identify the important parameters. The first of these is the application of the theory of dimensional analysis. The second approach is to create dimensionless forms of the governing conservation equations and their appropriate boundary conditions. The dimensionless parameters then appear as parameters in these dimensionless governing equations. The remainder of this chapter deals with this second approach and its application to certain problems in fluid mechanics and heat transfer. In spite of our focus on the use of dimensionless governing equations, a much more general approach invoking the Buckingham (or pi) theorem is frequently used to deal with complex problems where theory is incomplete or nonexistent.

Dimensionless Governing Equations

To illustrate how parameters can originate from a manipulation of the basic conservation equations, we choose to work with simplified versions of the momentum and energy-conservation principles. Using simplified equations allows us to focus on the method without being bogged down in a large number of terms. By choosing the simplified set of equations known as the boundary layer equations, we get the added benefit of dealing with a very important class of flows (i.e., boundary layer flows).

Figure 20–8

Progression of the velocity boundary layer leading to flow separation and reversal.

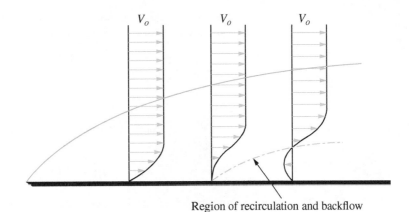

Region of recirculation and backflow

Dimensional Forms of the Boundary Layer Equations

The simplified boundary layer flow is that associated with the flow over a sharp-edged flat plate in a stream of a uniform velocity and a uniform temperature. Although understanding the details of this flow is not necessary for our current purposes, such understanding is quite useful in its own right because of the overwhelming importance of the boundary layer concept. Thus, we digress a bit to explore the flat-plate boundary layers.

Figure 20–8 illustrates the development of the hydrodynamic (or velocity) boundary layer. Upstream of the flat plate, the velocity field is uniform with a magnitude of V_o, and it is directed parallel to the plate surface. Upon encountering the flat plate, the fluid immediately adjacent to the plate comes to rest (the so-called nonslip boundary condition), creating a shearing action on the fluid layer above. The effect of shearing is to create a thin layer of fluid in which the magnitude of velocity is zero at the surface ($y = 0$) and rapidly approaches the free-stream value (V_o) with the perpendicular distance from the plate surface, y. The thickness of this layer, δ_H, grows with distance downstream, x, yet remains thin compared to this distance downstream. The existence of this thin layer adjacent to the surface was the insight of the famous fluid mechanician Ludwig Prandtl. That the effects of viscosity are confined to a thin layer provides a tremendous simplification to a wealth of flow problems, among them the calculation of lift and drag forces arising from airfoil surfaces.

Within a flat plate boundary layer, the velocity directed in the mean flow direction is much greater than that directed perpendicular to the plate. Furthermore, since the boundary layer is thin, gradients in the y-direction are much greater than those along the flat plate. As a consequence of these two conditions, the incompressible axial momentum equation can be greatly simplified. For steady flow in the absence of a pressure gradient, the axial momentum conservation equation can be written as follows:

$$V_x \frac{\partial V_x}{\partial x} + V_y \frac{\partial V_y}{\partial y} = \frac{\mu}{\rho} \frac{\partial^2 V_x}{\partial y^2}$$

where x is the distance along the flat plate. To fully describe this incompressible, two-dimensional flow, we must add the conservation of mass (or continuity) equation. This simplifies to:

$$\frac{\partial V_x}{\partial x} + \frac{\partial V_y}{\partial y} = 0$$

Subject to the application of the appropriate boundary conditions, the preceding two equations can be solved to provide the two velocity components:

$$V_x = V_x(x, y)$$

and

$$V_y = V_y(x, y)$$

Knowledge of these velocity components allows us to determine the shear stress at the wall:

$$\tau_w = \mu \left(\frac{\partial V_x}{\partial y} \right)_{y=0}$$

The engineering utility of this result is that the shear stress can be integrated over the plate surface to obtain the total drag force exerted by the fluid on the plate.

We now examine the corresponding thermal problem, that is, the development of a thermal boundary layer of thickness δ_T. Here, we see a flat plate maintained at a temperature T_s. For the temperature profiles sketched in Figure 20–8, the plate temperature T_s is greater than the free-stream temperature T_o. Thus, energy is transferred from the plate to the flowing fluid in a heat interaction situation. The thickness of the thermal boundary layer grows with distance downstream, similar to the growth of the hydrodynamic boundary layer. Assuming that the effects of temperature variation on the fluid density are small, we can use the incompressible energy equation to describe the temperature field. Treating the flow as steady and applying the same simplifying assumptions, we can express the energy conservation equation within the boundary layer as follows:

$$V_x \frac{\partial T}{\partial x} + V_y \frac{\partial T}{\partial y} = \frac{k}{\rho c_p} \frac{\partial^2 T}{\partial y^2}$$

where the viscous dissipation has been neglected and the thermophysical properties have been treated as constants. Solving this equation yields the temperature distribution,

$$T = T(x, y)$$

Knowing the temperature distribution allows us to compute the heat flux at the plate surface by applying Fourier's law to the stagnant fluid immediately adjacent to the surface ($y = 0$). This procedure is analogous to using the velocity distribution to find the shear stress. Equating this theoretical heat flux to the empirical definition of the convection heat flux introduces the convective heat transfer coefficient into the problem solution, i.e.,

$$\dot{Q}_H = -k \left(\frac{\partial T}{\partial y} \right)_{y=0} = h_{\text{conv},x}(T_s - T_o)$$

Because of their importance, these are called the boundary layer governing equations.

PROBLEMS

20–1   Determine the heat transfer surface required for a heat exchanger constructed from a 0.0254-m-OD tube to cool 6.93 kg/s of a 95 percent ethyl alcohol solution [$c_p = 3810$ J/(kg·K)] from 65.6°C to 39.4°C, using 6.30 kg/s of water available at 10°C. Assume that the overall coefficient

Figure 20–9

Input variables for problem 20–2.

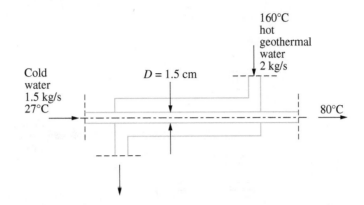

of heat transfer based on the outer-tube area is 568 W/(m². K), and consider each of the following arrangements:

a) Parallel-flow tube and shell

b) Counterflow tube and shell

c) Counterflow exchanger with two shell passes and 72 tube passes, the alcohol flowing through the shell and the water flowing through the tubes.

20–2 A counterflow double-pipe heat exchanger (see Figure 20–9) is to heat water from 27°C to 80°C at a rate of 1.2 kg/s. The heating is to be accomplished by geothermal water available at 160°C at a mass flow rate of 2.0 kg/s. The inner tube is thin-walled and has a diameter of 1.5 cm. If the overall heat transfer coefficient of the heat exchanger is 640 W/(m²·°C), determine the length of the heat exchanger required to achieve the desired heating. Assume the specific heat of the geothermal water to be 4.31 kJ/(kg·°C) and that of the water to be 4.18 kJ/(kg·°C).

20–3 A two-shell and four-tube-passes heat exchanger (Figure 20–10) is used to heat glycerin from 20°C to 50°C by hot water, which enters the thin-walled 2-cm-diameter tubes at 80°C and leaves at 40°C. The total length of the tubes in the heat exchanger is 60 m. The convection heat transfer coefficient is 25 W/(m²·°C) on the glycerin (shell) side, and 160 W/(m²·°C) on the water-tube side. Determine the rate of heat transfer in the heat exchanger assuming no fouling.

Figure 20–10

Input variables for problem 20–3.

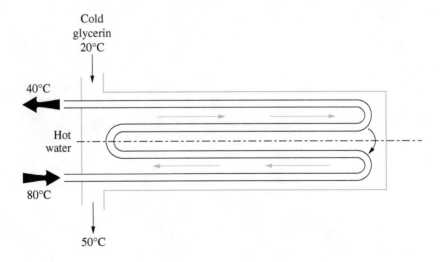

Figure 20–11

Input variables for
problem 20–4.

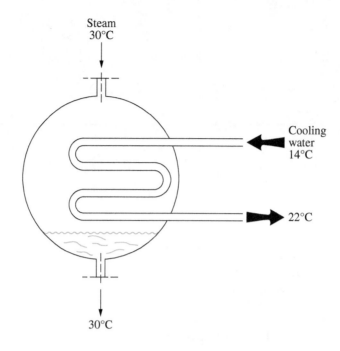

20–4   Steam in the condenser of a steam power plant is to be condensed at a temperature of 30°C, with
cooling water which enters the condenser at 14°C and leaves at 22°C (Figure 20–11). The surface
area of the tubes is 45 m², and the overall heat transfer coefficient is 2,100 W/(m²·°C). Determine
the mass flow rate of the cooling water needed and the rate of condensation of the steam in the
condenser.

21

Heat Radiation

Chapter Outline

21-1   Thermal Radiation: Blackbody   **344**

21-2   Planck's Law   **346**

21-3   Wien's Displacement Law   **346**

21-4   View Factor Relations   **348**

21-5   Radiation Functions   **348**

21-6   Radiation Properties   **350**

21-7   Total Radiation Properties   **350**

21-8   Radiation Shape Factor   **352**

21-9   Directional Radiation Properties   **352**

21–1 ## Thermal Radiation: Blackbody

A substance may be stimulated to emit electromagnetic radiation in a number of ways:

- An electric conductor carrying a high-frequency alternating current.
- A hot solid or liquid emits thermal radiation.
- A gas carrying an electric charge may emit visible or ultraviolet radiation.

The loss of energy due to the emission of thermal radiation may be compensated in a variety of ways. The emitting body may be:

- A metal plate bombarded by high-speed electrons emits x rays.
- A substance of which the atoms are radioactive may emit γ rays.

All of these radiations are electromagnetic waves, differing only in one aspect, their wavelength (Figure 21–1). We will be concerned only with thermal radiation, the radiation emitted by a solid or a liquid by virtue of which a body may receive energy by absorption of radiation from surrounding bodies. The radiation characteristics of gases require a special treatment. When thermal radiation is dispersed by a suitable prism, a continuous spectrum is obtained. The distribution of energy among the various wavelengths is such that at a temperature below about 500 °C, most of the energy is associated with infrared waves, whereas at high temperatures, some visible radiation is emitted. In general, the higher the temperature of a body, the greater is the total energy emitted.

As stated above, the loss of energy due to the emission of thermal radiation may be compensated in a variety of ways. The emitting body may be a source of energy itself,

Figure 21–1

Variation of the radiation energy coefficient.

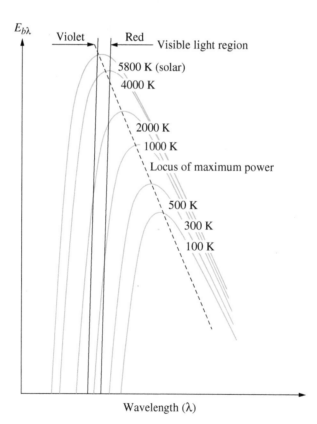

Figure 21–2

Absorption, reflection and transmission of the incident radiation.

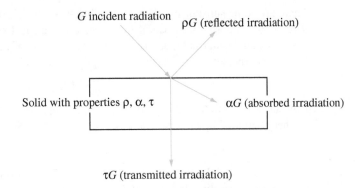

G incident radiation ρG (reflected irradiation)

Solid with properties ρ, α, τ αG (absorbed irradiation)

τG (transmitted irradiation)

as in the case of a filament of an electric light. Energy may be supplied also by heat conduction or by the performance of work on the emitting body. In the absence of these sources, the only other way in which the body may receive energy is by absorption of radiation from surrounding bodies. In the case of a body that is surrounded by other bodies, the internal energy that is emitted is equal to that at which it is absorbed.

Experiment shows that the rate at which a body emits thermal radiation depends on the temperature and on the surface nature. The total radiant energy emitted per second per unit area is called the *radiant emittance* of the body R. For example, the radiant emittance of tungsten that is capable of absorbing 17 °C is 50 W/cm². When radiant radiation is incident upon a body equally from all directions, the radiation is said to be isotropic. Some of the radiation may be absorbed, some reflected, and some transmitted. In general, the fraction of the incident isotropic radiation of all wavelengths that is absorbed depends on the temperature and the nature of the surface of the absorbing body. This fraction is called *absorptivity* (Figure 21–2). To summarize, we have:

- Radiant emittance $= R =$ total radiant energy emitted per second per cm²
- Absorptivity $= \alpha =$ fraction of the total energy of isotropic radiation that is absorbed

Figure 21–3

Situation used to derive Kirchhoff's law.

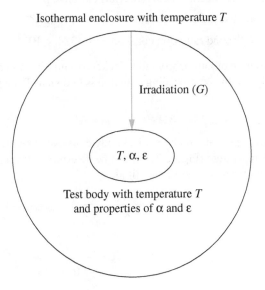

Isothermal enclosure with temperature T

Irradiation (G)

T, α, ε

Test body with temperature T
and properties of α and ε

There are some substances, such as lampblack, whose absorptivity is nearly unity. For theoretical purposes, it is useful to conceive of an ideal substance that is capable of absorbing all the thermal radiation falling on it. Such a body is referred to as a *blackbody*. If a blackbody is indicated by the subscript B, then we have:

$$\alpha_B = 1.0$$

Any surface will radiate energy if its absolute temperature is greater than zero. The rate of energy emitted by an ideal black surface is given by Stefan-Boltzmann law:

$$E_B = \sigma T^4$$

where T is the absolute temperature and σ is the Stefan-Boltzmann constant. The radiation leaving a real-life surface will depend on the surface radiation characteristics, such as whether it is polished, oxidized, or in some other condition and it will always be less than that leaving a blackbody at the same temperature.

The exchange of radiant energy between two or more surfaces is rather a complicated process. Since the electromagnetic waves travel in straight lines, the geometric orientation of the surface exchanging radiant energy must be considered. Techniques of heat transfer for simple configurations will be presented later in this section.

21-2 Planck's Law

When a blackbody is heated to a temperature T, photons are emitted from the surface of the body. The photons have a definite distribution of energy, depending upon the surface temperature T. It was Max Planck who showed that the energy emitted at a wavelength λ from a blackbody at a temperature T is:

$$E_{B\lambda}(T) = \frac{C_1}{\lambda^5 A}$$

where:

$$A = e^{\frac{C_2}{\lambda T}} - 1$$

where:

$E_{B\lambda} = $ Monochromatic (spectral) emissive power of a blackbody at T

$C_1 = $ First radiation constant $= 3.7418 \times 10^{-16}$

$C_2 = $ Second radiation constant $= 1.4388 \times 10^{-2}$

The variation of monochromatic blackbody emissive power with temperature and wavelength given by the preceding equation is known as Planck's law.

21-3 Wien's Displacement Law

The wavelength at which the blackbody emissive power reaches a maximum value for a given temperature (Figure 21–3) can be determined from Planck's law by satisfying the condition for a maximum magnitude:

$$\frac{dE_{B\lambda}}{d\lambda} = \frac{d}{d\lambda} A_{T=\text{constant}} = 0$$

where:

$$A = \frac{C_1}{\lambda^5 (e^{\frac{C_2}{\lambda T}} - 1)}$$

The result of this operation is:

$$\lambda_{max} T = 2.898 \times 10^{-3} \text{ m} \cdot \text{K}$$

where λ_{max} denotes the wavelength at which the maximum monochromatic emissive power occurs for a blackbody at a temperature T. The preceding equation is called Wien's displacement law.

The maximum magnitude for the monochromatic blackbody emissive power can be obtained by substitution in the preceding equation, which results in:

$$(E_{b\lambda})_{max} = 1.287 \times 10^{-5} T^5 \text{ W/m}^3$$

and then decreases with increasing wavelength.

We should already be familiar with the results of Wien's displacement law. Suppose an electric current is passed through a thin filament, causing the temperature to rise. At relatively low filament temperatures, below 900 K, the wavelength at which the emissive power reaches a maximum magnitude is around 3.2×10^{-6} m, which is in the infrared wavelength range. We can sense the radiant energy being emitted by the filament. However, our eyes are unable to detect visible radiation emitted by the filament because an insignificant amount of the energy falls in the visible wavelength range.

As the temperature of the filament is increased, the amount of radiant energy increases, and more of the energy is emitted at shorter wavelengths. Also at about 1000 K, a small portion of the energy falls in the long-wavelength or red end of the visible spectrum. Our eyes are able to detect this radiation, and the filament appears to be a dull red color. As the temperature increases further, more of the energy falls in the visible range, and above about 1600 K all visible wavelengths are included, so the filament appears "white-hot" at this temperature.

An example of an energy source that is at a high temperature of approximately 5800 K is the sun. According to Wien's law, the value of λ_{max} at this temperature is 5.2×10^{-7} m, which is near the center of the visible range. The human eye is perfectly adapted to sensing the maximum monochromatic energy that is emitted from the sun.

Our eyes are reliable detectors of radiant energy that falls in the visible wavelength range. An object that appears white when placed in the sun reflects nearly all and absorbs practically none of the radiation in the visible range. On the other hand, a blackbody absorbs all visible radiation while reflecting no radiation and therefore appears black to our eyes.

Since the human eye does not respond to radiant energy outside the visible range, it can only predict the surface behavior over a very small wavelength range.

Note that T is the absolute surface temperature, λ is the wavelength of the radiation emitted, and $k = 1.3805 \times 10^{-23}$ J/K. This relation is valid for a surface in a vacuum or a gas. For other media, it needs to be modified by replacing C_1 (above) with $\frac{C_1}{n^2}$, where n is the index of refraction of the medium. Note that the term *spectral* indicates dependence on the wavelength.

The variation of the blackbody emissive power with wavelength is plotted in Figure 21–1 for selected temperatures. Several observations can be made from this figure:

- The emitted radiation is a continuous function of the wavelength. At any specified temperature, it increases with wavelength, reaches a peak, and then decreases with increasing wavelength.

- At any wavelength, the amount of emitted radiation increases with increasing temperature.

- As the temperature increases, the curves get steeper and shift to the left of the shorter-wavelength region. Consequently, a large fraction of radiation is emitted at shorter wavelengths at higher temperatures.

- The radiation emitted by the sun, which is considered a blackbody at 5800 K, reaches its peak in the visible region of the spectrum. Therefore, the sun is in tune with our eyes. On the other hand, surfaces at $T \leq 800$ K emit almost entirely in the infrared region and thus are not visible to the eye unless they reflect light coming from other sources.

As the temperature increases, the peak of the curve in Figure 21–1 shifts toward shorter wavelengths. The wavelength at which the peak occurs for a specified temperature is given by Wien's displacement law as follows:

$$(\lambda T)_{\text{max-power}} = 2897.8 \ \mu\text{m} \cdot \text{K}$$

Once a sufficient number of view factors are available, the rest of them can be determined by utilizing some fundamental relationships. This relationship was originally developed by Willy Wien in 1894 using classic thermodynamics.

21–4   View Factor Relations

Radiation analysis on an enclosure consisting of N surfaces requires the evaluation of N^2 view factors, and this evaluation process is probably the most time-consuming part of the radiation analysis. However, it is neither practical nor necessary to evaluate all of the view factors available; the rest of them can be determined by utilizing some fundamental relations for view factors as discussed next.

The view factors F_{j-1} and F_{1-j} are not equal to one another, unless the areas of the two surfaces are equal.

Using the radiation intensity concept and going through some manipulations, it can be shown that the pair of the view factors F_{j-1} and F_{1-j} are related to each other as follows:

$$A_1 F_{1-j} = A_2 F_{j-1}$$

This relation is known as the *reciprocity rule*, and it enables us to determine the counterpart of a view factor from knowledge of the view factor itself and the areas of the two surfaces. When determining the pair of view factors (above), it makes sense to evaluate, first, the easier one directly and then the harder one by applying the reciprocity rule.

21–5   Radiation Functions

If the monochromatic blackbody emissive power given by Planck's law is integrated over the wavelength range from $\lambda = 0$ to λ_t, the result will be the total amount of radiative energy between the wavelengths of 0 and λ_t emitted from a black surface with a temperature T (Table 21–1). It can be shown by carrying out the integration that the result is only a function of the product $\lambda_t T$. The integral is denoted by:

$$\int_0^{\lambda_t} E_{B\lambda}(T) d\lambda = E_B(0 \rightarrow \lambda_t T)$$

Table 21–1

Blackbody
Radiation Functions

$\lambda T(\text{m K} \times 10^3)$	$\dfrac{E_b(0 \to \lambda T)}{\sigma T^4}$	$\lambda T(\text{m K} \times 10^3)$	$\dfrac{E_b(0 \to \lambda T)}{\sigma T^4}$
0.2	0.341796×10^{-26}	6.2	0.754187
0.4	0.186468×10^{-11}	6.4	0.769234
0.6	0.929299×10^{-7}	6.6	0.783248
0.8	0.164351×10^{-4}	6.8	0.796180
1.0	0.320780×10^{-3}	7.0	0.808160
1.2	0.213431×10^{-2}	7.2	0.819270
1.4	0.779084×10^{-2}	7.4	0.829580
1.6	0.197204×10^{-1}	7.6	0.839157
1.8	0.393449×10^{-1}	7.8	0.848060
2.0	0.667347×10^{-1}	8.0	0.856344
2.2	0.100897	8.5	0.874666
2.4	0.140268	9.0	0.890090
2.6	0.183135	9.5	0.903147
2.8	0.227908	10.0	0.914263
3.0	0.273252	10.5	0.923775
3.2	0.318124	11.0	0.931956
3.4	0.361760	11.5	0.939027
3.6	0.403633	12	0.945167
3.8	0.443411	13	0.955210
4.0	0.480907	14	0.962970
4.2	0.516046	15	0.969056
4.4	0.548830	16	0.973890
4.6	0.579316	18	0.980939
4.8	0.607597	20	0.985683
5.0	0.633786	25	0.992299
5.2	0.658011	30	0.995427
5.4	0.680402	40	0.998057
5.6	0.701090	50	0.999045
5.8	0.720203	75	0.999807
6.0	0.737864	100	1.000000

If we wish to determine the total amount of radiative energy emitted between two wavelengths λ_1 and λ_2 for a black surface at temperature T, we could take the difference between the two integrals:

$$\int_0^{\lambda_2} E_{B\lambda}(T)dA - \int_0^{\lambda_1} E_{B\lambda}(T)d\lambda = E_B(0 \to \lambda_2 T) - E_B(0 \to \lambda_1 T)$$

21–6   Radiation Properties

Radiative properties are those which quantitatively describe how radiant energy interacts with the surface of the material. Specifically, the radiative properties describe how the surface emits, reflects, absorbs, and transmits radiant energy.

In general, the radiative properties are functions of the wavelength. For example, a surface may be a good reflector in the visible wavelength range and a poor reflector in the infrared range. The properties that describe how a surface behaves as a function of the wavelength are called *monochromatic* or *spectral properties*. The radiative properties are also a function of the direction in which the radiation is incident upon the surface. Properties that describe how the distribution of energy varies with angle are called *directional properties*.

If we wish to perform an energy balance on a surface to determine its temperature, for example, we must know the radiative properties of the surface and all other surfaces that exchange energy with that surface. Even when the spectral directional properties of all these surfaces are known, the analysis is extremely involved. The complexity of the problem and more often the complete lack of the detailed properties suggest that we should search for a simplified approach. This involves using a single radiative property value that is an average magnitude over the wavelengths and all directions. The properties that are averaged over all wavelengths and angles are termed *total properties*. The use of total properties in a radiation-heat-transfer analysis often results in answers that are accurate enough for most engineering purposes, and they certainly reduce a very complex problem to a much simpler one.

Although we will almost exclusively use the total radiative properties in this text, it is important to be aware of the spectral and directional characteristics of surfaces so that their variations can be accounted for in problems where these effects are significant.

21–7   Total Radiation Properties

Consider a beam of radiant energy incident on a surface as shown in Figure 21–4. The total incident energy is referred to as the total irradiation and is given the symbol G. When the irradiation strikes a surface (Figure 21–4), a portion of the energy is absorbed within the material, a portion is reflected from the surface, and the remainder is transmitted through the body. Three of the radiative properties—the absorptivity, reflectivity, and transmissivity—describe how the incident energy is distributed into these categories.

The *absorptivity*, α, of the surface is the fraction of incident energy absorbed by the body. The *reflectivity*, ρ, of the surface is defined as the fraction of incident energy reflected from the surface. The *transmissivity*, τ, of the body is the fraction of incident energy that is transmitted through the body. We know that the irradiation must be absorbed by, reflected from, or transmitted through the body. The energy balance may be mathematically expressed as follows:

$$\alpha + \rho + \tau = 1$$

Often a surface is opaque. That is, it will not transmit any of the incident radiant energy. For an opaque surface, we have:

$$\tau = 0$$

which yields:

$$\alpha + \rho = 1$$

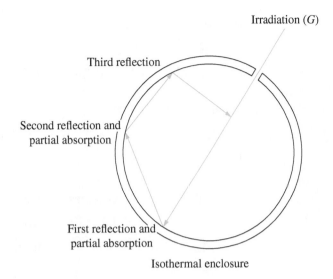

Isothermal enclosure

If a surface is said to be a perfect reflector, all radiation is reflected, or:

$$\rho = 1$$

A blackbody absorbs the maximum amount of incident energy, or:

$$\alpha = 1$$

Another important total property is called the *emissivity* of a body. The emissivity of a surface is defined as the total emitted energy divided by the total energy emitted by a blackbody at the same temperature. The mathematical definition of the total emissivity is then defined as follows:

$$\epsilon = \frac{E(T)}{E_B(T)} = \frac{E(T)}{\sigma T^4}$$

Since a blackbody emits the maximum amount of radiation at a given temperature, the emissivity of a surface is always between zero and unity. When a surface is a blackbody, we have:

$$\epsilon = \alpha = 1.0$$

Example 21-1

Assume that the sun ($T = 5800$ K) and a lightbulb are both blackbodies. Calculate, for these sources of radiant energy, the following variables:

- The total emissive power
- The maximum monochromatic blackbody emissive power
- The wavelength at which the maximum emissive power occurs
- The percent of total emitted energy that lies in the visible wavelength range

Solution

Part a
Applying the Stefan-Boltzmann law:

$$E_B(T) = \sigma T^4$$

For the sun and lightbulb:

$$E_B(T) = 6.42 \times 10^7 \text{ W/m}^2 \ldots \text{(for the sun where } T = 5800 \text{ K)}$$

$$E_B(T) = 3.49 \times 10^7 \text{ W/m}^2 \ldots \text{(for the lightbulb where } T = 2800 \text{ K)}$$

Part b
The maximum monochromatic blackbody emissive power is:

$$E_{B\lambda} = 1.287 \times 10^{-5}(5800)^5 \text{ W/m}^2 \ldots \text{(for the sun)}$$

$$E_{B|\lambda} = 2.21 \times 10^{12} \text{ W/m}^2 \ldots \text{(for the lightbulb)}$$

Part c
The value of $\lambda_{\max}$ is determined by Wien's law as follows:

$$\lambda_{\max} = 5.0 \times 10^{-7} \text{ m} \ldots \text{(for the sun)}$$

$$\lambda_{\max} = 1.04 \times 10^{-6} \text{ m} \ldots \text{(for the lightbulb)}$$

21-8 Radiation Shape Factor

To calculate radiative heat transfer rates between two surfaces, we must determine the percentage of total radiant energy that leaves one surface and arrives directly on the other surface. Let us define the *radiation shape factor* $F_{1\to2}$ as the fraction of total radiant energy from surface 1 which directly arrives on surface 2.

Figure 21–5 shows the shape factor between two coaxial cylinders.

21-9 Directional Radiation Properties

Thus far, we have assumed that the radiation properties are functions only of wavelength and conditions that describe the surface of the sending and receiving areas. To complicate matters further, the properties are also functions of the direction in which the energy is incident or leaves a surface. As was mentioned previously, the properties that describe the angular variation are called directional properties. Several examples of directional emissivities are shown in the polar plot of Figure 21–6. The angle θ in the figure is the angle between the normal to the surface and the directional energy emitted from that surface.

The directional emissivity for electrical conductors is characteristically higher for large θ angles than for small θ values. A conductor would, therefore, emit more energy at grazing angles than at angles more normal to its surface. Electrical nonconductors behave quite differently. They emit more strongly in directions close to the normal while their emissivity drops to zero as θ approaches $90°$.

Before we can relate directional properties to total properties, we need to discuss two quantities. The first is a solid angle, which is a measure of an angle in a solid geometry. Consider a differential area dA shown in Figure 21–7 which subtends a differential solid angle $d\omega$ at point O.

The solid angle is a dimensionless quantity defined as the normal projection of dA divided by the square of the radius between point O and the projected area. The solid angle $d\omega$ is then expressible in the following form:

$$d\omega = \frac{dA_N}{r^2} = \frac{da\cos\theta}{r^2}$$

Figure 21–5

Shape factor for two concentric cylinders.

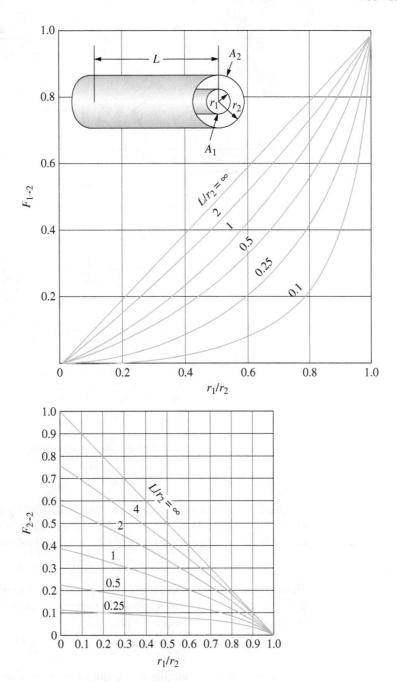

To calculate the radiative heat transfer rates between two surfaces, we must determine the percentage of the total radiant energy that leaves one surface and arrives directly on a second surface. Let us define the radiation shape factor F_{1-2} as the fraction of total radiant energy from surface 1 which directly arrives on surface 2. The shape factor is a dimensionless quantity. In some texts it is also called the view factor or the configuration factor.

Figure 21–6

Polar plot of directional emissivities.

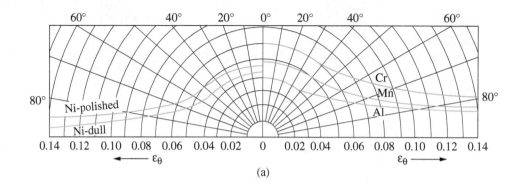

(a)

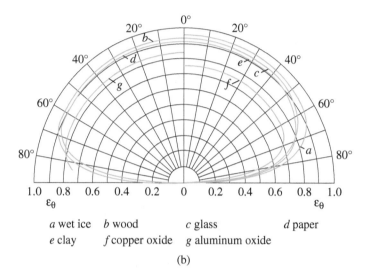

a wet ice   *b* wood   *c* glass   *d* paper

e clay   *f* copper oxide   *g* aluminum oxide

(b)

An expression for the shape factor can be derived by considering the geometry in Figure 21–8, where dA_1 is shown as the emitting area and dA_2 is the receiving area. The rate of radiant energy per unit area of dA_1 that leaves dA_1 and arrives on dA_2 is:

$$dq''_{1-2} = I_1 \cos\theta_1 d\omega_{1-2}$$

The subscript 1–2 refers to energy leaving surface 1 arriving on surface 2. The symbol $d\omega_{1-2}$ represents the solid angle subtended by the area dA_2 at dA_1. Substituting the expression for the solid angle, we get:

$$dq''_{1-2} = \frac{I_1 \cos\theta_1 \cos\theta_2 dA_2}{r^2}$$

If we assume that the emitting area is diffuse (definition provided later in this section), the intensity of radiation leaving dA_1 is independent of direction. The total radiant flux that leaves dA_1 must arrive on a hemisphere placed over dA_1, and the total flux leaving the diffuse area dA_1 is given by:

$$q''_{1-\text{hemisphere}} = \pi I_1$$

The radiation shape factor between the two differential areas dA_1 and dA_2 is:

$$F_{dA_1-dA_2} = \frac{\cos\theta_1 \cos\theta_2 dA_2}{\pi r^2}$$

Figure 21–7

Solid angle.

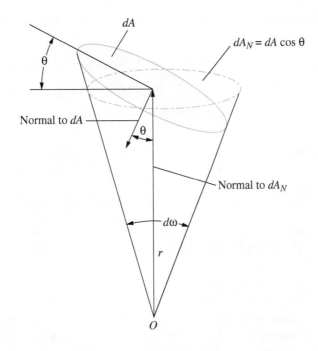

We very seldom wish to determine radiation exchange rates between two infinitesimal areas, because we are more concerned with radiation exchange between finite surfaces. Expressions for the shape factors for finite areas can be obtained by integrating the preceding equation over both the emitting and receiving areas.

Figure 21–8

Calculation of a special-geometry shape factor.

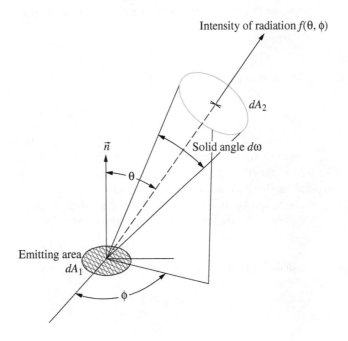

The shape factor between a diffuse differential emitting area and a finite receiving area is:

$$F_{dA_1-dA_2} = \int_{A_2} F_{dA_1-dA_2} = \int_{A_2} \frac{\cos\theta_1 \cos\theta_2 dA_2}{\pi r^2}$$

The shape factor for a finite diffusely emitting area A_1 and a finite receiving area A_2 is:

$$F_{A_1-A_2} = \frac{1}{A_1} \int_{A_1} \int_{A_2} \frac{\cos\theta_1 \cos\theta_2 dA_2 dA_1}{\pi r^2}$$

The unit of the solid angle is a steradian, abbreviated "sr."

We should note the similarity between a plane angle and a solid angle. The plane angle is an angle subtended by the normal projection of a line divided by the radius to the line. The plane angle is dimensionless and is measured in radians. There are 2π radians subtended by a closed line like a circle. The number of steradians subtended by a closed surface like a sphere can be determined by integrating the preceding expression over a sphere. Referring to Figure 21–8, we see that the number of steradians in a sphere is:

$$\omega_{\text{sphere}} = \int_{\text{sphere}} d\omega = \int_{\text{sphere}} \frac{dA_N}{r^2} = \int_{\phi=0}^{2\pi} \int_{\theta=0}^{\pi} \frac{r d\theta r \sin\theta d\phi}{r^2} = 4\pi$$

The second quantity that must be introduced when discussing the directional properties is the intensity of radiation. This is given the symbol I. The intensity is defined as the radiant energy per unit time, per unit solid angle, per unit area projected in a direction normal to the surface. The radiation intensity $I(\theta, \phi)$ is:

$$I(\theta, \phi) = \frac{dq''}{\cos\theta d\omega}$$

where the symbol q'' is used to denote energy per unit time and area. Units of intensity are W/m^2 · sr.

Once the intensity distribution is known, the emissive power which leaves a plane surface can be determined by integrating this equation over all solid angles subtended by a hemisphere placed over the surface. The emissive power of a surface is therefore:

$$E = q'' = \int_{\text{hemisphere}} I(\theta, \phi) \cos\theta d\omega$$

Through substitution of $d\omega$, we get:

$$E = \int_{\phi=0}^{2\pi} \int_{\theta=0}^{\pi/2} I(\theta, \phi) \sin\theta \cos\theta d\theta d\phi$$

The integration of this expression cannot be completed until we know the distribution of intensity over the angles θ and ϕ. The simplest distribution of intensity with angle would be to assume that the intensity is constant. A surface that emits with equal intensity over all angles is called a *diffuse surface* or, sometimes, a surface that abides by Lambert's cosine law because the energy leaving a diffuse surface in a given direction varies as the cosine of the angle to the surface normal. Therefore, a diffuse surface is one for which:

$$I(\theta, \phi) = \text{constant}$$

and the emissive power of a diffuse surface is:

$$E = I \int_{\phi=0}^{2\pi} \int_{\theta=0}^{\pi/2} \sin\theta \cos\theta d\theta d\phi = \pi I$$

A black surface is also a diffuse surface because if it did not emit with equal intensity in all directions, it would not be emitting with the maximum energy for its given temperature. Therefore, the blackbody emissive power and blackbody intensity are related by:

$$E_b = \pi I_b$$

Directional radiative properties cannot be defined in terms of emissive power because the emissive power is not dependent upon the direction from the surface. Directional radiative properties must be defined in terms of the intensity. For example, the directional emissivity $\epsilon(\theta, \phi)$ is defined as the intensity of radiation emitted from the surface in the direction specified by the angles θ and ϕ divided by the blackbody intensity in the same direction, i.e.,

$$\epsilon(\theta, \phi) = \frac{I(\theta, \phi)}{I_b}$$

The total emissivity of a surface is defined as follows:

$$\epsilon = \frac{E}{E_b}$$

If the surface is not diffuse, the emitted intensity is a function of direction. Through simple substitution, we obtain the following relationships:

$$\epsilon = \frac{\int_{\phi=0}^{2\pi} \int_{\theta=0}^{\pi/2} I(\theta, \phi) \sin\theta \cos\theta d\theta d\phi}{\pi I_b}$$

and

$$\epsilon = \frac{1}{\pi} \int_{\phi=0}^{2\pi} \int_{\theta=0}^{\pi/2} \epsilon(\theta, \phi) \sin\theta \cos\theta d\theta d\phi$$

Example 21–2

The directional emissivity of oxidized copper can be approximated by the following expression:

$$\epsilon = 0.70 \cos\theta$$

Determine the amount of radiant energy per unit time emitted by 0.5 m² of oxidized copper when heated to 800 K.

Solution

The total rate of energy emitted by the surface is:

$$q = EA = \epsilon E_b A = \epsilon \sigma T^4 A$$

where ϵ is the total emissivity of the surface. The directional emissivity is related to the total emissivity as follows:

$$\epsilon = 2\int_0^{\pi/2} 0.70 \cos^2\theta \sin\theta d\theta = \frac{-1.4}{3} \cos^2\theta = 0.467$$

where the start and end magnitudes of the angle θ of 0 and $\pi/2$ have already been substituted. Finally, the rate of emitted radiant energy is:

$$q = \epsilon A \sigma T^4 = 5,422 \text{ W}$$

Figure 21–9

Geometrical
configurations for
Example 21–3.

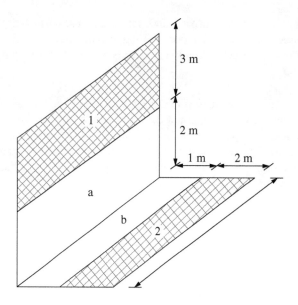

Example 21–3

Evaluate the shape factor $F_{1\to2}$ for the geometry in Figure 21–9. The surfaces are diffuse.

$$A_3 = A_a + A_1$$

$$A_5 = A_b + A_a$$

Solution

$$A_3 F_{3\to4} = A_a F_{a\to b} + A_a F_{a\to2} + A_1 F_{1\to b} + A_1 F_{1\to2}$$

$$A_3 F_{3\to4} = F_{a\to b} + A_1 F_{1\to b}$$

$$F_{a\to b} = F_{a\to b} + F_{a\to2}$$

Combining these three equations and solving for $F_{1\to2}$ gives:

$$F_{1\to2} = \frac{1}{A_1}(A_3 F_{3\to4} - A_a F_{a\to b} - A_a F_{a\to4} + A_a F_{a\to b})$$

These values are as follows:

$$F_{3\to4} = 0.19$$

$$F_{a\to4} = 0.32$$

$$F_{3\to b} = 0.08$$

$$F_{a\to b} = 0.19$$

Substituting these values yields:

$$F_{1\to2} = 0.097$$

The radiation shape factor between the two differential areas dA_1 and dA_2 can now be expressed as follows:

$$F_{dA_1\to dA_2} = \frac{\cos\theta_1 \cos\theta_2 dA_2}{\pi r^2}$$

The shape factor between a diffuse differential emitting area and a finite receiving area is:

$$F_{dA_1 \rightarrow dA_2} = \int_{A_2} F_{dA_1 \rightarrow dA_2} = \int_{A_2} \frac{\cos \theta_1 \cos \theta_2 dA_2}{\pi r^2}$$

The shape factor for a finite diffusely emitting area A_1 and a finite receiving area A_2 is:

$$F_{A_1 \rightarrow A_2} = \frac{1}{A_1} \int_{A_1} \int_{A_2} F_{dA_1 \rightarrow dA_2} dA_1$$

or:

$$F_{A_1 \rightarrow A_2} = \frac{1}{A_1} \int_{A_1} \int_{A_2} \frac{\cos \theta_1 \cos \theta_2 dA_2 dA_1}{\pi r^2}$$

The shape factor expressions (above) are restricted to diffusely emitting surfaces. The assumption of a diffuse surface is a great simplification, because for this condition the shape factors are only functions of the geometry and not of the distribution of radiation intensity.

If the subscripts 1 and 2 are interchanged on the terms of the last expression (above) by assuming that the surface denoted by 2 is the emitting surface and 1 the receiving area, the result is:

$$A_1 F_{1 \rightarrow 2} = A_2 F_{2 \rightarrow 1}$$

The subscript $A_1 \rightarrow A_2$ that has now been simplified to $1 \rightarrow 2$ in the preceding equation is called the *reciprocity relationship*. Reciprocity can be applied to any two surfaces i and j. The general form of the relationship can now be expressed as follows:

$$A_1 F_{i \rightarrow j} = A_2 F_{j \rightarrow i}$$

The simplified subscript will be used in future shape-factor expressions, and it implies a shape factor between two diffuse finite areas 1 and 2.

Example 21–4

A small spherical body is electrically heated and placed in a closed evacuated cylinder. The sphere is 10 cm in diameter with an emissivity of 0.8 and is maintained at a uniform temperature of 572 K. The inside surface of the cylinder, with a surface area of 0.5 m², has an emissivity of 0.2 and is maintained at a uniform temperature of 393 K. Determine the rate of heat transfer from the sphere.

Solution

The surface area of the sphere is 0.031 m². The radiation shape factor $F_{1 \rightarrow 2} = 1$, since all the radiation emitted by the sphere strikes the cylinder's inner surface. The rate of heat transfer is obtained as follows:

$$\dot{Q} = \frac{E_{B1} - E_{B2}}{\Sigma R} = 118.0 \text{ W}$$

Figure 21-10

Input variables for problem 21-4.

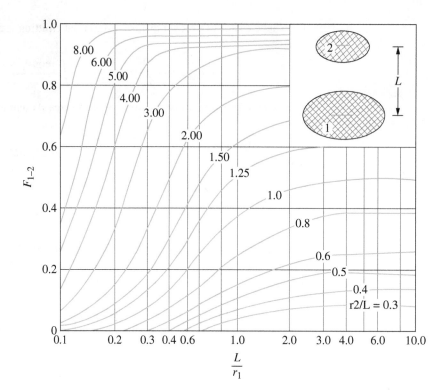

PROBLEMS

21-1 Estimate:
   a) The wavelength at which the monochromatic emissive power of a tungsten filament at 1400 K is a maximum.
   b) The monochromatic emissive power at that wavelength.
   c) The monochromatic emissive power at 5 μm.

21-2 Silica glass transmits 92 percent of the incident radiation in the wavelength range between 0.35 and 2.7 μm and is opaque at longer and shorter wavelengths. Estimate the percentage of solar radiation that the glass will transmit. The sun can be assumed to radiate as a blackbody at 5,800 K. (Use Table 21–1.)

21-3 Consider a 20-cm-diameter spherical ball at 800 K that is suspended in the air to closely represent a blackbody. Determine:
   a) The total blackbody emissive power
   b) The total amount of radiation emitted by the ball in a 5-min interval

21-4 Determine the geometric shape factor for a very small disk A_1 and a larger parallel disk A_2 (Figure 21–10) located at a distance L directly above the smaller one.

References

Baskharone, E. A., "Investigation of Different Cooling Categories of a Radial Inflow Turbine Rotor Using the Finite Element Technique," M.S. Thesis, University of Cincinnati, OH, 1975.

Baskharone, E. A., "F109 High Pressure Turbine First-Stage Rotor Redsign," Garrett Turbine Engine Co., Phoenix, AZ, March 1984.

Baskharone, E. A., "Optimization of the Three-Dimensional Flow Path in the Scroll-Nozzle Assembly of a Radial Inflow Turbine," *Journal for Gas Turbine and Power* (ASME Transactions), Vol. 106, No. 2, April 1984.

Baskharone, E. A., "Finite-Element Analysis of the Turbulent Flow in Annular Exhaust Diffusers for Gas Turbine Engines," *Journal of Fluids Engineering* (ASME Transactions), Vol. 113, No. 1, March 1991.

Baskharone, E. A., and Ghaly, A., "Theoretical versus Experimental Rotordynamic Coefficients of Incompressible-Flow Labyrinth Seals," *AIAA Journal of Propulsion and Power*, Vol. 10, No. 5, October, 1994, pp. 721–729.

Baskharone, E. A., "Principles of Turbomachinery in Air-Breathing Engines," Cambridge University Press, 2006.

Baskharone, E. A., and Hamed, A., "Flow in Non-Rotating Passages of Radial Inflow Turbines," NASA-CR-159679, September 1979.

Baskharone, E. A., and Hamed A., "A New Approach in Cascade Flow Analysis Using the Finite Element Method," *AIAA Journal*, Vol. 19, No. 1, January 1981, pp. 65-71.

Baskharone, E. A., and Hensel, S. J., A New Model for Leakage Prediction in Shrouded-Impeller Turbopumps," *Journal of Fluids Engineering* (ASME Transactions), Vol. 111, No. 2, June 1989.

Baskharone, E. A., and Hensel, S. J., "A Finite Element Perturbation Approach to the Fluid/Rotor Interaction in Turbomachinery Elements: Part 1 - Theory," *Journal of Fluids Engineering* (ASME Transactions), Vol. 113, No. 3, pp. 362-367, 1991

Baskharone, E. A., and Hensel, S. J., "Flow Field in the Secondary, Seal-Containing Passages of Centrifugal Pumps," *Journal of Fluids Engineering* (ASME Transactions), Vol. 115, No. 4, September 1993.

Baskharone, E. A., and McArthur, D. R., "Comprehensive Analysis of the Viscous Incompressible Flow in Quasi-Three-Dimensional Airfoil Cascades," *International Journal for Numerical Methods in Fluids*, Vol. 11, No. 2, July 1990.

Baskharone, E. A., and Wyman, N. J., "Primary/Leakage Flow Interaction in a Pump Stage," *Journal of Fluids Engineering* (ASME Transactions), Vol. 121, No. 1, March 1999.

Bathie, W. W., *Fundamentals of Gas Turbines*, New York: John Wiley and Sons, 1984.

Booth, T. C. "Low Aspect Ratio Turbine (LART)-Phase V Final Report," Design Report No. 75-211701 (5), AiResearch Manufacturing Co. of Arizona, Phoenix, AZ, May 1980.

Church, A. H., *Centrifugal Pumps and Blowers*, New York: John Wiley and Sons, 1944.

Cohen, H., Rogers, G. F. C., and Saravanamuttoo, H. I. H, *Gas Turbine Theory*, Essex, UK: Longman, 1972.

Galligab, J. E., "Advanced Technology Components for Model GTCP 305-2 Aircraft Auxiliary Power System," Design Report No. 31-2874, AiResearch Manufacturing Co. of Arizona, Phoenix, AZ, May 1979.

Glassman, A. J., "Turbine Design and Applications," NASA SP No. 290, Washington, D.C., 1973.

Gupta, S. K., and Tanji, K. K., "Computer Program for Solution of Large, Sparse, Unsymmetric Systems of Linear Equations," *International Journal for Numerical Methods in Engineering*, Vol. 11, 1977, pp. 1251-1259.

Hamed, A. and Baskharone, E. A., "Analysis of the Three-Dimensional Flow in a Turbine Scroll," *Journal of Fluids Engineering* (ASME Transactions), Vol. 102, No. 3, September 1980.

Hamed, A., Baskharone, E. A., and Tabakoff W., "Temperature Distribution Study in a Cooled Radial Turbine Rotor," *AIAA Journal of Aircraft*, Vol. 14, No. 2, February 1977.

Hill, P. G., and Peterson, C. R., *Mechanics and Thermodynamics of Propulsion*, Reading, MA: Addison-Wesley, 1992.

Hinch, D. V., "Axial Turbine Tip Clearance Study," Report No. 22-1648, Garrett Turbine Engine Co., Phoenix, AZ, April 1982.

Hlavaty, S. T., "A Finite Element Model of the Turbulent Flow Field in a Centrifugal Impeller," M. S. Thesis, Texas A&M University, August 1993.

Horlock, J. H., "Axial-Flow Turbines: Fluid Mechanics and Thermodynamics," Malabar, FL: Krieger, 1985.

Huebner, K. H., and Thornton, E. A., *The Finite Element Method for Engineers*, New York: John Wiley and Sons, 1974.

Katsanis, T., and McNally, W., "Fortran Program for Calculating Velocities and Streamlines on the Hub, Shroud, and Mid-Channel Flow Surface of Axial or Mixed-Flow Turbomachines," NASA TN D-7343, July 1973.

Kavanaough, P., and Ye, Z. Q., "Axial-Flow Turbine Design Procedure and Sample Design Cases," Technical Report No. TCRL-28, Iowa State University, Ames, IA, January 1984.

Khalil, I. M, Tabakoff, W., and Hamed, A. " Losses in Radial Inflow Turbines," *Journal of Fluids Engineering* (ASME Transactions), September 1976.

Kovats, A., *Design and Performance of Centrifugal and Axial-Flow Pumps and Compressors*, New York: MacMillan, 1964.

Kreith, F., and Black, W. Z., *Basic Heat Transfer*, New York: Harper and Row, 1980.

Kreith, F., and Bohn, M. S., Principles of Heat Transfer, Boston: PWS Publishing Co., 1993.

MacGregor, J. D., and Baskharone, E. A., "A Finite-Element Model of Cyclic Aerodynamic Loading of a Turbine Rotor Due to an Upstream Stator," *International Journal of Computational Fluid Dynamics*, Vol. 25, September 1977, pp. 291–306.

Mashimo, T., Watanabe, I., and Ariga, I., "Effects of Fluid Leakage on the Performance of a Centrifugal Compressor," *Journal of Engineering for Power* (ASME Transactions), Vol. 101, No. 3, July 1979.

Mattingly, J. D., *Elements of Gas Turbine Propulsion*, New York: McGraw-Hill, 1996.

McFarland, E. R., "A Rapid Blade-To-Blade Solution for Use in Turbomachinery Design," *Journal of Engineering for Gas Turbine and Power* (ASME Transactions), Vol. 106, No. 2, April 1984.

Obert, E. F., and Gaggioli, R. A., *Thermodynamics*, New York: McGraw-Hill, 1963.

Myers, G. E., *Analytical Methods in Conduction Heat Transfer*, Amsterdam, NY: Genium Publishing Co., 1987.

Schepherd, D. G., *Principles of Turbomachinery*, New York: McMillan, 1956.

Schlichting, H., *Boundary Layer Theory*, New York: McGraw-Hill, 1979.

Sczanka, E., Behning, F., and Shum, H., "Research Turbine for High Temperature Core Engine Application, Part II—Effect of Rotor Tip Clearance on Overall Performance," NASA TN D-7639, April 1974.

Sovran, G., and Klump, E. D., "Experimentally Determined Optimum Geometries for Diffusers with Rectangular, Conical or Annular Cross Sections," General Motors Research Publication HMR-511, General Motors Corporation, November 1965.

Vavra, M.H., *Aerothermodynamics and Flow in Turbomachines*, New York: John Wiley and Sons, 1960.

Zienkiewicz, O. C., *The Finite Element Method*, McGraw-Hill Book Company (UK) Limited, 1985.

Zucker, R. D., *Fundamentals of Gas Dynamics*, Beaverton, OR: Matrix Publishers, 1977.

Appendix Outline

Table A–1 Properties of Water: Saturated Liquid–Saturated Vapor—Temperature Table **365**

Table A–2 Properties of Water: Saturated Liquid–Saturated Vapor—Pressure Table **367**

Table A–3 Properties of Water: Superheated Vapor **369**

Table A–4 Properties of Refrigerant R-12: Saturated Liquid–Saturated Vapor—Temperature Table **370**

Table A–5 Properties of Refrigerant R-12: Superheated Vapor **372**

Table A–6 Saturated Refrigerant-134a: Temperature **380**

Table A–7 Saturated Refrigerant-134a: Pressure **381**

Table A–8 Superheated Refrigerant-134a **382**

Figure A–1 Generalized Compressibility Chart **385**

Table A–9 Equations of State **386**

Table A–10 Constant-Pressure Specific Heats for Several Ideal Gases **386**

Table A–11 Properties of Several Ideal Gases **387**

Table A–12 Thermophysical Properties of Air **388**

Table A–13 Thermophysical Properties of Saturated Water **390**

Table A–14 Thermophysical Properties of Oil **391**

Table A–15 Physical Properties of Common Liquids **391**

Figure A–2 Dynamic Viscosity of Common Fluids as a Function of Temperature **392**

Figure A–3 Kinematic Viscosity of Common Fluids as a Function of Temperature **393**

Table A–16 Thermophysical Properties of Selected Metallic Solids **394**

Table A–17 Thermophysical Properties of Common Materials **396**

Table A–18 Thermophysical Properties of Structural Building Materials **398**

Table A–19 Thermophysical Properties of Industrial Insulation **400**

Figure A–4 Friction Factor Chart **401**

Figure A–5 Oblique Shock Chart **402**

Table A–20 Isentropic Flow Parameters **403**

Table A–21 Normal Shock Parameters **416**

Table A–22   Fanno Flow Parameters   **426**

Table A–23   Rayleigh Flow Parameters   **439**

Table A–24   Oblique Shock Functions   **451**

Tables A–1 through A–19 and Figures A–1 and A–2 reprinted with permission from Schmidt, F. W., Henderson, R. E., and Wolgemuth, C. H., *Introduction to Thermal Sciences* (New York: John Wiley and Sons, 1984).

Tables A–20 through A–24 and Figures A–4 and A–5 reprinted with permission from Zucker, R. D., *Fundamentals of Gas Dynamics* (Champaign, Ill.: Matrix Publishers, 1977).

Table A–1

Properties of Water: Saturated Liquid–Saturated Vapor—Temperature Table

Temp. T, °C	Press. P, kPa	Specific volume, m³/kg Sat. liquid v_l	Sat. vapor v_g	Internal energy, kJ/kg Sat. liquid U_l	Evap. U_{lg}	Sat. vapor U_g	Enthalpy, kJ/kg Sat. liquid h_l	Evap. h_{lg}	Sat. vapor h_g	Entropy, kJ/(kg·K) Sat. liquid S_l	Evap. S_{lg}	Sat. vapor S_g
0.01	0.6113	0.001 000	206.14	.00	2375.3	2375.3	.01	2501.3	2501.4	.0000	9.1562	9.1562
5	0.8721	0.001 000	147.12	20.97	2361.3	2382.3	20.98	2489.6	2510.6	.0761	8.9496	9.0257
10	1.2276	0.001 000	106.38	42.00	2347.2	2389.2	42.01	2477.7	2519.8	.1510	8.7498	8.9008
15	1.7051	0.001 001	77.93	62.99	2333.1	2396.1	62.99	2465.9	2528.9	.2245	8.5569	8.7814
20	2.339	0.001 002	57.79	83.95	2319.0	2402.9	83.96	2454.1	2538.1	.2966	8.3706	8.6672
25	3.169	0.001 003	43.36	104.88	2304.9	2409.8	104.89	2442.3	2547.2	.3674	8.1905	8.5580
30	4.246	0.001 004	32.89	125.78	2290.8	2416.6	125.79	2430.5	2556.3	.4369	8.0164	8.4533
35	5.628	0.001 006	25.22	146.67	2276.7	2423.4	146.68	2418.6	2565.3	.5053	7.8478	8.3531
40	7.384	0.001 008	19.52	167.56	2262.6	2430.1	167.57	2406.7	2574.3	.5725	7.6845	8.2570
45	9.593	0.001 010	15.26	188.44	2248.4	2436.8	188.45	2394.8	2583.2	.6387	7.5261	8.1648
50	12.349	0.001 012	12.03	209.32	2234.2	2443.5	209.33	2382.7	2592.1	.7038	7.3725	8.0763
55	15.758	0.001 015	9.568	230.21	2219.9	2450.1	230.23	2370.7	2600.9	.7679	7.2234	7.9913
60	19.940	0.001017	7.671	251.11	2205.5	2456.6	251.13	2358.5	2609.6	.8312	7.0784	7.9096
65	25.03	0.001 020	6.197	272.02	2191.1	2463.1	272.06	2346.2	2618.3	.8935	6.9375	7.8310
70	31.19	0.001 023	5.042	292.95	2176.6	2469.6	292.98	2333.8	2626.8	.9549	6.8004	7.7553
75	38.58	0.001 026	4.131	313.90	2162.0	2475.9	313.93	2321.4	2635.3	1.0155	6.6669	7.6824
80	47.39	0.001 029	3.407	334.86	2147.4	2482.2	334.91	2308.8	2643.7	1.0753	6.5369	7.6122
85	57.83	0.001 033	2.828	355.84	2132.6	2488.4	355.90	2296.0	2651.9	1.1343	6.4102	7.5445
90	70.14	0.001 036	2.361	376.85	2117.7	2494.5	376.92	2283.2	2660.1	1.1925	6.2866	7.4791
95	84.55	0.001 040	1.982	397.88	2102.7	2500.6	3.97.96	2270.2	2668.1	1.2500	6.1659	7.4159
100	0.101 35	0.001 044	1.6729	418.94	2087.6	2506.5	419.04	2257.0	2676.1	1.3069	6.0480	7.3549
105	0.120 82	0.001 048	1.4194	440.02	2072.3	2512.4	440.15	2243.7	2683.8	1.3630	5.9328	7.2958
110	0.143 27	0.001 052	1.2102	461.14	2057.0	2518.1	461.30	2230.2	2691.5	1.4185	5.8202	7.2387
115	0.169 06	0.001 056	1.0366	482.30	2041.4	2523.7	482.48	2216.5	2699.0	1.4734	5.7100	7.1833
120	0.198 53	0.001 060	0.8919	503.50	2025.8	2529.3	503.71	2202.6	2706.3	1.5276	5.6020	7.1296
125	0.2321	0.001 065	0.7706	524.74	2009.9	2534.6	524.99	2188.5	2713.5	1.5813	5.4962	7.0775
130	0.2701	0.001 070	0.6685	546.02	1993.9	2539.9	546.31	2174.2	2720.5	1.6344	5.3925	7.0269
135	0.3130	0.001 075	0.5822	567.35	1977.7	2545.0	567.69	2159.6	2727.3	1.6870	5.2907	6.9777
140	0.3613	0.001 080	0.5089	588.74	1961.3	2550.0	589.13	2144.7	2733.9	1.7391	5.1908	6.9299
145	0.4154	0.001 085	0.4463	610.18	1944.7	2554.9	610.63	2129.6	2740.3	1.7907	5.0926	6.8833
150	0.4758	0.001 091	0.3928	631.68	1927.9	2559.5	632.20	2114.3	2746.5	1.8418	4.9960	6.8379
155	0.5431	0.001 096	0.3468	653.24	1910.8	2564.1	653.84	2098.6	2752.4	1.8925	4.9010	6.7935
160	0.6178	0.001 102	0.3071	674.87	1893.5	2568.4	675.55	2082.6	2758.1	1.9427	4.8075	6.7502
165	0.7005	0.001 108	0.2727	696.56	1876.0	2572.5	697.34	2066.2	2763.5	1.9925	4.7153	6.7078
170	0.7917	0.001 114	0.2428	718.33	1858.1	2576.5	719.21	2049.5	2768.7	2.0419	4.6244	6.6663

(continued)

Table A–1

Properties of Water: Saturated Liquid–Saturated Vapor—Temperature Table (Continued)

Temp. T, °C	Press. P, kPa	Specific volume, m³/kg		Internal energy, kJ/kg			Enthalpy, kJ/kg			Entropy, kJ/(kg·K)		
		Sat. liquid v_l	Sat. vapor v_g	Sat. liquid U_l	Evap. U_{lg}	Sat. vapor U_g	Sat. liquid h_l	Evap. h_{lg}	Sat. vapor h_g	Sat. liquid S_l	Evap. S_{lg}	Sat. vapor S_g
175	0.8920	0.001 121	0.2168	740.17	1840.0	2580.2	741.17	2032.4	2773.6	2.0909	4.5347	6.6256
180	1.0021	0.001 127	0.194 05	762.09	1821.6	2583.7	763.22	2015.0	2778.2	2.1396	4.4461	6.6857
185	1.1227	0.001 134	0.174 09	784.10	1802.9	2587.0	785.37	1997.1	2782.4	2.1879	4.3586	6.5465
190	1.2544	0.001 141	0.156 54	806.19	1783.8	2590.0	807.62	1978.8	2786.4	2.2359	4.2720	6.5079
195	1.3978	0.001 149	0.141 05	828.37	1764.4	2592.8	829.98	1960.0	2790.0	2.2835	4.1863	6.4698
200	1.5538	0.001 157	0.127 36	850.65	1744.7	2595.3	852.45	1940.7	2793.2	2.3309	4.1014	6.4323
205	1.7230	0.001 164	0.115 21	873.04	1724.5	2597.5	875.04	1921.0	2796.0	2.3780	4.0172	6.3952
210	1.9062	0.001 173	0.104 41	895.53	1703.9	2599.5	897.76	1900.7	2798.5	2.4248	3.9337	6.3585
215	2.104	0.001 181	0.094 79	918.14	1682.9	2601.1	920.62	1879.9	2800.5	2.4714	3.8507	6.3221
220	2.318	0.001 190	0.086 19	940.87	1661.5	2602.4	943.62	1858.5	2802.1	2.5178	3.7683	6.2861
225	2.548	0.001 199	0.078 49	963.73	1639.6	2603.3	966.78	1836.5	2803.3	2.5639	3.6863	6.2503
230	2.795	0.001 209	0.071 58	986.74	1617.2	2603.9	990.12	1813.8	2804.0	2.6099	3.6047	6.2146
235	3.060	0.001 219	0.065 37	1009.89	1594.2	2604.1	1013.62	1790.5	2804.2	2.6558	3.5233	6.1791
240	3.344	0.001 229	0.059 76	1033.21	1570.8	2604.0	1037.32	1766.5	2803.8	2.7015	3.4422	6.1437
245	3.648	0.001 240	0.054 71	1056.71	1546.7	2603.4	1061.23	1741.7	2803.0	2.7472	3.3612	6.1083
250	3.973	0.001 251	0.050 13	1080.39	1522.0	2602.4	1085.36	1716.2	2801.5	2.7927	3.2802	6.0730
255	4.319	0.001 263	0.045 98	1104.28	1496.7	2600.9	1109.73	1689.8	2799.5	2.8383	3.1992	6.0375
260	4.688	0.001 276	0.042 21	1128.39	1470.6	2599.0	1134.37	1662.5	2796.9	2.8838	3.1181	6.0019
265	5.081	0.001 289	0.038 77	1152.74	1443.9	2596.6	1159.28	1634.4	2793.6	2.9294	3.0368	5.9662
270	5.499	0.001 302	0.035 64	1177.36	1416.3	2593.7	1184.51	1605.2	2789.7	2.9751	2.9551	5.9301
275	5.942	0.001 317	0.032 79	1202.25	1387.9	2590.2	1210.07	1574.9	2785.0	3.0208	2.8730	5.8938
280	6.412	0.001 332	0.030 17	1227.46	1358.7	2586.1	1235.99	1 543.6	2779.6	3.0668	2.7903	5.8571
285	6.909	0.001 348	0.027 77	1253.00	1328.4	2581.4	1262.31	1511.0	2773.3	3.1130	2.7070	5.8199
290	7.436	0.001 366	0.025 57	1278.92	1297.1	2576.0	1289.07	1477.1	2766.2	3.1594	2.6227	5.7821
295	7.993	0.001 384	0.023 54	1305.2	1264.7	2569.9	1316.3	1441.8	2758.1	3.2062	2.5375	5.7437
300	8.581	0.001 404	0.021 67	1332.0	1231.0	2563.0	1344.0	1404.9	2749.0	3.2534	2.4511	5.7045
305	9.202	0.001 425	0.019 948	1359.3	1195.9	2555.2	1372.4	1366.4	2738.7	3.3010	2.3633	5.6643
310	9.856	0.001 447	0.018 350	1387.1	1159.4	2546.4	1401.3	1326.0	2727.3	3.3493	2.2737	5.6230
315	10.547	0.001 472	0.016 867	1415.5	1121.1	2536.6	1431.0	1283.5	2714.5	3.3982	2.1821	5.5804
320	11.274	0.001 499	0.015 488	1444.6	1080.9	2525.5	1461.5	1238.6	2700.1	3.4480	2.0882	5.5362
330	12.845	0.001 561	0.012 996	1505.3	993.7	2498.9	1525.3	1140.6	2665.9	3.5507	1.8909	5.4417
340	14.586	0.001 638	0.010 797	1570.3	894.3	2464.6	1594.2	1027.9	2622.0	3.6594	1.6763	5.3357
350	16.513	0.001 740	0.008 813	1641.9	776.6	2418.4	1670.6	893.4	2563.9	3.7777	1.4335	5.2112
360	18.651	0.001 893	0.006 945	1725.2	626.3	2351.5	1760.5	720.5	2481.0	3.9147	1.1379	5.0526
370	21.03	0.002 213	0.004 925	1844.0	384.5	2228.5	1890.5	441.6	2332.1	4.1106	0.6865	4.7971
374.14	22.09	0.003 155	0.003 155	2029.6	0	2029.6	2099.3	0	2099.3	4.4298	0	4.4298

Table A–2

Properties of Water: Saturated Liquid–Saturated Vapor—Pressure Table

Press. P, kPa	Temp. T,°C	Specific volume, m³/kg Sat. liquid v_l	Sat. vapor v_g	Internal energy, kJ/kg Sat. liquid U_l	Evap. U_{lg}	Sat. vapor U_g	Enthalpy, kJ/kg Sat. liquid h_l	Evap. h_{lg}	Sat. vapor h_g	Entropy, kJ/(kg·K) Sat. liquid S_l	Evap. S_{lg}	Sat. vapor S_g
0.6113	0.01	0.001 000	206.14	.00	2375.3	2375.3	.01	2501.3	2501.4	.0000	9.1562	9.1562
1.0	6.98	0.001 000	129.21	29.30	2355.7	2385.0	29.30	2484.9	2514.2	.1059	8.8697	8.9756
1.5	13.03	0.001 001	87.98	54.71	2338.6	2393.6	54.71	2470.6	2525.3	.1957	8.6322	8.8279
2.0	17.50	0.001 001	67.00	73.48	2326.0	2399.5	73.48	2460.0	2533.5	.2607	8.4629	8.7237
2.5	21.08	0.001 002	54.25	88.48	2315.9	2404.4	88.49	2451.6	2540.0	.3120	8.3311	8.6432
3.0	24.08	0.001 003	45.67	101.04	2307.5	2408.5	101.05	2444.5	2545.5	.3545	8.2231	8.5776
4.0	28.96	0.001 004	34.80	121.45	2293.7	2415.2	121.46	2432.9	2554.4	.4226	8.0520	8.4746
5.0	32.88	0.001 005	28.19	137.81	2282.7	2420.5	137.82	2423.7	2561.5	.4764	7.9187	8.3951
7.5	40.29	0.001 008	19.24	168.78	2261.7	2430.5	168.79	2406.0	2574.8	.5764	7.6750	8.2515
10	45.81	0.001 010	14.67	191.82	2246.1	2437.9	191.83	2392.8	2584.7	.6493	7.5009	8.1502
15	53.97	0.001 014	10.02	225.92	2222.8	2448.7	225.94	2373.1	2599.1	.7549	7.2536	8.0085
20	60.06	0.001 017	7.649	251.38	2205.4	2456.7	251.40	2358.3	2609.7	.8320	7.0766	7.9085
25	64.97	0.001 020	6.204	271.90	2191.2	2463.1	271.93	2346.3	2618.2	.8931	6.9383	7.8314
30	69.10	0.001 022	5.229	289.20	2179.2	2468.4	289.23	2336.1	2625.3	.9439	6.8247	7.7686
40	75.87	0.001 027	3.993	317.53	2159.5	2477.0	317.58	2319.2	2636.8	1.0259	6.6441	7.6700
50	81.33	0.001 030	3.240	340.44	2143.4	2483.9	340.49	2305.4	2645.9	1.0910	6.5029	7.5939
75	91.78	0.001 037	2.217	384.31	2112.4	2496.7	384.39	2278.6	2663.0	1.2130	6.2434	7.4564
MPa												
0.100	99.63	0.001 043	1.6940	417.36	2088.7	2506.1	417.46	2258.0	2675.5	1.3026	6.0568	7.3594
0.125	105.99	0.001 048	1.3749	444.19	2069.3	2513.5	444.32	2241.0	2685.4	1.3740	5.9104	7.2844
0.150	111.37	0.001 053	1.1593	466.94	2052.7	2519.7	467.11	2226.5	2693.6	1.4336	5.7897	7.2233
0.175	116.06	0.001 057	1.0036	486.80	2038.1	2524.9	486.99	2213.6	2700.6	1.4849	5.6868	7.1717
0.200	120.23	0.001 061	0.8857	504.49	2025.0	2529.5	504.70	2201.9	2706.7	1.5301	5.5970	7.1271
0.225	124.00	0.001 064	0.7933	520.47	2013.1	2533.6	520.72	2191.3	2712.1	1.5706	5.5173	7.0878
0.250	127.44	0.001 067	0.7187	535.10	2002.1	2537.2	535.37	2181.5	2716.9	1.6072	5.4455	7.0527
0.275	130.60	0.001 070	0.6573	548.59	1991.9	2540.5	548.89	2172.4	2721.3	1.6408	5.3801	7.0209
0.300	133.55	0.001 073	0.6058	561.15	1982.4	2543.6	561.47	2163.8	2725.3	1.6718	5.3201	6.9919
0.325	136.30	0.001 076	0.5620	572.90	1973.5	2546.4	573.25	2155.8	2729.0	1.7006	5.2646	6.9652
0.350	138.88	0.001 079	0.5243	583.95	1965.0	2548.9	584.33	2148.1	2732.4	1.7275	5.2130	6.9405
0.375	141.32	0.001 081	0.4914	594.40	1956.9	2551.3	594.81	2140.8	2735.6	1.7528	5.1647	6.9175
0.40	143.63	0.001 084	0.4625	604.31	1949.3	2553.6	604.74	2133.8	2738.6	1.7766	5.1193	6.8959
0.45	147.93	0.001 088	0.4140	622.77	1934.9	2557.6	623.25	2120.7	2743.9	1.8207	5.0359	6.8565
0.50	151.86	0.001 093	0.3749	639.68	1921.6	2561.2	640.23	2108.5	2748.7	1.8607	4.9606	6.8213
0.55	155.48	0.001 097	0.3427	655.32	1909.2	2564.5	655.93	2097.0	2753.0	1.8973	4.8920	6.7893

(*continued*)

Table A–2

Properties of Water: Saturated Liquid–Saturated Vapor—Pressure Table (Continued)

		Specific volume, m³/kg		Internal energy, kJ/kg			Enthalpy, kJ/kg			Entropy, kJ/(kg·K)		
Press. P, kPa	Temp. T,°C	Sat. liquid v_l	Sat. vapor v_g	Sat. liquid U_l	Evap. U_{lg}	Sat. vapor U_g	Sat. liquid h_l	Evap. h_{lg}	Sat. vapor h_g	Sat. liquid S_l	Evap. S_{lg}	Sat. vapor S_g
0.60	158.85	0.001 101	0.3157	669.90	1897.5	2567.4	670.56	2086.3	2756.8	1.9312	4.8288	6.7600
0.65	162.01	0.001 104	0.2927	683.56	1886.5	2570.1	684.28	2076.0	2760.3	1.9627	4.7703	6.7331
0.70	164.97	0.001 108	0.2729	696.44	1876.1	2572.5	697.22	2066.3	2763.5	1.9922	4.7158	6.7080
0.75	167.78	0.001 112	0.2556	708.64	1866.1	2574.7	709.47	2057.0	2766.4	2.0200	4.6647	6.6847
0.80	170.43	0.001 115	0.2404	720.22	1856.6	2576.8	721.11	2048.0	2769.1	2.0462	4.6166	6.6628
0.85	172.96	0.001 118	0.2270	731.27	1847.4	2578.7	732.22	2039.4	2771.6	2.0710	4.5711	6.6421
0.90	175.38	0.001 121	0.2150	741.83	1838.6	2580.5	742.83	2031.1	2773.9	2.0946	4.5280	6.6226
0.95	177.69	0.001 124	0.2042	751.95	1830.2	2582.1	753.02	2023.1	2776.1	2.1172	4.4869	6.6041
1.00	179.91	0.001 127	0.194 44	761.68	1822.0	2583.6	762.81	2015.3	2778.1	2.1387	4.4478	6.5865
1.10	184.09	0.001 133	0.177 53	780.09	1806.3	2586.4	781.34	2000.4	2781.7	2.1792	4.3744	6.5536
1.20	187.99	0.001 139	0.163 33	797.29	1791.5	2588.8	798.65	1986.2	2784.8	2.2166	4.3067	6.5233
1.30	191.64	0.001 144	0.151 25	813.44	1775.5	2591.0	814.93	1972.7	2787.6	2.2515	4.2438	6.4953
1.40	195.07	0.001 149	0.140 84	828.70	1764.1	2592.8	830.30	1959.7	2790.0	2.2842	4.1850	6.4693
1.50	198.32	0.001 154	0.131 77	843.16	1751.3	2594.5	844.89	1947.3	2792.2	2.3150	4.1298	6.4448
1.75	205.76	0.001 166	0.113 49	876.46	1721.4	2597.8	878.50	1917.9	2796.4	2.3851	4.0044	6.3896
2.00	212.42	0.001 177	0.099 63	906.44	1693.8	2600.3	908.79	1890.7	2799.5	2.4474	3.8935	6.3409
2.25	218.45	0.001 187	0.088 75	933.83	1668.2	2602.0	936.49	1865.2	2801.7	2.5035	3.7937	6.2972
2.5	223.99	0.001197	0.079 98	959.11	1644.0	2603.1	962.11	1841.0	2803.1	2.5547	3.7028	6.2575
3.0	233.90	0.001 217	0.066 68	1004.78	1599.3	2604.1	1008.42	1795.7	2804.2	2.6457	3.5412	6.1869

Table A–3

Properties of Water: Superheated Vapor

T	P = 0.010 MPa (45.81)				P = 0.050 MPa (81.33)				P = 0.10 MPa (99.63)			
	v	u	h	s	v	u	h	s	v	u	h	s
Sat.	14.674	2437.9	2584.7	8.1502	3.240	2483.9	2645.9	7.5939	1.6940	2506.1	2675.5	7.3594
50	14.869	2443.9	2592.6	8.1749								
100	17.196	2515.5	2687.5	8.4479	3.418	2511.6	2682.5	7.6947	1.6958	2506.7	2676.2	7.3614
150	19.512	2587.9	2783.0	8.6882	3.889	2585.6	2780.1	7.9401	1.9364	2582.8	2776.4	7.6134
200	21.825	2661.3	2879.5	8.9038	4.356	2659.9	2877.7	8.1580	2.172	2658.1	2875.3	7.8343
250	24.136	2736.0	2977.3	9.1002	4.820	2735.0	2976.0	8.3556	2.406	2733.7	2974.3	8.0333
300	26.445	2812.1	3076.5	9.2813	5.284	2811.3	3075.5	8.5373	2.639	2810.4	3074.3	8.2158
400	31.063	2968.9	3279.6	9.6077	6.209	2968.5	3278.9	8.8642	3.103	2967.9	3278.2	8.5435
500	35.679	3132.3	3489.1	9.8978	7.134	3132.0	3488.7	9.1546	3.565	3131.6	3488.1	8.8342
600	40.295	3302.5	3705.4	10.1608	8.057	3302.2	3705.1	9.4178	4.028	3301.9	3704.7	9.0976
700	44.911	3479.6	3928.7	10.4028	8.981	3479.4	3928.5	9.6599	4.490	3479.2	3928.2	9.3398
800	49.526	3663.8	4159.0	10.6281	9.904	3663.6	4158.9	9.8852	4.952	3663.5	4158.6	9.5652
900	54.141	3855.0	4396.4	10.8396	10.828	3854.9	4396.3	10.0967	5.414	3854.8	4396.1	9.7767
1000	58.757	4053.0	4640.6	11.0393	11.751	4052.9	4640.5	10.2964	5.875	4052.8	4640.3	9.9764
1100	63.372	4257.5	4891.2	11.2287	12.674	4257.4	4891.1	10.4859	6.337	4257.3	4891.0	10.1659
1200	67.987	4467.9	5147.8	11.4091	13.597	4467.8	5147.7	10.6662	6.799	4467.7	5147.6	10.3463
1300	72.602	4683.7	5409.7	11.5811	14.521	4683.6	5409.6	10.8382	7.260	4683.5	5409.5	10.5183
	P = 0.20 MPa (120.23)				P = 0.30 MPa (133.55)				P = 0.40 MPa (143.63)			
Sat.	0.8857	2529.5	2706.7	7.1272	0.6058	2543.6	2725.3	6.9919	0.4625	2553.6	2738.5	6.8959
150	0.9596	2576.9	2768.8	7.2795	0.6339	2570.8	2761.0	7.0778	0.4708	2564.5	2752.8	6.9299
200	1.0803	2654.4	2870.5	7.5066	0.7163	2650.7	2865.6	7.3115	0.5342	2646.8	2860.5	7.1706
250	1.1988	2731.2	2971.0	7.7086	0.7964	2728.7	2967.6	7.5166	0.5951	2726.1	2964.2	7.3789
300	1.3162	2808.6	3071.8	7.8926	0.8753	2806.7	3069.3	7.7022	0.6548	2804.8	3066.8	7.5662
400	1.5493	2966.7	3276.6	8.2218	1.0315	2965.6	3275.0	8.0330	0.7726	2964.4	3273.4	7.8985
500	1.7814	3130.8	3487.1	8.5133	1.1867	3130.0	3486.0	8.3251	0.8893	3129.2	3484.9	8.1913
600	2.013	3301.4	3704.0	8.7770	1.3414	3300.8	3703.2	8.5892	1.0055	3300.2	3702.4	8.4558
700	2.244	3478.8	3927.6	9.0194	1.4957	3478.4	3927.1	8.8319	1.1215	3477.9	3926.5	8.6987
800	2.475	3663.1	4158.2	9.2449	1.6499	3662.9	4157.8	9.0576	1.2372	3662.4	4157.3	8.9244
900	2.706	3854.5	4395.8	9.4566	1.8041	3854.2	4395.4	9.2692	1.3529	3853.9	4395.1	9.1362

Table A–4

Properties of Refrigerant R-12: Saturated Liquid–Saturated Vapor?—Temperature Table

Press. P, MPa	Temp. T, °C	Specific volume, m³/kg		Internal energy, kJ/kg			Enthalpy, kJ/kg			Entropy, kJ/(kg·K)		
		Sat. liquid v_f	Sat. vapor v_g	Sat. liquid U_f	Evap. U_{fg}	Sat. vapor U_g	Sat. liquid h_f	Evap. h_{fg}	Sat. vapor h_g	Sat. liquid S_f	Evap. S_{tg}	Sat. vapor S_g
3.5	242.60	0.001 235	0.057 07	1045.43	1558.3	2603.7	1049.75	1753.7	2803.4	2.7253	3.4000	6.1253
4	250.40	0.001 252	0.049 78	1082.31	1520.0	2602.3	1087.31	1714.1	2801.4	2.7964	3.2737	6.0701
5	263.99	0.001 286	0.039 44	1147.81	1449.3	2597.1	1154.23	1640.1	2794.3	2.9202	3.0532	5.9734
6	275.64	0.001 319	0.032 44	1205.44	1384.3	2589.7	1213.35	1571.0	2784.3	3.0267	2.8625	5.8892
7	285.88	0.001 351	0.027 37	1257.55	1323.0	2580.5	1267.00	1505.1	2772.1	3.1211	2.6922	5.8133
8	295.06	0.001 384	0.023 52	1305.57	1264.2	2569.8	1316.64	1441.3	2758.0	3.2068	2.5364	5.7432
9	303.40	0.001 418	0.020 48	1350.51	1207.3	2557.8	1363.26	1378.9	2742.1	3.2858	2.3915	5.6772
10	311.06	0.001 452	0.018 026	1393.04	1151.4	2544.4	1407.56	1317.1	2724.7	3.3596	2.2544	5.6141
11	318.15	0.001 489	0.015 987	1433.7	1096.0	2529.8	1450.1	1255.5	2705.6	3.4295	2.1233	5.5527
12	324.75	0.001 527	0.014 263	1473.0	1040.7	2513.7	1491.3	1193.6	2684.9	3.4962	1.9962	5.4924
13	330.93	0.001 567	0.012 780	1511.1	985.0	2496.1	1531.5	1130.7	2662.2	3.5606	1.8718	5.4323
14	336.75	0.001 611	0.011 485	1548.6	928.2	2476.8	1571.1	1066.5	2637.6	3.6232	1.7485	5.3717
15	342.24	0.001 658	0.010 337	1585.6	869.8	2455.5	1610.5	1000.0	2610.5	3.6848	1.6249	5.3098
16	347.44	0.001 711	0.009 306	1622.7	809.0	2431.7	1650.1	930.6	2580.6	3.7461	1.4994	5.2455
17	352.37	0.001 770	0.008 364	1660.2	744.8	2405.0	1690.3	856.9	2547.2	3.8079	1.3698	5.1777
18	357.06	0.001 840	0.007 489	1598.9	675.4	2374.3	1732.0	777.1	2509.1	3.8715	1.2329	5.1044
19	361.54	0.001 924	0.006 657	1739.9	598.1	2338.1	1776.5	688.0	2464.5	3.9388	1.0839	5.0228
20	365.81	0.002 036	0.005 834	1785.6	507.5	2293.0	1826.3	583.4	2409.7	4.0139	0.9130	4.9269
21	369.89	0.002 207	0.004 952	1842.1	388.5	2230.6	1884.4	446.2	2334.6	4.1075	0.6938	4.8013
22	373.80	0.002 742	0.003 568	1961.9	125.2	2087.1	2022.2	143.4	2165.6	4.3110	0.2216	4.5327
22.09	374.14	0.003 155	0.003 155	2029.6	0	2029.6	2099.3	0	2099.3	4.4298	0	4.4298

Table A–4

Properties of Refrigerant R-12: Saturated Liquid–Saturated Vapor—Temperature Table (Continued)

Temp., °C	Abs. press. P, MPa	Specific volume, m³/kg Sat. liquid v_f	Evap. v_{fg}	Sat. vapor v_g	Enthalpy, kJ/kg Sat. liquid h_f	Evap. h_{fg}	Sat. vapor h_g	Entropy, kJ/(kg·K) Sat. liquid s_f	Evap. s_{fg}	Sat. vapor s_g
−90	0.0028	0.000 608	4.414 937	4.415 545	−43.243	189.618	146.375	−0.2084	1.0352	0.8268
−85	0.0042	0.000 612	3.036 704	3.037 316	−38.968	187.608	148.640	−0.1854	0.9970	0.8116
−80	0.0062	0.000 617	2.137 728	2.138 345	−34.688	185.612	150.924	−0.1630	0.9609	0.7979
−75	0.0088	0.000 622	1.537 030	1.537 651	−30.401	183.625	153.224	−0.1411	0.9266	0.7855
−70	0.0123	0.000 627	1.126 654	1.127 280	−26.103	181.640	155.536	−0.1197	0.8940	0.7744
−65	0.0168	0.000 632	0.840 534	0.841 166	−21.793	179.651	157.857	−0.0987	0.8630	0.7643
−60	0.0226	0.000 637	0.637 274	0.637 910	−17.469	177.653	160.184	−0.0782	0.8334	0.7552
−55	0.0300	0.000 642	0.490 358	0.491 000	−13.129	175.641	162.512	−0.0581	0.8051	0.7470
−50	0.0391	0.000 648	0.382 457	0.383 105	−8.772	173.611	164.840	−0.0384	0.7779	0.7396
−45	0.0504	0.000 654	0.302 029	0.302 682	−4.396	171.558	167.163	−0.0190	0.7519	0.7329
−40	0.0642	0.000 659	0.241 251	0.241 910	−0.000	169.479	169.479	−0.0000	0.7269	0.7269
−35	0.0807	0.000 666	0.194 732	0.195 398	4.416	167.368	171.784	0.0187	0.7027	0.7214
−30	0.1004	0.000 672	0.158 703	0.159 375	8.854	165.222	174.076	0.0371	0.6795	0.7165
−25	0.1237	0.000 679	0.130 487	0.131 166	13.315	163.037	176.352	0.0552	0.6570	0.7121
−20	0.1508	0.000 686	0.108 162	0.108 847	17.800	160.810	178.610	0.0730	0.6352	0.7082
−15	0.1826	0.000 693	0.090 326	0.091 018	22.312	158.534	180.846	0.0906	0.6141	0.7046
−10	0.2191	0.000 700	0.075 946	0.076 646	26.851	156.207	183.058	0.1079	0.5936	0.7014
−5	0.2610	0.000 708	0.064 255	0.064 963	31.420	153.823	185.243	0.1250	0.5736	0.6986
0	0.3086	0.000 716	0.054 673	0.055 389	36.022	151.376	187.397	0.1418	0.5542	0.6960
5	0.3626	0.000 724	0.046 761	0.047 485	40.659	148.859	189.518	0.1585	0.5351	0.6937
10	0.4233	0.000 733	0.040 180	0.040 914	45.337	146.265	191.602	0.1750	0.5165	0.6916
15	0.4914	0.000 743	0.034 671	0.035 413	50.058	143.586	193.644	0.1914	0.4983	0.6897
20	0.5673	0.000 752	0.030 028	0.030 780	54.828	140.812	195.641	0.2076	0.4803	0.6879
25	0.6516	0.000 763	0.026 091	0.026 854	59.653	137.933	197.586	0.2237	0.4626	0.6863
30	0.7449	0.000 774	0.022 734	0.023 508	64.539	134.936	199.475	0.2397	0.4451	0.6848
35	0.8477	0.000 786	0.019 855	0.020 641	69.494	131.805	201.299	0.2557	0.4277	0.6834
40	0.9607	0.000 798	0.017 373	0.018 171	74.527	128.525	203.051	0.2716	0.4104	0.6820
45	1.0843	0.000 811	0.015 220	0.016 032	79.647	125.074	204.722	0.2875	0.3931	0.6806
50	1.2193	0.000 826	0.013 344	0.014 170	84.868	121.430	206.298	0.3034	0.3758	0.6792
55	1.3663	0.000 841	0.011 701	0.012 542	90.201	117.565	207.766	0.3194	0.3582	0.6777
60	1.5259	0.000 858	0.010 253	0.011 111	95.665	113.443	209.109	0.3355	0.3405	0.6760
65	1.6988	0.000 877	0.008 971	0.009 847	101.279	109.024	210.303	0.3518	0.3224	0.6742
70	1.8858	0.000 897	0.007 828	0.008 725	107.067	104.255	211.321	0.3683	0.3038	0.6721
75	2.0874	0.000 920	0.006 802	0.007 723	113.058	99.068	212.126	0.3851	0.2845	0.6697
80	2.3046	0.000 946	0.005 875	0.006 821	119.291	93.373	212.665	0.4023	0.2644	0.6667
85	2.5380	0.000 976	0.005 029	0.006 005	125.818	87.047	212.865	0.4201	0.2430	0.6631
90	2.7885	0.001 012	0.004 246	0.005 258	132.708	79.907	212.614	0.4385	0.2200	0.6585
95	3.0569	0.001 056	0.003 508	0.004 563	140.068	71.658	211.726	0.4579	0.1946	0.6526
100	3.3440	0.001 113	0.002 790	0.003 903	148.076	61.768	209.843	0.4788	0.1655	0.6444

Table A–5

Properties of Refrigerant R-12: Superheated Vapor

T	P = 0.20 MPa (120.23) v	u	h	s	P = 0.30 MPa (133.55) v	u	h	s	P = 0.40 MPa (143.63) v	u	h	s
1000	2.937	4052.5	4640.0	9.6563	1.9581	4052.3	4639.7	9.4690	1.4685	4052.0	4639.4	9.3360
1100	3.168	4257.0	4890.7	9.8458	2.1121	4256.8	4890.4	9.6585	1.5840	4256.5	4890.2	9.5256
1200	3.399	4467.5	5147.3	10.0262	2.2661	4467.2	5147.1	9.8389	1.6996	4467.0	5146.8	9.7060
1300	3.630	4683.2	5409.3	10.1982	2.4201	4683.0	5409.0	10.0110	1.8151	4682.8	5408.8	9.8780

T	P = 0.50 MPa (151.86) v	u	h	s	P = 0.60 MPa (158.85) v	u	h	s	P = 0.80 MPa (170.43) v	u	h	s
Sat.	0.3749	2561.2	2748.7	6.8213	0.3157	2567.4	2756.8	6.7600	0.2404	2576.8	2769.1	6.6628
200	0.4249	2642.9	2855.4	7.0592	0.3520	2638.9	2850.1	6.9665	0.2608	2630.6	2839.3	6.8158
250	0.4744	2723.5	2960.7	7.2709	0.3938	2720.9	2957.2	7.1816	0.2931	2715.5	2950.0	7.0384
300	0.5226	2802.9	3064.2	7.4599	0.4344	2801.0	3061.6	7.3724	0.3241	2797.2	3056.5	7.2328
350	0.5701	2882.6	3167.7	7.6329	0.4742	2881.2	3165.7	7.5464	0.3544	2878.2	3161.7	7.4089
400	0.6173	2963.2	3271.9	7.7938	0.5137	2962.1	3270.3	7.7079	0.3843	2959.7	3267.1	7.5716
500	0.7109	3128.4	3483.9	8.0873	0.5920	3127.6	3482.8	8.0021	0.4433	3126.0	3480.6	7.8673
600	0.8041	3299.6	3701.7	8.3522	0.6697	3299.1	3700.9	8.2674	0.5018	3297.9	3699.4	8.1333
700	0.8969	3477.5	3925.9	8.5952	0.7472	3477.0	3925.3	8.5107	0.5601	3476.2	3924.2	8.3770
800	0.9896	3662.1	4156.9	8.8211	0.8245	3661.8	4156.5	8.7367	0.6181	3661.1	4155.6	8.6033
900	1.0822	3653.6	4394.7	9.0329	0.9017	3853.4	4394.4	8.9486	0.6761	3852.8	4393.7	8.8153
1000	1.1747	4051.8	4639.1	9.2328	0.9788	4051.5	4638.8	9.1485	0.7340	4051.0	4638.2	9.0153
1100	1.2672	4256.3	4889.9	9.4224	1.0559	4256.1	4889.6	9.3381	0.7919	4255.6	4889.1	9.2050
1200	1.3596	4466.8	5146.6	9.6029	1.1330	4466.5	5146.3	9.5185	0.8497	4466.1	5145.9	9.3855
1300	1.4521	4682.5	5408.6	9.7749	1.2101	4682.3	5408.3	9.6906	0.9076	4681.8	5407.9	9.5575

T	P = 1.00 MPa (179.91) v	u	h	s	P = 1.20 MPa (187.99) v	u	h	s	P = 1.40 MPa (195.07) v	u	h	s
Sat.	0.194 44	2583.6	2778.1	6.5865	0.163 33	2588.8	2784.8	6.5233	0.140 84	2592.8	2790.0	6.4693
200	0.2060	2621.9	2827.9	6.6940	0.169 30	2612.8	2815.9	6.5898	0.143 02	2603.1	2803.3	6.4975
250	0.2327	2709.9	2942.6	6.9247	0.192 34	2704.2	2935.0	6.8294	0.163 50	2698.3	2927.2	6.7467
300	0.2579	2793.2	3051.2	7.1229	0.2138	2789.2	3045.8	7.0317	0.182 28	2785.2	3040.4	6.9534
350	0.2825	2875.2	3157.7	7.3011	0.2345	2872.2	3153.6	7.2121	0.2003	2869.2	3149.5	7.1360
400	0.3066	2957.3	3263.9	7.4651	0.2548	2954.9	3260.7	7.3774	0.2178	2952.5	3257.5	7.3026
500	0.3541	3124.4	3478.5	7.7622	0.2946	3122.8	3476.3	7.6759	0.2521	3121.1	3474.1	7.6027
600	0.4011	3296.8	3697.9	8.0290	0.3339	3295.6	3696.3	7.9435	0.2860	3294.4	3694.8	7.8710
700	0.4478	3475.3	3923.1	8.2731	0.3729	3474.4	3922.0	8.1881	0.3195	3473.6	3920.8	8.1160
800	0.4943	3660.4	4154.7	8.4996	0.4118	3659.7	4153.8	8.4148	0.3528	3659.0	4153.0	8.3431
900	0.5407	3852.2	4392.9	8.7118	0.4505	3851.6	4392.2	8.6272	0.3861	3851.1	4391.5	8.5556
1000	0.5871	4050.5	4637.6	8.9119	0.4892	4050.0	4637.0	8.8274	0.4192	4049.5	4636.4	8.7559
1100	0.6335	4255.1	4888.6	9.1017	0.5278	4254.6	4888.0	9.0172	0.4524	4254.1	4887.5	8.9457
1200	0.6798	4465.6	5145.4	9.2822	0.5665	4465.1	5144.9	9.1977	0.4855	4464.7	5144.4	9.1262
1300	0.7261	4681.3	5407.4	9.4543	0.6051	4680.9	5407.0	9.3698	0.5186	4680.4	5406.5	9.2984

Table A–5

Properties of Refrigerant R-12: Superheated Vapor (Continued)

T	v	u	h	s	v	u	h	s	v	u	h	s
	P = 1.60 MPa (201.41)				**P = 1.80 MPa (207.15)**				**P = 2.00 MPa (212.24)**			
Sat.	0.123 80	2596.0	2794.0	6.4218	0.110 42	2598.4	2797.1	6.3794	0.099 63	2600.3	2799.5	6.3409
225	0.132 87	2644.7	2857.3	6.5518	0.116 73	2636.6	2846.7	6.4808	0.103 77	2628.3	2835.8	6.4147
250	0.141 84	2692.3	2919.2	6.6732	0.124 97	2686.0	2911.0	6.6066	0.111 44	2679.6	2902.5	6.5453
300	0.158 62	2781.1	3034.8	6.8844	0.140 21	2776.9	3029.2	6.8226	0.125 47	2772.6	3023.5	6.7664
350	0.174 56	2866.1	3145.4	7.0694	0.154 57	2863.0	3141.2	7.0100	0.138 57	2859.8	3137.0	6.9563
400	0.190 05	2950.1	3254.2	7.2374	0.168 47	2947.7	3250.9	7.1794	0.151 20	2945.2	3247.6	7.1271
500	0.2203	3119.5	3472.0	7.5390	0.195 50	3117.9	3469.8	7.4825	0.175 68	3116.2	3467.6	7.4317
600	0.2500	3293.3	3693.2	7.8080	0.2220	3292.1	3691.7	7.7523	0.199 60	3290.9	3690.1	7.7024
700	0.2794	3472.7	3919.7	8.0535	0.2482	3471.8	3918.5	7.9983	0.2232	3470.9	3917.4	7.9487
800	0.3086	3658.3	4152.1	8.2808	0.2742	3657.6	4151.2	8.2258	0.2467	3657.0	4150.3	8.1765
900	0.3377	3850.5	4390.8	8.4935	0.3001	3849.9	4390.1	8.4386	0.2700	3849.3	4389.4	8.3895
1000	0.3668	4049.0	4635.8	8.6938	0.3260	4048.5	4635.2	8.6391	0.2933	4048.0	4634.6	8.5901
1100	0.3958	4253.7	4887.0	8.8837	0.3518	4253.2	4886.4	8.8290	0.3166	4252.7	4885.9	8.7800
1200	0.4248	4464.2	5143.9	9.0643	0.3776	4463.7	5143.4	9.0096	0.3398	4463.3	5142.9	8.9607
1300	0.4538	4679.9	5406.0	9.2364	0.4034	4679.5	5405.6	9.1818	0.3631	4679.0	5405.1	9.1329
	P = 2.50 MPa (223.99)				**P = 3.00 MPa (233.90)**				**P = 3.50 MPa (242.60)**			
Sat.	0.079 98	2603.1	2803.1	6.2575	0.066 68	2604.1	2804.2	6.1869	0.057 07	2603.7	2803.4	6.1253
225	0.080 27	2605.6	2806.3	6.2639								
250	0.087 00	2662.6	2880.1	6.4085	0.070 58	2644.0	2855.8	6.2872	0.058 72	2623.7	2829.2	6.1749
300	0.098 90	2761.6	3008.8	6.6438	0.081 14	2750.1	2993.5	6.5390	0.068 42	2738.0	2977.5	6.4461
350	0.109 76	2851.9	3126.3	6.8403	0.090 53	2843.7	3115.3	6.7428	0.076 78	2835.3	3104.0	6.6579
400	0.120 10	2939.1	3239.3	7.0148	0.099 36	2932.8	3230.9	6.9212	0.084 53	2926.4	3222.3	6.8405
450	0.130 14	3025.5	3350.8	7.1746	0.107 87	3020.4	3344.0	7.0834	0.091 96	3015.3	3337.2	7.0052
500	0.139 98	3112.1	3462.1	7.3234	0.116 19	3108.0	3456.5	7.2338	0.099 18	3103.0	3450.9	7.1572
600	0.159 30	3288.0	3686.3	7.5960	0.132 43	3285.0	3682.3	7.5085	0.113 24	3282.1	3678.4	7.4339
	P = 2.50 MPa (223.99)				**P = 3.00 MPa (233.90)**				**P = 3.50 MPa (242.60)**			
700	0.178 32	3468.7	3914.5	7.8435	0.148 38	3466.5	3911.7	7.7571	0.126 99	3464.3	3908.8	7.6837
800	0.197 16	3655.3	4148.2	8.0720	0.164 14	3653.5	4145.9	7.9862	0.140 56	3651.8	4143.7	7.9134
900	0.215 90	3847.9	4387.6	8.2853	0.179 80	3846.5	4385.9	8.1999	0.154 02	3845.0	4384.1	8.1276
1000	0.2346	4046.7	4633.1	8.4861	0.195 41	4045.4	4631.6	8.4009	0.167 43	4044.1	4630.1	8.3288
1100	0.2532	4251.5	4884.6	8.6762	0.210 98	4250.3	4883.3	8.5912	0.180 80	4249.2	4881.9	8.5192
1200	0.2718	4462.1	5141.7	8.8569	0.226 52	4460.9	5140.5	8.7720	0.194 15	4459.8	5139.3	8.7000
1300	0.2905	4677.8	5404.0	9.0291	0.242 06	4676.6	5402.8	8.9442	0.207 49	4675.5	5401.7	8.8723

(*continued*)

Table A–5

Properties of Refrigerant R-12: Superheated Vapor (Continued)

	P = 4.0 MPa (250.40)				P = 4.5 MPa (257.49)				P = 5.0 MPa (263.99)			
T	v	u	h	s	v	u	h	s	v	u	h	s
Sat.	0.049 78	2602.3	2801.4	6.0701	0.044 06	2600.1	2798.3	6.0198	0.039 44	2597.1	2794.3	5.9734
275	0.054 57	2667.9	2886.2	6.2285	0.047 30	2650.3	2863.2	6.1401	0.041 41	2631.3	2838.2	6.0544
300	0.058 84	2725.3	2960.7	6.3615	0.051 35	2712.0	2943.1	6.2828	0.045 32	2698.0	2924.5	6.2084
350	0.066 45	2826.7	3092.5	6.5821	0.058 40	2817.8	3080.6	6.5131	0.051 94	2808.7	3068.4	6.4493
400	0.073 41	2919.9	3213.6	6.7690	0.064 75	2913.3	3204.7	6.7047	0.057 81	2906.6	3195.7	6.6459
450	0.080 02	3010.2	3330.3	6.9363	0.070 74	3005.0	3323.3	6.8746	0.063 30	2999.7	3316.2	6.8186
500	0.086 43	3099.5	3445.3	7.0901	0.076 51	3095.3	3439.6	7.0301	0.068 57	3091.0	3433.8	6.9759
600	0.098 85	3279.1	3674.4	7.3688	0.087 65	3276.0	3670.5	7.3110	0.078 69	3273.0	3666.5	7.2589
700	0.110 95	3462.1	3905.9	7.6198	0.098 47	3459.9	3903.0	7.5631	0.088 49	3457.6	3900.1	7.5122
800	0.122 87	3650.0	4141.5	7.8502	0.109 11	3648.3	4139.3	7.7942	0.098 11	3646.6	4137.1	7.7440
900	0.134 69	3843.6	4382.3	8.0647	0.119 65	3842.2	4380.6	8.0091	0.107 62	3840.7	4378.8	7.9593
1000	0.146 45	4042.9	4628.7	8.2662	0.130 13	4041.6	4627.2	8.2108	0.117 07	4040.4	4625.7	8.1612
1100	0.158 17	4248.0	4880.6	8.4567	0.140 56	4246.8	4879.3	8.4015	0.126 48	4245.6	4878.0	8.3520
1200	0.169 87	4458.6	5138.1	8.6376	0.150 98	4457.5	5136.9	8.5825	0.135 87	4456.3	5135.7	8.5331
1300	0.181 56	4674.3	5400.5	8.8100	0.161 39	4673.1	5399.4	8.7549	0.145 26	4672.0	5398.2	8.7055

	P = 6.0 MPa (275.64)				P = 7.0 MPa (285.88)				P = 8.0 MPa (295.06)			
Sat.	0.032 44	2589.7	2784.3	5.8892	0.027 37	2580.5	2772.1	5.8133	0.023 52	2569.8	2758.0	5.7432
300	0.036 16	2667.2	2884.2	6.0674	0.029 47	2632.2	2838.4	5.9305	0.024 26	2590.9	2785.0	5.7906
350	0.042 23	2789.6	3043.0	6.3335	0.035 24	2769.4	3016.0	6.2283	0.029 95	2747.7	2987.3	6.1301
400	0.047 39	2892.9	3177.2	6.5408	0.039 93	2878.6	3158.1	6.4478	0.034 32	2863.8	3138.3	6.3634
450	0.052 14	2988.9	3301.8	6.7193	0.044 16	2978.0	3287.1	6.6327	0.038 17	2966.7	3272.0	6.5551
500	0.056 65	3082.2	3422.2	6.8803	0.048 14	3073.4	3410.3	6.7975	0.041 75	3064.3	3398.3	6.7240
550	0.061 01	3174.6	3540.6	7.0288	0.051 95	3167.2	3530.9	6.9486	0.045 16	3159.8	3521.0	6.8778
600	0.065 25	3266.9	3658.4	7.1677	0.055 65	3260.7	3650.3	7.0894	0.048 45	3254.4	3642.0	7.0206
700	0.073 52	3453.1	3894.2	7.4234	0.062 83	3448.5	3888.3	7.3476	0.054 81	3443.9	3882.4	7.2812
800	0.081 60	3643.1	4132.7	7.6566	0.069 81	3639.5	4128.2	7.5822	0.060 97	3636.0	4123.8	7.5173
900	0.089 58	3837.8	4375.3	7.8727	0.076 69	3835.0	4371.8	7.7991	0.067 02	3832.1	4368.3	7.7351
1000	0.097 49	4037.8	4622.7	8.0751	0.083 50	4035.3	4619.8	8.0020	0.073 01	4032.8	4616.9	7.9384
1100	0.105 36	4243.3	4875.4	8.2661	0.090 27	4240.9	4872.8	8.1933	0.078 96	4238.6	4870.3	8.1300
1200	0.113 21	4454.0	5133.3	8.4474	0.097 03	4451.7	5130.9	8.3747	0.084 89	4449.5	5128.5	8.3115
1300	0.121 06	4669.6	5396.0	8.6199	0.103 77	4667.3	5393.7	8.5473	0.090 80	4665.0	5391.5	8.4842

Table A–5

Properties of Refrigerant R-12: Superheated Vapor (Continued)

T	v	u	h	s	v	u	h	s	v	u	h	s
	P = 9.0 MPa (303.40)				*P* = 10.0 MPa (311.06)				*P* = 12.5 MPa (327.89)			
Sat.	0.020 48	2557.8	2742.1	5.6772	0.018 026	2544.4	2724.7	5.6141	0.013 495	2505.1	2673.8	5.4624
325	0.023 27	2646.6	2856.0	5.8712	0.019 861	2610.4	2809.1	5.7568				
350	0.025 80	2724.4	2956.6	6.0361	0.022 42	2699.2	2923.4	5.9443	0.016 126	2624.5	2826.2	5.7118
400	0.029 93	2848.4	3117.8	6.2854	0.026 41	2832.4	3096.5	6.2120	0.020 00	2789.3	3039.3	6.0417
450	0.033 50	2955.2	3256.6	6.4844	0.029 75	2943.4	3240.9	6.4190	0.022 99	2912.5	3199.8	6.2719
500	0.036 77	3055.2	3386.1	6.6576	0.032 79	3045.8	3373.7	6.5966	0.025 60	3021.7	3341.8	6.4618
550	0.039 87	3152.2	3511.0	6.8142	0.035 64	3144.6	3500.9	6.7561	0.028 01	3125.0	3475.2	6.6290
600	0.042 85	3243.1	3633.7	6.9589	0.038 37	3241.7	3625.3	6.9029	0.030 29	3225.4	3604.0	6.7810
650	0.045 74	3343.6	3755.3	7.0943	0.041 01	3338.2	3748.2	7.0398	0.032 48	3324.4	3730.4	6.9218
700	0.048 57	3439.3	3876.5	7.2221	0.043 58	3434.7	3870.5	7.1687	0.034 60	3422.9	3855.3	7.0536
800	0.054 09	3632.5	4119.3	7.4596	0.048 59	3628.9	4114.8	7.4077	0.038 69	3620.0	4103.6	7.2965
900	0.059 50	3829.2	4364.8	7.6783	0.053 49	3826.3	4361.2	7.6272	0.042 67	3819.1	4352.5	7.5182
1000	0.064 85	4030.3	4614.0	7.8821	0.058 32	4027.8	4611.0	7.8315	0.046 58	4021.6	4603.8	7.7237
1100	0.070 16	4236.3	4867.7	8.0740	0.063 12	4234.0	4865.1	8.0237	0.050 45	4228.2	4858.8	7.9165
1200	0.075 44	4447.2	5126.2	8.2556	0.067 89	4444.9	5123.8	8.2055	0.054 30	4439.3	5118.0	8.0987
1300	0.080 72	4662.7	5389.2	8.4284	0.072 65	4460 5	5387.0	8.3783	0.058 13	4654.8	5381.4	8.2717
	P = 15.0 MPa (342.24)				*P* = 17.5 MPa (354.75)				*P* = 20.0 MPa (365.81)			
Sat.	0.010 337	2455.5	2610.5	5.3098	0.007 920	2390.2	2528.8	5.1419	0.005 834	2293.0	2409.7	4.9269
350	0.011 470	2520.4	2692.4	5.4421								
400	0.015 649	2740.7	2975.5	5.8811	0.012 447	2685.0	2902.9	5.7213	0.009 942	2619.3	2818.1	5.5540
450	0.018 445	2879.5	3156.2	6.1404	0.015 174	2844.2	3109.7	6.0184	0.012 695	2806.2	3060.1	5.9017
500	0.020 80	2996.6	3308.6	6.3443	0.017 358	2970.3	3274.1	6.2383	0.014 768	2942.9	3238.2	6.1401
550	0.022 93	3104.7	3448.6	6.5199	0.019 288	3083.9	3421.4	6.4230	0.016 555	3062.4	3393.5	6.3348
600	0.024 91	3208.6	3582.3	6.6776	0.021 06	3191.5	3560.1	6.5866	0.018 178	3174.0	3537.6	6.5048
650	0.026 80	3310.3	3712.3	6.8224	0.022 74	3296.0	3693.9	6.7357	0.019 693	3281.4	3675.3	6.6582
700	0.028 61	3410.9	3840.1	6.9572	0.024 34	3398.7	3824.6	6.8736	0.021 13	3386.4	3809.0	6.7993
800	0.032 10	3610.9	4092.4	7.2040	0.027 38	3601.8	4081.1	7.1244	0.023 85	3592.7	4069.7	7.0544
900	0.035 46	3811.9	4343.8	7.4279	0.030 31	3804.7	4335.1	7.3507	0.026 45	3797.5	4326.4	7.2830
1000	0.038 75	4015.4	4596.6	7.6348	0.033 16	4009.3	4589.5	7.5589	0.028 97	4003.1	4582.5	7.4925
1100	0.042 00	4222.6	4852.6	7.8283	0.035 97	4216.9	4846.4	7.7531	0.031 45	4211.3	4840.2	7.6874
1200	0.045 23	4433.8	5112.3	8.0108	0.038 76	4428.3	5106.6	7.9360	0.033 91	4422.8	5101.0	7.8707
1300	0.048 45	4649.1	5376.0	8.1840	0.041 54	4643.5	5370.5	8.1093	0.036 36	4638.0	5365.1	8.0442
	P = 25.0 MPa				*P* = 30.0 MPa				*P* = 35.0 MPa			
375	0.001 973	1798.7	1848.0	4.0320	0.001 789	1737.8	1791.5	3.9305	0.001 700	1702.9	1762.4	3.8722
400	0.006 004	2430.1	2580.2	5.1418	0.002 790	2067.4	2151.1	4.4728	0.002 100	1914.1	1987.6	4.2126
425	0.007 881	2609.2	2806.3	5.4723	0.005 303	2455.1	2614.2	5.1504	0.003 428	2253.4	2373.4	4.7747
450	0.009 162	2720.7	2949.7	5.6744	0.006 735	2619.3	2821.4	5.4424	0.004 961	2498.7	2672.4	5.1962

(*continued*)

Table A–5

Properties of Refrigerant R-12: Superheated Vapor (Continued)

	P = 25.0 MPa				P = 30.0 MPa				P = 35.0 MPa		
500	0.011 123	2884.3	3162.4	5.9592	0.008 678	2820.7	3081.1	5.7905	0.006 927	2751.9	2994.4	5.6282
550	0.012 724	3017.5	3335.6	6.1765	0.010 168	2970.3	3275.4	6.0342	0.008 345	2921.0	3213.0	5.9026
600	0.014 137	3137.9	3491.4	6.3602	0.011 446	3100.5	3443.9	6.2331	0.009 527	3062.0	3395.5	6.1179
650	0.015 433	3251.6	3637.4	3.5229	0.012 596	3221.0	3598.9	6.4058	0.010 575	3189.8	3559.9	6.3010
700	0.016 646	3361.3	3777.5	6.6707	0.013 661	3335.8	3745.6	6.5606	0.011 533	3309.8	3713.5	6.4631
800	0.018 912	3574.3	4047.1	6.9345	0.015 623	3555.5	4024.2	6.8332	0.013 278	3536.7	4001.5	6.7450
900	0.021 045	3783.0	4309.1	7.1680	0.017 448	3768.5	4291.9	7.0718	0.014 883	3754.0	4274.9	6.9886
1000	0.023 10	3990.9	4568.5	7.3802	0.019 196	3978.8	4554.7	7.2867	0.016 410	3966.7	4541.1	7.2064
1100	0.025 12	4200.2	4828.2	7.5765	0.020 903	4189.2	4816.3	7.4845	0.017 895	4178.3	4804.6	7.4057
1200	0.027 11	4412.0	5089.9	7.7605	0.022 589	4401.3	5079.0	7.6692	0.019 360	4390.7	5068.3	7.5910
1300	0.029 10	4626.9	5354.4	7.9342	0.024 266	4616.0	5344.0	7.8432	0.020 815	4605.1	5333.6	7.7653
	P = 40.0 MPa				P = 50.0 MPa				P = 60.0 MPa		
375	0.001 641	1677.1	1742.8	3.8290	0.001 559	1638.6	1716.6	3.7639	0.001 503	1609.4	1699.5	3.7141
400	0.001 908	1854.6	1930.9	4.1135	0.001 730	1788.1	1874.6	4.0031	0.001 634	1745.4	1843.4	3.9318
425	0.002 532	2096.9	2198.1	4.5029	0.002 007	1959.7	2060.0	4.2734	0.001 817	1892.7	2001.7	4.1626
450	0.003 693	2365.1	2512.8	4.9459	0.002 486	2159.6	2284.0	4.5884	0.002 085	2053.9	2179.0	4.4121
500	0.005 622	2678.4	2903.3	5.4700	0.003 892	2525.5	2720.1	5.1726	0.002 956	2390.6	2567.9	4.9321
550	0.006 984	2869.7	3149.1	5.7785	0.005 118	2763.6	3019.5	5.5485	0.003 956	2658.8	2896.2	5.3441
600	0.008 094	3022.6	3346.4	6.0114	0.006 112	2942.0	3247.6	5.8178	0.004 834	2861.1	3151.2	5.6452
650	0.009 063	3158.0	3520.6	6.2054	0.006 966	3093.5	3441.8	6.0342	0.005 595	3028.8	3364.5	5.8829
700	0.009 941	3283.6	3681.2	6.3750	0.007 727	3230.5	3616.8	6.2189	0.006 272	3177.2	3553.5	6.0824
800	0.011 523	3517.8	3978.7	6.6662	0.009 076	3479.8	3933.6	6.5290	0.007 459	3441.5	3889.1	6.4109
900	0.012 962	3739.4	4257.9	6.9150	0.010 283	3710.3	4224.4	6.7882	0.008 508	3681.0	4191.5	6.6805
1000	0.014 324	3954.6	4527.6	7.1356	0.011 411	3930.5	4501.1	7.0146	0.009 480	3906.4	4475.2	6.9127
1100	0.015 642	4167.4	4793.1	7.3364	0.012 496	4145.7	4770.5	7.2184	0.010 409	4124.1	4748.6	7.1195
1200	0.016 940	4380.1	5057.7	7.5224	0.013 561	4359.1	5037.2	7.4058	0.011 317	4338.2	5017.2	7.3083
1300	0.018 229	4594.3	5323.5	7.6969	0.014 616	4572.8	5303.6	7.5808	0.012 215	4551.4	5284.3	7 4837

Table A–5

Properties of Refrigerant R-12: Superheated Vapor (Continued)

Temp., °C	v, m³/kg	h, kJ/kg	s, kJ/ (kg·K)	v, m³/kg	h, kJ/kg	s, kJ/ (kg·K)	v, m³/kg	h, kJ/kg	s, kJ/ (kg·K)
	0.05 MPa			0.10 MPa			0.15 MPa		
−20.0	0.341 857	181.042	0.7912	0.167 701	179.861	0.7401			
−10.0	0.356 227	186.757	0.8133	0.175 222	185.707	0.7628	0.114 716	184.619	0.7318
0.0	0.370 508	192.567	0.8350	0.182 647	191.628	0.7849	0.119 866	190.660	0.7543
10.0	0.384 716	198.471	0.8562	0.189 994	197.628	0.8064	0.124 932	196.762	0.7763
20.0	0.398 863	204.469	0.8770	0.197 277	203.707	0.8275	0.129 930	202.927	0.7977

Table A–5

Properties of Refrigerant R-12: Superheated Vapor (Continued)

Temp., °C	v, m³/kg	h, kJ/kg	s, kJ/(kg·K)	v, m³/kg	h, kJ/kg	s, kJ/(kg·K)	v, m³/kg	h, kJ/kg	s, kJ/(kg·K)
		0.05 MPa			0.10 MPa			0.15 MPa	
30.0	0.412 959	210.557	0.8974	0.204 506	209.866	0.8482	0.134 873	209.160	0.8186
40.0	0.427 012	216.733	0.9175	0.211 691	216.104	0.8684	0.139 768	215.463	0.8390
50.0	0.441 030	222.997	0.9372	0.218 839	222.421	0.8883	0.144 625	221.835	0.8591
60.0	0.455 017	229.344	0.9565	0.225 955	228.815	0.9078	0.149 450	228.277	0.8787
70.0	0.468 978	235.774	0.9755	0.233 044	235.285	0.9269	0.154 247	234.789	0.8980
80.0	0.482 917	242.282	0.9942	0.240 111	241.829	0.9457	0.159 020	241.371	0.9169
90.0	0.496 838	248.868	1.0126	0.247 159	248.446	0.9642	0.163 774	248.020	0.9354
		0.20 MPa			0.25 MPa			0.30 MPa	
0.0	0.088 608	189.669	0.7320	0.069 752	188.644	0.7139	0.057 150	187.583	0.6984
10.0	0.092 550	195.878	0.7543	0.073 024	194.969	0.7366	0.059 984	194.034	0.7216
20.0	0.096 418	202.135	0.7760	0.076 218	201.322	0.7587	0.062 734	200.490	0.7440
30.0	0.100 228	208.446	0.7972	0.079 350	207.715	0.7801	0.065 418	206.969	0.7658
40.0	0.103 989	214.814	0.8178	0.082 431	214.153	0.8010	0.068 049	213.480	0.7869
50.0	0.107 710	221.243	0.8381	0.085 470	220.642	0.8214	0.070 635	220.030	0.8075
60.0	0.111 397	227.735	0.8578	0.088 474	227.185	0.8413	0.073 185	226.627	0.8276
70.0	0.115 055	234.291	0.8772	0.091 449	233.785	0.8608	0.075 705	233.273	0.8473
80.0	0.118 690	240.910	0.8962	0.094 398	240.443	0.8800	0.078 200	239.971	0.8665
90.0	0.122 304	247.593	0.9149	0.097 327	247.160	0.8987	0.080 673	246.723	0.8853
100.0	0.125 901	254.339	0.9332	0.100 238	253.936	0.9171	0.083 127	253.530	0.9038
110.0	0.129 483	261.147	0.9512	0.103 134	260.770	0.9352	0.085 566	260.391	0.9220
		0.40 MPa			0.50 MPa			0.60 MPa	
20.0	0.045 836	198.762	0.7199	0.035 646	196.935	0.6999			
30.0	0.047 971	205.428	0.7423	0.037 464	203.814	0.7230	0.030 422	202.116	0.7063
40.0	0.050 046	212.095	0.7639	0.039 214	210.656	0.7452	0.031 966	209.154	0.7291
50.0	0.052 072	218.779	0.7849	0.040 911	217.484	0.7667	0.033 450	216.141	0.7511
60.0	0.054 059	225.488	0.8054	0.042 565	224.315	0.7875	0.034 887	223.104	0.7723
70.0	0.056 014	232.230	0.8253	0.044 184	231.161	0.8077	0.036 285	230.062	0.7929
80.0	0.057 941	239.012	0.8448	0.045 774	238.031	0.8275	0.037 653	237.027	0.8129
90.0	0.059 846	245.837	0.8638	0.047 340	244.932	0.8467	0.038 995	244.009	0.8324
100.0	0.061 731	252.707	0.8825	0.048 886	251.869	0.8656	0.040 316	251.016	0.8514
110.0	0.063 600	259.624	0.9008	0.050 415	258.845	0.8840	0.041 619	258.053	0.8700
120.0	0.065 455	266.590	0.9187	0.051 929	265.862	0.9021	0.042 907	265.124	0.8882
130.0	0.067 298	273.605	0.9364	0.053 430	272.923	0.9198	0.044 181	272.231	0.9061

(continued)

Table A–5

Properties of Refrigerant R-12: Superheated Vapor (Continued)

Temp., °C	v, m³/kg	h, kJ/kg	s, kJ/(kg·K)	v, m³/kg	h, kJ/kg	s, kJ/(kg·K)	v, m³/kg	h, kJ/kg	s, kJ/(kg·K)
		0.70 MPa			0.80 MPa			0.90 MPa	
40.0	0.026 761	207.580	0.7148	0.022 830	205.924	0.7016	0.019 744	204.170	0.6982
50.0	0.028 100	214.745	0.7373	0.024 068	213.290	0.7248	0.020 912	211.765	0.7131
60.0	0.029 387	221.854	0.7590	0.025 247	220.558	0.7469	0.022 012	219.212	0.7358
70.0	0.030 632	228.931	0.7799	0.026 380	227.766	0.7682	0.023 062	226.564	0.7575
80.0	0.031 843	235.997	0.8002	0.027 477	234.941	0.7888	0.024 072	233.856	0.7785
90.0	0.033 027	243.066	0.8199	0.028 545	242.101	0.8088	0.025 051	241.113	0.7987
100.0	0.034 189	250.146	0.8392	0.029 588	249.260	0.8283	0.026 005	248.355	0.8184
110.0	0.035 332	257.247	0.8579	0.030 612	256.428	0.8472	0.026 937	255.593	0.8376
120.0	0.036 458	264.374	0.8763	0.031 619	263.613	0.8657	0.027 851	262.839	0.8562
130.0	0.037 572	271.531	0.8943	0.032 612	270.820	0.8838	0.028 751	270.100	0.8745
140.0	0.038 673	278.720	0.9119	0.033 592	278.055	0.9016	0.029 639	277.381	0.8923
150.0	0.039 764	285.946	0.9292	0.034 563	285.320	0.9189	0.030 515	284.687	0.9098
		1.00 MPa			1.20 MPa			1.40 MPa	
50.0	0.018 366	210.162	0.7021	0.014 483	206.661	0.6812			
60.0	0.019 410	217.810	0.7254	0.015 463	214.805	0.7060	0.012 579	211.457	0.6876
70.0	0.020 397	225.319	0.7476	0.016 368	222.687	0.7293	0.013 448	219.822	0.7123
80.0	0.021 341	232.739	0.7689	0.017 221	230.398	0.7514	0.014 247	227.891	0.7355
90.0	0.022 251	240.101	0.7895	0.018 032	237.995	0.7727	0.014 997	235.766	0.7575
100.0	0.023 133	247.430	0.8094	0.018 812	245.518	0.7931	0.015 710	243.512	0.7785
110.0	0.023 993	254.743	0.8287	0.019 567	252.993	0.8129	0.016 393	251.170	0.7988
120.0	0.024 835	262.053	0.8475	0.020 301	260.441	0.8320	0.017 053	258.770	0.8183
130.0	0.025 661	269.369	0.8659	0.021 018	267.875	0.8507	0.017 695	266.334	0.8373
140.0	0.026 474	276.699	0.8839	0.021 721	275.307	0.8689	0.018 321	273.877	0.8558
150.0	0.027 275	284.047	0.9015	0.022 412	282.745	0.8867	0.018 934	281.411	0.8738
160.0	0.028 068	291.419	0.9187	0.023 093	290.195	0.9041	0.019 535	288.946	0.8914
		1.60 MPa			1.80 MPa			2.00 MPa	
70.0	0.011 208	216.650	0.6959	0.009 406	213.049	0.6794			
80.0	0.011 984	225.177	0.7204	0.010 187	222.198	0.7057	0.008 704	218.859	0.6909
90.0	0.012 698	233.390	0.7433	0.010 884	230.835	0.7298	0.009 406	228.056	0.7166
100.0	0.013 366	241.397	0.7651	0.011 526	239.155	0.7524	0.010 035	236.760	0.7402
110.0	0.014 000	249.264	0.7859	0.012 126	247.264	0.7739	0.010 615	245.154	0.7624
120.0	0.014 608	257.035	0.8059	0.012 697	255.228	0.7944	0.011 159	253.341	0.7835
130.0	0.015 195	264.742	0.8253	0.013 244	263.094	0.8141	0.011 676	261.384	0.8037
140.0	0.015 765	272.406	0.8440	0.013 772	270.891	0.8332	0.012 172	269.327	0.8232

Table A–5

Properties of Refrigerant R-12: Superheated Vapor (Continued)

Temp., °C	v, m³/kg	h, kJ/kg	s, kJ/ (kg·K)	v, m³/kg	h, kJ/kg	s, kJ/ (kg·K)	v, m³/kg	h, kJ/kg	s, kJ/ (kg·K)
	1.60 MPa			**1.80 MPa**			**2.00 MPa**		
150.0	0.016 320	280.044	0.8623	0.014 284	278.642	0.8518	0.012 651	277.201	0.8420
160.0	0.016 864	287.669	0.8801	0.014 784	286.384	0.8698	0.013 116	285.027	0.8603
170.0	0.017 398	295.290	0.8975	0.015 272	294.069	0.8874	0.013 570	292.822	0.8781
180.0	0.017 923	302.914	0.9145	0.015 752	301.767	0.9046	0.014 013	300.598	0.8955
	2.50 MPa			**3.00 MPa**			**3.50 MPa**		
90.0	0.006 595	219.562	0.6823						
100.0	0.007 264	229.852	0.7103	0.005 231	220.529	0.6770			
110.0	0.007 837	239.271	0.7352	0.005 886	232.068	0.7075	0.004 324	222.121	0.6750
120.0	0.008 351	248.192	0.7582	0.006 419	242.208	0.7336	0.004 959	234.875	0.7078
130.0	0.008 827	256.794	0.7798	0.006 887	251.632	0.7573	0.005 456	245.661	0.7349
140.0	0.009 273	265.180	0.8003	0.007 313	260.620	0.7793	0.005 884	255.524	0.7591
150.0	0.009 697	273.414	0.8200	0.007 709	269.319	0.8001	0.006 270	264.846	0.7814
160.0	0.010 104	281.540	0.8390	0.008 083	277.817	0.8200	0.006 626	273.817	0.8023
170.0	0.010 497	289.589	0.8574	0.008 439	286.171	0.8391	0.006 961	282.545	0.8222
180.0	0.010 879	297.583	0.8752	0.008 782	294.422	0.8575	0.007 279	291.100	0.8413
190.0	0.011 250	305.540	0.8926	0.009 114	302.597	0.8753	0.007 584	299.528	0.8597
200.0	0.011 614	313.472	0.9095	0.009 436	310.718	0.8927	0.007 878	307.864	0.8775
	4.00 MPa								
120.0	0.003 736	224.863	0.6771						
130.0	0.004 325	238.443	0.7111						
140.0	0.004 781	249.703	0.7386						
150.0	0.005 172	259.904	0.7630						
160.0	0.005 522	269.492	0.7854						
170.0	0.005 845	278.684	0.8063						
180.0	0.006 147	287.602	0.8262						
190.0	0.006 434	296.326	0.8453						
200.0	0.006 708	304.906	0.8636						
210.0	0.006 972	313.380	0.8813						
220.0	0.007 228	321.774	0.8985						
230.0	0.007 477	330.108	0.9152						

Table A–6

Saturated Refrigerant-134a: Temperature

Temp. T °C	Press. P_{sat} MPa	Specific volume m³/kg Sat. liquid v_f	Sat. vapor v_g	Internal Energy kJ/kg Sat. liquid u_f	Sat. vapor u_g	Enthalpy kJ/kg Sat. liquid h_f	Evap. h_{fg}	Sat. vapor h_g	Entropy kJ/(kg·k) Sat. liquid s_f	Sat. vapor s_g
−40	0.05164	0.0007055	0.3569	−0.04	204.45	0.00	222.88	222.88	0.0000	0.9560
−36	0.06332	0.0007113	0.2947	4.68	206.73	4.73	220.67	225.40	0.0201	0.9506
−32	0.07704	0.0007172	0.2451	9.47	209.01	9.52	218.37	227.90	0.0401	0.9456
−28	0.09305	0.0007233	0.2052	14.31	211.29	14.37	216.01	230.38	0.0600	0.9411
−26	0.10199	0.0007265	0.1882	16.75	212.43	16.82	214.80	231.62	0.0699	0.9390
−24	0.11160	0.0007296	0.1728	19.21	213.57	19.29	213.57	232.85	0.0798	0.9370
−22	0.12192	0.0007328	0.1590	21.68	214.70	21.77	212.32	234.08	0.0897	0.9351
−20	0.13299	0.0007361	0.1464	24.17	215.84	24.26	211.05	235.31	0.0996	0.9332
−18	0.14483	0.0007395	0.1350	26.67	216.97	26.77	209.76	236.53	0.1094	0.9315
−16	0.15748	0.0007428	0.1247	29.18	218.10	29.30	208.45	237.74	0.1192	0.9298
−12	0.18540	0.0007498	0.1068	34.25	220.36	34.39	205.77	240.15	0.1388	0.9267
−8	0.21704	0.0007569	0.0919	39.38	222.60	39.54	203.00	242.54	0.1583	0.9239
−4	0.25274	0.0007644	0.0794	44.56	224.84	44.75	200.15	244.90	0.1777	0.9213
0	0.29282	0.0007721	0.0689	49.79	227.06	50.02	197.21	247.23	0.1970	0.9190
4	0.33765	0.0007801	0.0600	55.08	229.27	55.35	194.19	249.53	0.2162	0.9169
8	0.38756	0.0007884	0.0525	60.43	231.46	60.73	191.07	251.80	0.2354	0.9150
12	0.44294	0.0007971	0.0460	65.83	233.63	66.18	187.85	254.03	0.2545	0.9132
16	0.50416	0.0008062	0.0405	71.29	235.78	71.69	184.52	256.22	0.2735	0.9116
20	0.57160	0.0008157	0.0358	76.80	237.91	77.26	181.09	258.36	0.2924	0.9102
24	0.64566	0.0008257	0.0317	82.37	240.01	82.90	177.55	260.45	0.3113	0.9089
26	0.68530	0.0008309	0.0298	85.18	241.05	85.75	175.73	261.48	0.3208	0.9082
28	0.72675	0.0008362	0.0281	88.00	242.08	88.61	173.89	262.50	0.3302	0.9076
30	0.77006	0.0008417	0.0265	90.84	243.10	91.49	172.00	263.50	0.3396	0.9070
32	0.81528	0.0008473	0.0250	93.70	244.12	94.39	170.09	264.48	0.3490	0.9064
34	0.86247	0.0008530	0.0236	96.58	245.12	97.31	168.14	265.45	0.3584	0.9058
36	0.91168	0.0008590	0.0223	99.47	246.11	100.25	166.15	266.40	0.3678	0.9053
38	0.96298	0.0008651	0.0210	102.38	247.09	103.21	164.12	267.33	0.3772	0.9047
40	1.0164	0.0008714	0.0199	105.30	248.06	106.19	162.05	268.24	0.3866	0.9041
42	1.0720	0.0008780	0.0188	108.25	249.02	109.19	159.94	269.14	0.3960	0.9035
44	1.1299	0.0008847	0.0177	111.22	249.96	112.22	157.79	270.01	0.4054	0.9030
48	1.2526	0.0008989	0.0159	117.22	251.79	118.35	153.33	271.68	0.4243	0.9017
52	1.3851	0.0009142	0.0142	123.31	253.55	124.58	148.66	273.24	0.4432	0.9004
56	1.5278	0.0009308	0.0127	129.51	255.23	130.93	143.75	274.68	0.4622	0.8990
60	1.6813	0.0009488	0.0114	135.82	256.81	137.42	138.57	275.99	0.4814	0.8973
70	2.1162	0.0010027	0.0086	152.22	260.15	154.34	124.08	278.43	0.5302	0.8918
80	2.6324	0.0010766	0.0064	169.88	262.14	172.71	106.41	279.12	0.5814	0.8827
90	3.2435	0.0011949	0.0046	189.82	261.34	193.69	82.63	276.32	0.6380	0.8655
100	3.9742	0.0015443	0.0027	218.60	248.49	224.74	34.40	259.13	0.7196	0.8117

Table A–7

Saturated Refrigerant-134a: Pressure

Press. P MPa	Temp. T_{sat} °C	Specific volume m³/kg		Internal Energy kJ/kg		Enthalpy kJ/kg			Entropy kJ/(kg · k)	
		Sat. liquid v_f	Sat. vapor v_g	Sat. liquid u_f	Sat. vapor u_g	Sat. liquid h_f	Evap. h_{fg}	Sat. vapor h_g	Sat. liquid s_f	Sat. vapor s_g
0.06	−37.07	0.0007097	0.3100	3.41	206.12	3.46	221.27	224.72	0.0147	0.9520
0.08	−31.21	0.0007184	0.2366	10.41	209.46	10.47	217.92	228.39	0.0440	0.9447
0.10	−26.43	0.0007258	0.1917	16.22	212.18	16.29	215.06	231.35	0.0678	0.9395
0.12	−22.36	0.0007323	0.1614	21.23	214.50	21.32	212.54	233.86	0.0879	0.9354
0.14	−18.80	0.0007381	0.1395	25.66	216.52	25.77	210.27	236.04	0.1055	0.9322
0.16	−15.62	0.0007435	0.1229	29.66	218.32	29.78	208.18	237.97	0.1211	0.9295
0.18	−12.73	0.0007485	0.1098	33.31	219.94	33.45	206.26	239.71	0.1352	0.9273
0.20	−10.09	0.0007532	0.0993	36.69	221.43	36.84	204.46	241.30	0.1481	0.9253
0.24	−5.37	0.0007618	0.0834	42.77	224.07	42.95	201.14	244.09	0.1710	0.9222
0.28	−1.23	0.0007697	0.0719	48.18	226.38	48.39	198.13	246.52	0.1911	0.9197
0.32	2.48	0.0007770	0.0632	53.06	228.43	53.31	195.35	248.66	0.2089	0.9177
0.36	5.84	0.0007839	0.0564	57.54	230.28	57.82	192.76	250.58	0.2251	0.9160
0.4	8.93	0.0007904	0.0509	61.69	231.97	62.00	190.32	252.32	0.2399	0.9145
0.5	15.74	0.0008056	0.0409	70.93	235.64	71.33	184.74	256.07	0.2723	0.9117
0.6	21.58	0.0008196	0.0341	78.99	238.74	79.48	179.71	259.19	0.2999	0.9097
0.7	26.72	0.0008328	0.0292	86.19	241.42	86.78	175.07	261.85	0.3242	0.9080
0.8	31.33	0.0008454	0.0255	92.75	243.78	93.42	170.73	264.15	0.3459	0.9066
0.9	35.53	0.0008576	0.0226	98.79	245.88	99.56	166.62	266.18	0.3656	0.9054
1.0	39.39	0.0008695	0.0202	104.42	247.77	105.29	162.68	267.97	0.3838	0.9043
1.2	46.32	0.0008928	0.0166	114.69	251.03	115.76	155.23	270.99	0.4164	0.9023
1.4	52.43	0.0009159	0.0140	123.98	253.74	125.26	148.14	273.40	0.4453	0.9003
1.6	57.92	0.0009392	0.0121	132.52	256.00	134.02	141.31	275.33	0.4714	0.8982
1.8	62.91	0.0009631	0.0105	140.49	257.88	142.22	134.60	276.83	0.4954	0.8959
2.0	67.49	0.0009878	0.0093	148.02	259.41	149.99	127.95	277.94	0.5178	0.8934
2.5	77.59	0.0010562	0.0069	165.48	261.84	168.12	111.06	279.17	0.5687	0.8854
3.0	86.22	0.0011416	0.0053	181.88	262.16	185.30	92.71	278.01	0.6156	0.8735

Table A–8

Superheated Refrigerant-134a

T	v	u	h	s	v	u	h	s	v	u	h	s
				kJ/				kJ/				kJ/
°C	m³/kg	kJ/kg	kJ/kg	(kg · K)	m³/kg	kJ/kg	kJ/kg	(kg · k)	m³/kg	kJ/kg	kJ/kg	(kg · k)
	P = 0.06 MPa (T_{sat} = −37.07°C)				P = 0.10 MPa (T_{sat} = −26.43°C)				P = 0.14 MPa (T_{sat} = −18.80°C)			
Sat.	0.31003	206.12	224.72	0.9520	0.19170	212.18	231.35	0.9395	0.13945	216.52	236.04	0.9322
−20	0.33536	217.86	237.98	1.0062	0.19770	216.77	236.54	0.9602				
−10	0.34992	224.97	245.96	1.0371	0.20686	224.01	244.70	0.9918	0.14549	223.03	243.40	0.9606
0	0.36433	232.24	254.10	1.0675	0.21587	231.41	252.99	1.0227	0.15219	230.55	251.86	0.9922
10	0.37861	239.69	262.41	1.0973	0.22473	238.96	261.43	1.0531	0.15875	238.21	260.43	1.0230
20	0.39279	247.32	270.89	1.1267	0.23349	246.67	270.02	1.0829	0.16520	246.01	269.13	1.0532
30	0.40688	255.12	279.53	1.1557	0.24216	254.54	278.76	1.1122	0.17155	253.96	277.97	1.0828
40	0.42091	263.10	288.35	1.1844	0.25076	262.58	287.66	1.1411	0.17783	262.06	286.96	1.1120
50	0.43487	271.25	297.34	1.2126	0.25930	270.79	296.72	1.1696	0.18404	270.32	296.09	1.1407
60	0.44879	279.58	306.51	1.2405	0.26779	279.16	305.94	1.1977	0.19020	278.74	305.37	1.1690
70	0.46266	288.08	315.84	1.2681	0.27623	287.70	315.32	1.2254	0.19633	287.32	314.80	1.1969
80	0.47650	296.75	325.34	1.2954	0.28464	296.40	324.87	1.2528	0.20241	296.06	324.39	1.2244
90	0.49031	305.58	335.00	1.3224	0.29302	305.27	334.57	1.2799	0.20846	304.95	334.14	1.2516
100									0.21449	314.01	344.04	1.2785
	P = 0.18 MPa (T_{sat} = −12.73°C)				P = 0.20 MPa (T_{sat} = −10.09°C)				P = 0.24 MPa (T_{sat} = −5.37°C)			
Sat.	0.10983	219.94	239.71	0.9273	0.09933	221.43	241.30	0.9253	0.08343	224.07	244.09	0.9222
−10	0.11135	222.02	242.06	0.9362	0.09938	221.50	241.38	0.9256				
0	0.11678	229.67	250.69	0.9684	0.10438	229.23	250.10	0.9582	0.08574	228.31	248.89	0.9399
10	0.12207	237.44	259.41	0.9998	0.10922	237.05	258.89	0.9898	0.08993	236.26	257.84	0.9721
20	0.12723	245.33	268.23	1.0304	0.11394	244.99	267.78	1.0206	0.09399	244.30	266.85	1.0034
30	0.13230	253.36	277.17	1.0604	0.11856	253.06	276.77	1.0508	0.09794	252.45	275.95	1.0339
40	0.13730	261.53	286.24	1.0898	0.12311	261.26	285.88	1.0804	0.10181	260.72	285.16	1.0637
50	0.14222	269.85	295.45	1.1187	0.12758	269.61	295.12	1.1094	0.10562	269.12	294.47	1.0930
60	0.14710	278.31	304.79	1.1472	0.13201	278.10	304.50	1.1380	0.10937	277.67	303.91	1.1218
70	0.15193	286.93	314.28	1.1753	0.13639	286.74	314.02	1.1661	0.11307	286.35	313.49	1.1501
80	0.15672	295.71	323.92	1.2030	0.14073	295.53	323.68	1.1939	0.11674	295.18	323.19	1.1780
90	0.16148	304.63	333.70	1.2303	0.14504	304.47	333.48	1.2212	0.12037	304.15	333.04	1.2055
100	0.16622	313.72	343.63	1.2573	0.14932	313.57	343.43	1.2483	0.12398	313.27	343.03	1.2326
	P = 0.28 MPa (T_{sat} = −1.23°C)				P = 0.32 MPa (T_{sat} = 2.48°C)				P = 0.40 MPa (T_{sat} = 8.93°C)			
Sat.	0.07193	226.38	246.52	0.9197	0.06322	228.43	248.66	0.9177	0.05089	231.97	252.32	0.9145
0	0.07240	227.37	247.64	0.9238								
10	0.07613	235.44	256.76	0.9566	0.06576	234.61	255.65	0.9427	0.05119	232.87	253.35	0.9182
20	0.07972	243.59	265.91	0.9883	0.06901	242.87	264.95	0.9749	0.05397	241.37	262.96	0.9515
30	0.08320	251.83	275.12	1.0192	0.07214	251.19	274.28	1.0062	0.05662	249.89	272.54	0.8937

Table A–8

Superheated Refrigerant-134a (Continued)

T	v	u	h	s	v	u	h	s	v	u	h	s
				kJ/				kJ/				kJ/
°C	m³/kg	kJ/kg	kJ/kg	(kg · K)	m³/kg	kJ/kg	kJ/kg	(kg · k)	m³/kg	kJ/kg	kJ/kg	(kg · k)
	P = 0.28 MPa (T_sat = −1.23°C)				**P = 0.32 MPa (T_sat = 2.48°C)**				**P = 0.40 MPa (T_sat = 8.93°C)**			
40	0.08660	260.17	284.42	1.0494	0.07518	259.61	283.67	1.0367	0.05917	258.47	282.14	1.0148
50	0.08992	268.64	293.81	1.0789	0.07815	268.14	293.15	1.0665	0.06164	267.13	291.79	1.0452
60	0.09319	277.23	303.32	1.1079	0.08106	276.79	302.72	1.0957	0.06405	275.89	301.51	1.0748
70	0.09641	285.96	312.95	1.1364	0.08392	285.56	312.41	1.1243	0.06641	284.75	311.32	1.1038
80	0.09960	294.82	322.71	1.1644	0.08674	294.46	322.22	1.1525	0.06873	293.73	321.23	1.1322
90	0.10275	303.83	332.60	1.1920	0.08953	303.50	332.15	1.1802	0.07102	302.84	331.25	1.1602
100	0.10587	312.98	342.62	1.2193	0.09229	312.68	342.21	1.2076	0.07327	312.07	341.38	1.1878
110	0.10897	322.27	352.78	1.2461	0.09503	322.00	352.40	1.2345	0.07550	321.44	351.64	1.2149
120	0.11205	331.71	363.08	1.2727	0.09774	331.45	362.73	1.2611	0.07771	330.94	362.03	1.2417
130									0.07991	340.58	372.54	1.2681
140									0.08208	350.35	383.18	1.2941
	P = 0.50 MPa (T_sat = 15.74°C)				**P = 0.60 MPa (T_sat = 21.58°C)**				**P = 0.70 MPa (T_sat = 26.72°C)**			
Sat.	0.04086	235.64	256.07	0.9117	0.03408	238.74	259.19	0.9097	0.02918	241.42	261.85	0.9080
20	0.04188	239.40	260.34	0.9264								
30	0.04416	248.20	270.28	0.9597	0.03581	246.41	267.89	0.9388	0.02979	244.51	265.37	0.9197
40	0.04633	256.99	280.16	0.9918	0.03774	255.45	278.09	0.9719	0.03157	253.83	275.93	0.9539
50	0.04842	265.83	290.04	1.0229	0.03958	264.48	288.23	1.0037	0.03324	263.08	286.35	0.9867
60	0.05043	274.73	299.95	1.0531	0.04134	273.54	298.35	1.0346	0.03482	272.31	296.69	1.0182
70	0.05240	283.72	309.92	1.0825	0.04304	282.66	308.48	1.0645	0.03634	281.57	307.01	1.0487
80	0.05432	292.80	319.96	1.1114	0.04469	291.86	318.67	1.0938	0.03781	290.88	317.35	1.0784
90	0.05620	302.00	330.10	1.1397	0.04631	301.14	328.93	1.1225	0.03924	300.27	327.74	1.1074
100	0.05805	311.31	340.33	1.1675	0.04790	310.53	339.27	1.1505	0.04064	309.74	338.19	1.1358
110	0.05988	320.74	350.68	1.1949	0.04946	320.03	349.70	1.1781	0.04201	319.31	348.71	1.1637
120	0.06168	330.30	361.14	1.2218	0.05099	329.64	360.24	1.2053	0.04335	328.98	359.33	1.1910
130	0.06347	339.98	371.72	1.2484	0.05251	339.38	370.88	1.2320	0.04468	338.76	370.04	1.2179
140	0.06524	349.79	382.42	1.2746	0.05402	349.23	381.64	1.2584	0.04599	348.66	380.86	1.2444
150					0.05550	359.21	392.52	1.2844	0.04729	358.68	391.79	1.2706
160					0.05698	369.32	403.51	1.3100	0.04857	368.82	402.82	1.2963
	P = 0.80 MPa (T_sat = 31.33°C)				**P = 0.90 MPa (T_sat = 35.53°C)**				**P = 1.00 MPa (T_sat = 39.39°C)**			
Sat.	0.02547	243.78	264.15	0.9066	0.02255	245.88	266.18	0.9054	0.02020	247.77	267.97	0.9043
40	0.02691	252.13	273.66	0.9374	0.02325	250.32	271.25	0.9217	0.02029	248.39	268.68	0.9066
50	0.02846	261.62	284.39	0.9711	0.02472	260.09	282.34	0.9566	0.02171	258.48	280.19	0.9428
50	0.02992	271.04	294.98	1.0034	0.02609	269.72	293.21	0.9897	0.02301	268.35	291.36	0.9768

(continued)

Table A–8

Superheated Refrigerant-134a (Continued)

T	v	u	h	s	v	u	h	s	v	u	h	s
				kJ/				kJ/				kJ/
°C	m³/kg	kJ/kg	kJ/kg	(kg · K)	m³/kg	kJ/kg	kJ/kg	(kg · k)	m³/kg	kJ/kg	kJ/kg	(kg · k)
	$P = 0.80$ MPa ($T_{sat} = 31.33°C$)				$P = 0.90$ MPa ($T_{sat} = 35.53°C$)				$P = 1.00$ MPa ($T_{sat} = 39.39°C$)			
70	0.03131	280.45	305.50	1.0345	0.02738	279.30	303.94	1.0214	0.02423	278.11	302.34	1.0093
80	0.03264	289.89	316.00	1.0647	0.02861	288.87	314.62	1.0521	0.02538	287.82	313.20	1.0405
90	0.03393	299.37	326.52	1.0940	0.02980	298.46	325.28	1.0819	0.02649	297.53	324.01	1.0707
100	0.03519	308.93	337.08	1.1227	0.03095	308.11	335.96	1.1109	0.02755	307.27	334.82	1.1000
110	0.03642	318.57	347.71	1.1508	0.03207	317.82	346.68	1.1392	0.02858	317.06	345.65	1.1286
120	0.03762	328.31	358.40	1.1784	0.03316	327.62	357.47	1.1670	0.02959	326.93	356.52	1.1567
130	0.03881	338.14	369.19	1.2055	0.03423	337.52	368.33	1.1943	0.03058	336.88	367.46	1.1841
140	0.03997	348.09	380.07	1.2321	0.03529	347.51	379.27	1.2211	0.03154	346.92	378.46	1.2111
150	0.04113	358.15	391.05	1.2584	0.03633	357.61	390.31	1.2475	0.03250	357.06	389.56	1.2376
160	0.04227	368.32	402.14	1.2843	0.03736	367.82	401.44	1.2735	0.03344	367.31	400.74	1.2638
170	0.04340	378.61	413.33	1.3098	0.03838	378.14	412.68	1.2992	0.03436	377.66	412.02	1.2893
180	0.04452	339.02	424.63	1.3351	0.03939	388.57	424.02	1.3245	0.03528	388.12	423.40	1.3149
	$P = 1.20$ MPa ($T_{sat} = 46.32°C$)				$P = 1.40$ MPa ($T_{sat} = 52.43°C$)				$P = 1.60$ MPa ($T_{sat} = 57.92°C$)			
Sat.	0.01663	251.03	270.99	0.9023	0.01405	253.74	273.40	0.9003	0.01208	256.00	275.33	0.8982
50	0.01712	254.98	275.52	0.9164								
60	0.01835	265.42	287.44	0.9527	0.01495	262.17	283.10	0.9297	0.01233	258.48	278.20	0.9069
70	0.01947	275.59	298.96	0.9868	0.01603	272.87	295.31	0.9658	0.01340	269.89	291.33	0.9457
80	0.02051	285.62	310.24	1.0192	0.01701	283.29	307.10	0.9997	0.01435	280.78	303.74	0.9813
90	0.02150	295.59	321.39	1.0503	0.01792	293.55	318.63	1.0319	0.01521	291.39	315.72	1.0146
100	0.02244	305.54	332.47	1.0804	0.01878	303.73	330.02	1.0628	0.01601	301.84	327.46	1.0467
110	0.02335	315.50	343.52	1.1096	0.01960	313.88	341.32	1.0927	0.01677	312.20	339.04	1.0773
120	0.02423	325.51	354.58	1.1381	0.02039	324.05	352.59	1.1218	0.01750	322.53	350.53	1.1069
130	0.02508	335.58	365.68	1.1660	0.02115	334.25	363.86	1.1501	0.01820	332.87	361.99	1.1357
140	0.02592	345.73	376.83	1.1933	0.02189	344.50	375.15	1.1777	0.01887	343.24	373.44	1.1638
150	0.02674	355.95	388.04	1.2201	0.02262	354.82	386.49	1.2048	0.01953	353.66	384.91	1.1912
160	0.02754	366.27	399.33	1.2465	0.02333	365.22	397.89	1.2315	0.02017	364.15	396.43	1.2181
170	0.02834	376.69	410.70	1.2724	0.02403	375.71	409.36	1.2576	0.02080	374.71	407.99	1.2445
180	0.02912	387.21	422.16	1.2980	0.02472	386.29	420.90	1.2834	0.02142	385.35	419.62	1.2704
190					0.02541	396.96	432.53	1.3088	0.02203	396.08	431.33	1.2960
200					0.02608	407.73	444.24	1.3338	0.02263	406.90	443.11	1.3212

Figure A–1

Generalized Compressibility Chart

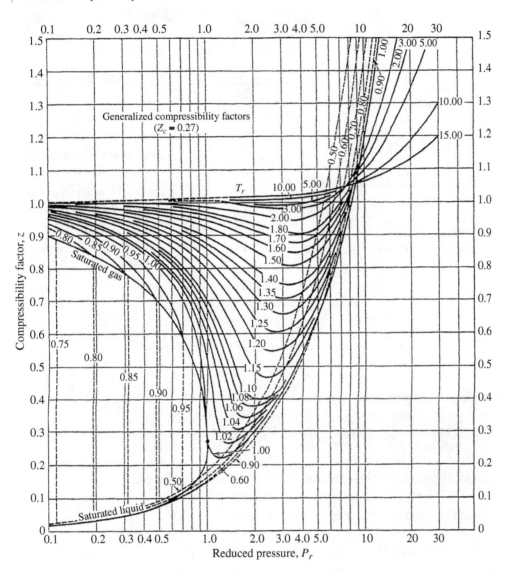

Table A-9

Equations of State

Name of equation	Equation	Constants	Comments
van der Waals	$\left(P + \dfrac{a}{v^2}\right)(v - b) = RT$	$a = \dfrac{27}{64}\dfrac{R^2 T_{CR}^2}{P_{CR}}$ $b = \dfrac{RT_{CR}}{8 P_{CR}}$	Lacks accuracy—mainly of historical interest
Redlich–Kwong	$P = \dfrac{RT}{v - b} - \dfrac{a}{T^{1/2}v(v - b)}$	$a = 0.4278\dfrac{R^2 T_{CR}^{2.5}}{P_{CR}}$ $b = 0.0867\dfrac{RT_{CR}}{P_{CR}}$	Good at high pressures and temperatures near and above the critical magnitude
Beattie–Bridgeman	$P = \dfrac{RT(1 - e)(v + B)}{v^2} - \dfrac{A}{v^2}$	$e = \dfrac{c}{vT^3}$ $A = A_0\left(1 - \dfrac{a}{v}\right)$ $B = B_0\left(1 - \dfrac{b}{v}\right)$	High accuracy but requires that five constants (A_0, B_0, a, b, and c) be determined experimentally for each substance
Virial	$Pv = a + bP + cP^2 + dP^3 + \cdots$	$a, b, c, d, \cdots$ are functions of temperature and can be determined from statistical mechanics for each substance	High accuracy but coefficients, a, b, c, etc. are functions of temperature (not constants)

Table A-10

Constant-Pressure Specific Heats for Several Ideal Gases

Gas		Range, K	Max. error, %
N_2	$\bar{C}_P = 39.060 - 512.79\theta^{-1.5} + 1072.7\theta^{-2} - 820.40\theta^{-3}$	300–3500	0.43
O_2	$\bar{C}_P = 37.432 + 0.020102\theta^{1.5} - 178.57\theta^{-1.5} + 236.88\theta^{-2}$	300–3500	0.30
H_2	$\bar{C}_P = 56.505 - 702.74\theta^{-0.75} + 1165.0\theta^{-1} - 560.70\theta^{-1.5}$	300–3500	0.60
CO	$\bar{C}_P = 69.145 - 0.70463\theta^{0.75} - 200.77\theta^{-0.5} + 176.76\theta^{-0.75}$	300–3500	0.42
OH	$\bar{C}_P = 81.546 - 59.350\theta^{0.25} + 17.329\theta^{0.75} - 4.2660\theta$	300–3500	0.43
NO	$\bar{C}_P = 59.283 - 1.7096\theta^{0.5} - 70.613\theta^{-0.5} + 74.889\theta^{-1.5}$	300–3500	0.34
H_2O	$\bar{C}_P = 143.05 - 183.54\theta^{0.25} + 82.751\theta^{0.5} - 3.6989\theta$	300–3500	0.43
CO_2	$\bar{C}_P = -3.7357 + 30.529\theta^{0.5} - 4.1034\theta + 0.024198\theta^2$	300–3500	0.19
NO_2	$\bar{C}_P = 46.045 + 216.10\theta^{-0.5} - 363.66\theta^{-0.75} + 232.550\theta^{-2}$	300–3500	0.26
CH_4	$\bar{C}_P = -672.87 + 439.74\theta^{0.25} - 24.875\theta^{0.75} + 323.88\theta^{-0.5}$	300–2000	0.15
C_2H_4	$\bar{C}_P = -95.395 + 123.15\theta^{0.5} - 35.641\theta^{0.75} + 182.77\theta^{-3}$	300–2000	0.07
C_2H_6	$\bar{C}_P = 6.895 + 17.26\theta - 0.6402\theta^2 + 0.00728\theta^3$	300–1500	0.83
C_3H_8	$\bar{C}_P = -4.042 + 30.46\theta - 1.571\theta^2 + 0.03171\theta^3$	300–1500	0.40
C_4H_{10}	$\bar{C}_P = 3.954 + 37.12\theta - 1.833\theta^2 + 0.03498\theta^3$	300–1500	0.54

Table A–11

Properties of Several Ideal Gases

Gas	Chemical formula	Molecular weight	R kJ/(kg·K)	c_p^* kJ/(kg · K)	c_v kJ/(kg · K)	y	Critical temperature, T_{CR} K	Critical pressure P_{CR} MPa
Air	—	28.97	0.287 00	1.0035	0.7165	1.400	—	—
Argon	Ar	39.948	0.208 13	0.5203	0.3122	1.667	151	4.86
Butane	C_4H_{10}	58.124	0.143 04	1.7164	1.5734	1.091	425.2	3.80
Carbon dioxide	CO_2	44.01	0.188 92	0.8418	0.6529	1.289	304.2	7.39
Carbon monoxide	CO	28.01	0.296 83	1.0413	0.7445	1.400	133	3.50
Ethane	C_2H_8	30.07	0.276 50	1.7662	1.4897	1.186	305.5	4.88
Ethylene	C_2H_4	28.054	0.296 37	1.5482	1.2518	1.237	282.4	5.12
Helium	He	4.003	2.077 03	5.1926	3.1156	1.667	5.3	0.23
Hydrogen	H_2	2.016	4.124 18	14.2091	10.0849	1.409	33.3	1.30
Methane	CH_4	16.04	0.518 35	2.2537	1.7354	1.299	191.1	4.64
Neon	Ne	20.183	0.411 95	1.0299	0.6179	1.667	44.5	2.73
Nitrogen	N_2	28.013	0.296 80	1.0416	0.7448	1.400	126.2	3.39
Octane	C_8H_{18}	114.23	0.072 79	1.7113	1.6385	1.044	—	—
Oxygen	O_2	31.999	0.259 83	0.9216	0.6618	1.393	154.8	5.08
Propane	C_3H_8	44.097	0.188 55	1.6794	1.4909	1.126	370	4.26

Table A–12

Thermophysical Properties of Air

T °C	c_p kJ/(kg·°C)	ρ kg/m³	$\mu \times 10^6$ kg/(m·s)	$\nu \times 10^6$ m²/s	$k \times 10^3$ W/(m·°C)	Pr
−50	1.0064	1.5819	14.63	9.25	20.04	0.735
−40	1.0060	1.5141	15.17	10.02	20.86	0.731
−30	1.0058	1.4518	15.69	10.81	21.68	0.728
−20	1.0057	1.3944	16.20	11.62	22.49	0.724
−10	1.0056	1.3414	16.71	12.46	23.29	0.721
0	1.0057	1.2923	17.20	13.31	24.08	0.718
10	1.0058	1.2467	17.69	14.19	24.87	0.716
20	1.0061	1.2042	18.17	15.09	25.64	0.713
30	1.0064	1.1644	18.65	16.01	26.38	0.712
40	1.0068	1.1273	19.11	16.96	27.10	0.710
50	1.0074	1.0924	19.57	17.92	27.81	0.709
60	1.0080	1.0596	20.03	18.90	28.52	0.708
70	1.0087	1.0287	20.47	19.90	29.22	0.707
80	1.0095	0.9996	20.92	20.92	29.91	0.706
90	1.0103	0.9721	21.35	21.96	30.59	0.705
100	1.0113	0.9460	21.78	23.02	31.27	0.704
110	1.0123	0.9213	22.20	24.10	31.94	0.704
120	1.0134	0.8979	22.62	25.19	32.61	0.703
130	1.0146	0.8756	23.03	26.31	33.28	0.702
140	1.0159	0.8544	23.44	27.44	33.94	0.702
150	1.0172	0.8342	23.84	28.58	34.59	0.701
160	1.0186	0.8150	24.24	29.75	35.25	0.701
170	1.0201	0.7966	24.63	30.93	35.89	0.700
180	1.0217	0.7790	25.03	32.13	36.54	0.700
190	1.0233	0.7622	25.41	33.34	37.18	0.699
200	1.0250	0.7461	25.79	34.57	37.81	0.699
210	1.0268	0.7306	26.17	35.82	38.45	0.699
220	1.0286	0.7158	26.54	37.08	39.08	0.699
230	1.0305	0.7016	26.91	38.36	39.71	0.698
240	1.0324	0.6879	27.27	39.65	40.33	0.698
250	1.0344	0.6748	27.64	40.96	40.95	0.698
260	1.0365	0.6621	27.99	42.28	41.57	0.698
270	1.0386	0.6499	28.35	43.62	42.18	0.698
280	1.0407	0.6382	28.70	44.97	42.79	0.698
290	1.0429	0.6268	29.05	46.34	43.40	0.698

Table A–12

Thermophysical Properties of Air (Continued)

T °C	c_p kJ/(kg·°C)	ρ kg/m³	$\mu \times 10^6$ kg/(m·s)	$v \times 10^6$ m²/s	$k \times 10^3$ W/(m·°C)	Pr
300	1.0452	0.6159	29.39	47.72	44.01	0.698
310	1.0475	0.6053	29.73	49.12	44.61	0.698
320	1.0499	0.5951	30.07	50.53	45.21	0.698
330	1.0523	0.5853	30.41	51.95	45.84	0.698
340	1.0544	0.5757	30.74	53.39	46.38	0.699
350	1.0568	0.5665	31.07	54.85	46.92	0.700
360	1.0591	0.5575	31.40	56.31	47.47	0.701
370	1.0615	0.5489	31.72	57.79	48.02	0.701
380	1.0639	0.5405	32.04	59.29	48.58	0.702
390	1.0662	0.5323	32.36	60.79	49.15	0.702
400	1.0686	0.5244	32.68	62.31	49.72	0.702
410	1.0710	0.5167	32.99	63.85	50.29	0.703
420	1.0734	0.5093	33.30	65.39	50.86	0.703
430	1.0758	0.5020	33.61	66.95	51.44	0.703
440	1.0782	0.4950	33.92	68.52	52.01	0.703
450	1.0806	0.4882	34.22	70.13	52.59	0.703
460	1.0830	0.4815	34.52	71.70	53.16	0.703
470	1.0854	0.4750	34.82	73.31	53.73	0.703
480	1.0878	0.4687	35.12	74.93	54.31	0.704
490	1.0902	0.4626	35.42	76.57	54.87	0.704
500	1.0926	0.4566	35.71	78.22	55.44	0.704
510	1.0949	0.4508	36.00	79.87	56.01	0.704
520	1.0973	0.4451	36.29	81.54	56.57	0.704
530	1.0996	0.4395	36.58	83.23	57.13	0.704
540	1.1020	0.4341	36.87	84.92	57.68	0.704
550	1.1043	0.4288	37.15	86.63	58.24	0.704
560	1.1066	0.4237	37.43	88.35	58.79	0.705
570	1.1088	0.4187	37.71	90.07	59.33	0.705
580	1.1111	0.4138	37.99	91.82	59.87	0.705
590	1.1133	0.4090	38.27	93.57	60.41	0.705
600	1.1155	0.4043	38.54	95.33	60.94	0.705
610	1.1177	0.3997	38.81	97.11	61.47	0.706
620	1.1198	0.3952	39.09	98.89	62.00	0.706
630	1.1219	0.3908	39.36	100.69	62.52	0.706
640	1.1240	0.3866	39.62	102.50	63.03	0.707
650	1.1260	0.3824	39.89	104.32	63.55	0.707

Table A–13

Thermophysical Properties of Saturated Water (Liquid)

T °C	c_p kJ/(kg·°C)	ρ kg/m·s³	$\mu \times 10^3$ kg/(m·s)	$v \times 10^6$ m²/s	k W/(m·°C)	$\alpha \times 10^7$ M²/s	$\beta \times 10^3$ 1/°K	Pr
0	4.218	999.8	1.791	1.792	0.5619	1.332	−0.0853	13.45
5	4.203	1000.0	1.520	1.520	0.5723	1.362	0.0052	11.16
10	4.193	999.8	1.308	1.308	0.5820	1.389	0.0821	9.42
15	4.187	999.2	1.139	1.140	0.5911	1.413	0.148	8.07
20	4.182	998.3	1.003	1.004	0.5996	1.436	0.207	6.99
25	4.180	997.1	0.8908	0.8933	0.6076	1.458	0.259	6.13
30	4.180	995.7	0.7978	0.8012	0.6150	1.478	0.306	5.42
35	4.179	994.1	0.7196	0.7238	0.6221	1.497	0.349	4.83
40	4.179	992.3	0.6531	0.6582	0.6286	1.516	0.389	4.34
45	4.182	990.2	0.5962	0.6021	0.6347	1.533	0.427	3.93
50	4.182	998.0	0.5471	0.5537	0.6405	1.550	0.462	3.57
55	4.184	985.7	0.5043	0.5116	0.6458	1.566	0.496	3.27
60	4.186	983.1	0.4668	0.4748	0.6507	1.581	0.529	3.00
65	4.187	980.5	0.4338	0.4424	0.6553	1.596	0.560	2.77
70	4.191	977.7	0.4044	0.4137	0.6594	1.609	0.590	2.57
75	4.191	974.7	0.3783	0.3881	0.6633	1.624	0.619	2.39
80	4.195	971.6	0.3550	0.3653	0.6668	1.636	0.647	2.23
85	4.201	968.4	0.3339	0.3448	0.6699	1.647	0.675	2.09
90	4.203	965.1	0.3150	0.3264	0.6727	1.659	0.702	1.97
95	4.210	961.7	0.2978	0.3097	0.6753	1.668	0.728	1.86
100	4.215	958.1	0.2822	0.2945	0.6775	1.677	0.755	1.76
120	4.246	942.8	0.2321	0.2461	0.6833	1.707	0.859	1.44
140	4.282	925.9	0.1961	0.2118	0.6845	1.727	0.966	1.23
160	4.339	907.3	0.1695	0.1869	0.6815	1.731	1.084	1.08
180	4.411	886.9	0.1494	0.1684	0.6745	1.724	1.216	0.98
200	4.498	864.7	0.1336	0.1545	0.6634	1.706	1.372	0.91
220	4.608	840.4	0.1210	0.1439	0.6483	1.674	1.563	0.86
240	4.770	813.6	0.1105	0.1358	0.6292	1.622	1.806	0.84
260	4.991	783.9	0.1015	0.1295	0.6059	1.549	2.130	0.84
280	5.294	750.5	0.0934	0.1245	0.5780	1.455	2.589	0.86
300	5.758	712.2	0.0858	0.1205	0.5450	1.329	3.293	0.91
320	6.566	666.9	0.0783	0.1174	0.5063	1.156	4.511	1.02
340	8.234	610.2	0.0702	0.1151	0.4611	0.918	7.170	1.25
360	16.138	526.2	0.0600	0.1139	0.4115	0.485	21.28	2.35

Table A–14

Thermophysical Properties of Oil

T K	ρ kg/m³	c_p kJ/(kg·°C)	$\mu \times 10^2$ N·s/m²	$v \times 10^6$ m²/s	$k \times 10^3$ W/(m·°C)	$\alpha \times 10^7$ m²/s	Pr	$\beta \times 10^3$ K⁻¹
Engine oil (unused)								
273	899.1	1.796	385	4,280	147	0.910	47,000	0.70
280	895.3	1.827	217	2,430	144	0.880	27,500	0.70
290	890.0	1.868	99.9	1,120	145	0.872	12,900	0.70
300	884.1	1.909	48.6	550	145	0.859	6,400	0.70
310	877.9	1.951	25.3	288	145	0.847	3,400	0.70
320	871.8	1.993	14.1	161	143	0.823	1,965	0.70
330	865.8	2.035	8.36	96.6	141	0.800	1,205	0.70
340	859.9	2.076	5.31	61.7	139	0.779	793	0.70
350	853.9	2.118	3.56	41.7	138	0.763	546	0.70
360	847.8	2.161	2.52	29.7	138	0.753	395	0.70
370	841.8	2.206	1.86	22.0	137	0.738	300	0.70
380	836.0	2.250	1.41	16.9	136	0.723	233	0.70
390	830.6	2.294	1.10	13.3	135	0.709	187	0.70
400	825.1	2.337	0.874	10.6	134	0.695	152	0.70
410	818.9	2.381	0.698	8.52	133	0.682	125	0.70
420	812.1	2.427	0.564	6.94	133	0.675	103	0.70
430	806.5	2.471	0.470	5.83	132	0.662	88	0.70

Table A–15

Physical Properties of Common Liquids

Liquid	Specific gravity (–)
Benzene	0.879
Carbon tetrachloride	1.595
Castor oil	0.969
Gasoline	0.72
Glycerine	1.26
Heptane	0.684
Kerosene	0.82
Lubricating oil	0.88
Mercury	13.55
Octane	0.702
Sea water	1.025
Fresh water	1.000

Figure A–2

Dynamic Viscosity of Common Fluids as a Function of Temperature

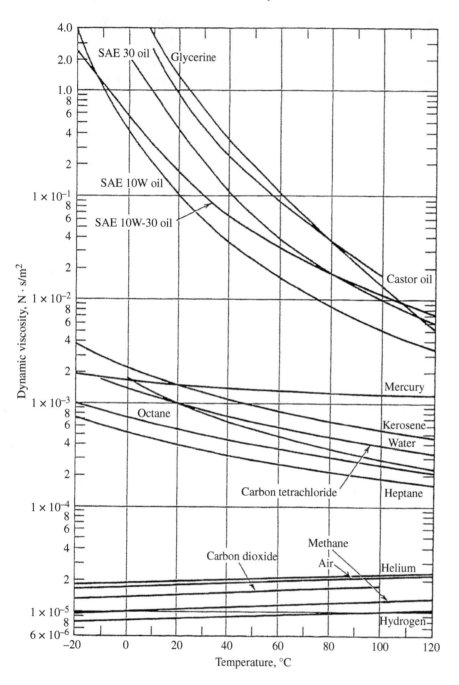

Figure A–3

Kinematic Viscosity of Common Fluids as a Function of Temperature

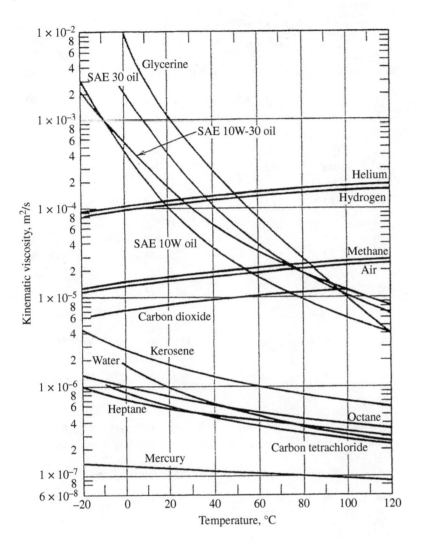

Table A–16

Thermophysical Properties of Selected Metallic Solids

| Composition | Melting point, K | ρ, kg/m³ | Properties at 300 K | | | Properties at various temperatures, K | | | | | | | |
| | | | c_p, J/(kg·°C) | k, W/(m·°C) | $\alpha \times 10^6$, m²/s | k, W/(m·°C) | | | | c_p, J/(kg·°C) | | | |
						100	200	400	600	100	200	400	600
Aluminum													
Pure	933	2,702	903	237	97.1	302	237	240	231	482	796	949	1,033
Alloy 2024–T6 (4.5% Cu, 1.5% Mg, 0.6% Mn)	775	2,770	875	177	73.0	65	163	186	186	473	787	925	1,042
Alloy 195, cast (4.5% Cu)	—	2,790	883	168	68.2	—	—	174	185	—	—	—	—
Chromium	2,118	7,160	449	93.7	29.1	159	111	90.9	80.7	192	384	484	542
Copper													
Pure	1,358	8,933	385	401	117	482	413	393	379	252	356	397	417
Commercial bronze (90% Cu, 10% Al)	1,293	8,800	420	52	14	—	42	52	59	—	785	460	545
Phosphor gear bronze (89% Cu, 11% Sn)	1,104	8,780	355	54	17	—	41	65	74	—	—	—	—
Cartridge brass (70% Cu, 30% Zn)	1,188	8,530	380	110	33.9	75	95	137	149	—	360	395	425
Constantan (55% Cu, 45% Ni)	1,493	8,920	384	23	6.71	17	19	—	—	237	362	—	—
Iron													
Pure	1,810	7,870	447	80.2	23.1	134	94.0	69.5	54.7	216	384	490	574
Armco (99.75% pure)	—	7,870	447	72.7	20.7	95.6	80.6	65.7	53.1	215	384	490	574
Carbon steels													
Plain carbon (Mn ≤ 1%, Si ≤ 0.1%)	—	7,854	434	60.5	17.7	—	—	56.7	48.0	—	—	487	559
AISI 1010	—	7,832	434	63.9	18.8	—	—	58.7	48.8	—	—	487	559
Carbon-silicon (Mn≤1%, 0.1% < Si≤ 0.6%)	—	7,817	446	51.9	14.9	—	—	49.8	44.0	—	—	501	582
Carbon—manganese—silicon (1%< Mn≤ 1.65%, 0.1% < Si≤ 0.6%)	—	8,131	434	41.0	11.6	—	—	42.2	39.7	—	—	487	559

Table A–16

Thermophysical Properties of Selected Metallic Solids (Continued)

Composition	Melting point, K	ρ, kg/m³	c_p, J/(kg·°C)	k, W/(m·°C)	$\alpha \times 10^6$, m²/s	k, W/(m·°C) 100	200	400	600	c_p, J/(kg·°C) 100	200	400	600
Chromium (low) steels													
½ Cr–¼ Mo–Si (0.18% C, 0.65% Cr), 0.23% Mo, 0.6% Si)	—	7,822	444	37.7	10.9	—	—	38.2	36.7	—	—	492	575
1 Cr–½ Mo (0.16% C, 1% Cr, 0.54% Mo, 0.39% Si)	—	7,858	442	42.3	12.2	—	—	42.0	39.1	—	—	492	575
1 Cr–V (0.2% C, 1.02% Cr, 0.15% V)	—	7,836	443	48.9	14.1	—	—	46.8	42.1	—	—	492	575
Stainless steels													
AISI 302	—	8,055	480	15.1	3.91	—	—	17.3	20.0	—	—	512	559
AISI 304	1,670	7,900	477	14.9	3.95	9.2	12.6	16.6	19.8	272	402	515	557
AISI 316	—	8,238	468	13.4	3.48	—	—	15.2	18.3	—	—	504	550
AISI 347	—	7,978	480	14.2	3.71	—	—	15.8	18.9	—	—	513	559
Lead	601	11,340	129	35.3	24.1	39.7	36.7	34.0	31.4	118	125	132	142
Magnesium	923	1,740	1,024	156	87.6	169	159	153	149	649	934	1,074	1,170
Molybdenum	2,894	10,240	251	138	53.7	179	143	134	126	141	224	261	275
Nickel													
Pure	1,728	8,900	444	90.7	23.0	164	107	80.2	65.6	232	383	485	592
Nichrome (80% Ni, 20% Cr)	1,672	8,400	420	12	3.4	—	—	14	16	—	—	480	525
Inconel X–750 (73% Ni, 15% Cr, 6.7% Fe)	1,665	8,510	439	11.7	3.1	8.7	10.3	13.5	17.0	—	372	473	510
Platinum													
Pure	2045	21450	133	71.6	25.1	77.5	72.6	71.8	73.2	100	125	136	141
Alloy 60 Pt–40 Rh (60% Pt, 40% Rh)	1800	16630	162	47	17.4	—	—	52	59	—	—	—	—
Silicon	1685	2330	712	148	89.2	884	264	98.9	61.9	259	556	790	867
Silver	1235	10500	235	429	174	444	430	425	412	187	225	239	250
Tin	505	7310	227	66.6	40.1	85.2	73.3	62.2	—	188	215	243	—
Titanium	1953	4500	522	21.9	9.32	30.5	24.5	20.4	19.4	300	465	551	591
Tungsten	3660	19300	132	174	68.3	208	186	159	137	87	122	137	142
Uranium	1406	19070	116	27.6	12.5	21.7	25.1	29.6	34.0	94	108	125	146
Zinc	693	7140	389	116	41.8	117	118	111	103	297	367	402	436

Table A–17

Thermophysical Properties of Common Materials

Description/composition	Temperature K	Density ρ, kg/m³	Thermal conductivity k, W/(m·°C).	Specific heat c_p, J/(Kg·°C)
Asphalt	300	2115	0.062	920
Bakelite	300	1300	1.4	1465
Brick, refractory				
Carborundum	872	—	18.5	—
	1672	—	11.0	—
Chrome brick	473	3010	2.3	835
	823	—	2.5	—
	1173	—	2.0	—
Diatomaceous silica, fired	478	—	0.25	—
	1145	—	0.30	—
Fire clay, burnt 1600 K	773	2050	1.0	960
	1073	—	1.1	—
	1373	—	1.1	—
Fire clay, burnt 1725 K	773	2325	1.3	960
	1073	—	1.4	—
	1373	—	1.4	—
Fire clay brick	478	2645	1.0	960
	922	—	1.5	—
	1478	—	1.8	—
Magnesite	478	—	3.8	1130
	922	—	2.8	—
	1478	—	1.9	—
Clay	300	1460	1.3	880
Coal, anthracite	300	1350	0.26	1260
Concrete (stone mix)	300	2300	1.4	880
Cotton	300	80	0.06	1300
Foodstuffs				
Banana (75.7% water content)	300	980	0.481	3350
Apple, red (75% water content)	300	840	0.513	3600
Cake, batter	300	720	0.223	—
Cake, fully baked	300	280	0.121	—
Chicken meat, white (74.4% water content)	233	—	1.49	—
	273	—	0.476	—
	293	—	0.489	—

Table A–17

Thermophysical Properties of Common Materials (Continued)

Description/composition	Temperature K	Density ρ, kg/m^3	Thermal conductivity k, W/(m·°C).	Specific heat c_p, J/(Kg·°C)
Glass				
Plate (soda lime)	300	2500	1.4	750
Pyrex	300	2225	1.4	835
Ice	273	920	0.188	2040
	253	—	0.203	1945
Leather (sole)	300	998	0.013	—
Paper	300	930	0.011	1340
Paraffin	300	900	0.020	2890
Rock				
Granite, Barre	300	2630	2.79	775
Limestone, Salem	300	2320	21.5	810
Marble, Halston	300	2680	2.80	830
Quartzite, Sioux	300	2640	5.38	1105
Sandstone, Berea	300	2150	2.90	745
Rubber, vulcanized				
Soft	300	1100	0.012	2010
Hard	300	1190	0.013	—
Sand	300	1515	0.027	800
Soil	300	2050	0.52	1840
Snow	273	110	0.049	—
	273	500	0.190	—
Teflon	300	2200	0.35	—
	400	—	0.45	—
Tissue, human				
Skin	300	—	0.37	—
Fat layer (adipose)	300	—	0.2	—
Muscle	300	—	0.41	—
Wood, cross grain				
Balsa	300	140	0.055	—
Cypress	300	465	0.097	—
Fir	300	415	0.11	2720
Oak	300	545	0.17	2385
Yellow pine	300	640	0.15	2805
White pine	300	435	0.11	—
Wood, radial				
Oak	300	545	0.19	2385
Fir	300	420	0.14	2720

Table A–18

Table A–18

Thermophysical Properties of Structural Building Materials

Description/composition	Density ρ, kg/m³	Typical properties at 300 K Thermal conductivity k, W/(m·°C)	Specific heat c_p, J/(kg·°C)
Blanket and batt			
Glass fiber, paper faced	16	0.046	—
	28	0.038	—
	40	0.035	—
Glass fiber, coated, duct liner	32	0.038	835
Board and slab			
Cellular glass	145	0.058	1000
Glass fiber, organic bonded	105	0.036	795
Polystrene, expanded			
Extruded (R-12)	55	0.027	1210
Molded beads	16	0.040	1210
Mineral fiberboard, roofing material	265	0.049	—
Wood, shredded/cemented	350	0.087	1590
Cork	120	0.039	1800
Loose fill			
Cork, granulated	160	0.045	—
Diatomaceous silica, coarse powder	350	0.069	—
	400	0.091	—
Diatomaceous silica, fine powder	200	0.052	—
	275	0.061	—
Glass fiber, poured or blown	16	0.043	835
Vermiculite, flakes	80	0.068	835
	160	0.063	1000
Formed/foamed-in-place			
Mineral wood granules with asbestos/inorganic binders, sprayed	190	0.046	—
Polyvinyl acetate cork mastic, sprayed or troweled	—	0.100	—
Urethane, two-part mixture, rigid foam	70	0.026	1045
Reflective			
Aluminium foil separating fluffy glass mats, 10–12 layers, evacuated for cryogenic application (150 °C)	40	0.00016	—
Aluminum foil and glass paper laminate, 75–150 layers, evacuated, for cryogenic application (150 °C)	120	0.000017	—
Typical silica powder, evacuated	160	0.0017	—

Table A–18

Thermophysical Properties of Structural Building Materials (Continued)

| | Typical properties at 300 K | | |
| | Density ρ, | Thermal conductivity k, | Specific heat c_p, |
Description/composition	kg/m³	W/(m·°C)	J/(kg·°C)
Building boards			
Asbestos–cement board	1920	0.58	—
Gypsum or plaster board	800	0.17	—
Plywood	545	0.12	1215
Sheathing, regular density	290	0.055	1300
Acoustic tile	290	0.058	1340
Hardboard, siding	640	0.094	1170
Hardboard, high-density	1010	0.15	1380
Particle board, low-density	590	0.078	1300
Particle board, high-density	1000	0.170	1300
Woods			
Hardwoods (oak, maple)	720	0.16	1255
Softwards (fir, pine)	510	0.12	1380
Masonry materials			
Cement mortar	1860	0.72	780
Brick, common	1920	0.72	835
Brick, face	2083	1.3	—
Clay tile, hollow			
one cell deep, 10 cm thick	—	0.52	—
three cell deep, 30 cm thick	—	0.69	—
Concrete block, three oval cores			
sand/gravel, 20 cm thick	—	1.0	—
cinder aggregate, 20 cm thick	—	0.67	—
Concrete block, rectangular core			
two-core, 20 cm thick, 16 kg	—	1.1	—
same with filled cores	—	0.60	—
Plastering materials			
Cement plaster, sand aggregate	1860	0.72	—
Gypsum plaster, sand aggregate	1860	0.22	1085
Gypsum plaster, vermiculite aggregate	720	0.25	—

Table A–19

Thermophysical Properties of Industrial Insulation

Description/ Composition	Max. service temp., °C	Typical density, ρ kg/m³	200	215	230	240	255	270	285	300	310	365	420	530	645	750
Blankets																
Blanket, mineral fiber, metal	920	96–192								0.038	0.046	0.056	0.078			
reinforced	815	40–96								0.035	0.045	0.058	0.088			
Blanket, mineral fiber, glass, fine																
fiber,	450	10			0.036	0.038	0.040	0.043	0.048	0.052	0.076					
organic bonded		12			0.035	0.036	0.039	0.042	0.046	0.049	0.069					
		16			0.033	0.035	0.036	0.039	0.042	0.046	0.062					
		24			0.030	0.032	0.033	0.036	0.039	0.040	0.053					
		32			0.029	0.030	0.032	0.033	0.036	0.038	0.048					
		48			0.027	0.029	0.030	0.032	0.033	0.035	0.045					
Blanket, alumina-silica fiber	1530	48												0.071	0.105	0.150
		64												0.059	0.087	0.125
		96												0.052	0.076	0.100
		128												0.049	0.068	0.091
Felt, semirigid,	480	50–125						0.035	0.036	0.038	0.039	0.051	0.063			
organic bonded	730	50	0.023	0.025	0.026	0.027	0.029	0.030	0.032	0.033	0.035	0.051	0.079			
Felt, laminated, no binder	920	120												0.051	0.065	0.087
Blocks, boards, and pipe insulations																
Asbestos paper, laminated and corrugated																
four-ply	420	190									0.078	0.082	0.098			
six-ply	420	255									0.071	0.074	0.085			
eight-ply	420	300									0.068	0.071	0.082			
Magnesia, 85%	590	185										0.051	0.055	0.061		
Calcium silicate	920	190									0.055	0.059	0.063	0.075	0.089	0.104
Cellular glass	700	145			0.046	0.048	0.051	0.052	0.055	0.058	0.062	0.069	0.079			
Diatomaceous silica	1145	345												0.092	0.098	0.104
	1310	385												0.101	0.100	0.115
Polystyrene rigid																
Extruded (R–12)	350	56	0.023	0.023	0.023	0.023	0.023	0.025	0.026	0.027	0.029					
Extruded (R–12)	350	35	0.023	0.023	0.025	0.025	0.025	0.026	0.027	0.029						
Molded beads	350	16	0.026	0.029	0.030	0.033	0.035	0.036	0.038	0.040						
Rubber, rigid foamed	340	70							0.029	0.030	0.032	0.033				
Loose fill																
Cellulose, wood, or Paper pulp		45								0.038	0.039	0.042				
Perlite, expanded		105	0.036	0.039	0.042	0.043	0.046	0.049	0.051	0.053	0.056					
Vermiculite,		122			0.056	0.058	0.061	0.063	0.065	0.068	0.071					
expanded		80			0.049	0.051	0.055	0.058	0.061	0.063	0.066					

Figure A–4

Friction Factor Chart

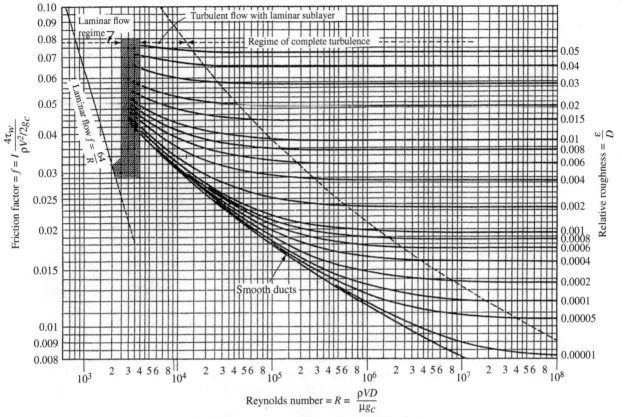

Moody diagram for determination of friction factor

(Adapted with permission from *Friction Factors for Pipe Flow*
by L.F. Moody, transactions of ASME. vol. 66, 1944.)

Figure A–5

Oblique Shock Chart

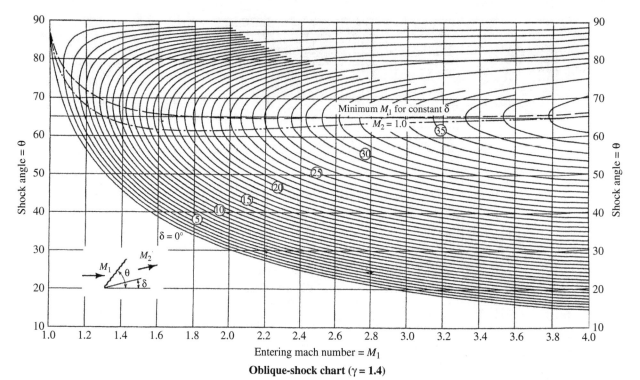

Oblique-shock chart ($\gamma = 1.4$)

(Adapted with permission from *Gas Dynamics* by A.B. Cambel
and B.H. Jennings, McGraw-Hill, New York, 1958.)

Table A–20

Isentropic Flow Parameters

M	p/p_t	T/T_t	A/A^*	pA/p_tA^*	ν	μ
0.0	1.00000	1.00000	∞	∞		
0.01	0.99993	0.99998	57.87384	57.86979		
0.02	0.99972	0.99992	28.94213	28.93403		
0.03	0.99937	0.99982	19.30054	19.28839		
0.04	0.99888	0.99968	14.48149	14.46528		
0.05	0.99825	0.99950	11.59144	11.57118		
0.06	0.99748	0.99928	9.66591	9.64159		
0.07	0.99658	0.99902	8.29153	8.26315		
0.08	0.99553	0.99872	7.26161	7.22917		
0.09	0.99435	0.99838	6.46134	6.42484		
0.10	0.99303	0.99800	5.82183	5.78126		
0.11	0.99158	0.99759	5.29923	5.25459		
0.12	0.98998	0.99713	4.86432	4.81560		
0.13	0.98826	0.99663	4.49686	4.44406		
0.14	0.98640	0.99610	4.18240	4.12552		
0.15	0.98441	0.99552	3.91034	3.84937		
0.16	0.98228	0.99491	3.67274	3.60767		
0.17	0.98003	0.99425	3.46351	3.39434		
0.18	0.97765	0.99356	3.27793	3.20465		
0.19	0.97514	0.99283	3.11226	3.03487		
0.20	0.97250	0.99206	2.96352	2.88201		
0.21	0.96973	0.99126	2.82929	2.74366		
0.22	0.96685	0.99041	2.70760	2.61783		
0.23	0.96383	0.98953	2.59681	2.50290		
0.24	0.96070	0.98861	2.49556	2.39750		
0.25	0.95745	0.98765	2.40271	2.30048		
0.26	0.95408	0.98666	2.31729	2.21089		
0.27	0.95060	0.98563	2.23847	2.12789		
0.28	0.94700	0.98456	2.16555	2.05078		
0.29	0.94329	0.98346	2.09793	1.97896		
0.30	0.93947	0.98232	2.03507	1.91188		
0.31	0.93554	0.98114	1.97651	1.84910		
0.32	0.93150	0.97993	1.92185	1.79021		
0.33	0.92736	0.97868	1.87074	1.73486		
0.34	0.92312	0.97740	1.82288	1.68273		

(*continued*)

Table A–20

Isentropic Flow Parameters (Continued)

M	P/P_t	T/T_t	A/A^*	PA/P_tA^*	ν	μ
0.35	0.91877	0.97609	1.77797	1.63355		
0.36	0.91433	0.97473	1.73578	1.58707		
0.37	0.90979	0.97335	1.69609	1.54308		
0.38	0.90516	0.97193	1.65870	1.50138		
0.39	0.90043	0.97048	1.62343	1.46179		
0.40	0.89561	0.96899	1.59014	1.42415		
0.41	0.89071	0.96747	1.55867	1.38833		
0.42	0.88572	0.96592	1.52890	1.35419		
0.43	0.88065	0.96434	1.50072	1.32161		
0.44	0.87550	0.96272	1.47401	1.29049		
0.45	0.87027	0.96108	1.44867	1.26073		
0.46	0.86496	0.95940	1.42463	1.23225		
0.47	0.85958	0.95769	1.40180	1.20495		
0.48	0.85413	0.95595	1.38010	1.17878		
0.49	0.84861	0.95418	1.35947	1.15365		
0.50	0.84302	0.95238	1.33984	1.12951		
0.51	0.83737	0.95055	1.32117	1.10630		
0.52	0.83165	0.94869	1.30339	1.08397		
0.53	0.82588	0.94681	1.28645	1.06246		
0.54	0.82005	0.94489	1.27032	1.04173		
0.55	0.81417	0.94295	1.25495	1.02173		
0.56	0.80823	0.94098	1.24029	1.00244		
0.57	0.80224	0.93898	1.22633	0.98381		
0.58	0.79621	0.93696	1.21301	0.96580		
0.59	0.79013	0.93491	1.20031	0.94840		
0.60	0.78400	0.93284	1.18820	0.93155		
0.61	0.77784	0.93073	1.17665	0.91525		
0.62	0.77164	0.92861	1.16565	0.89946		
0.63	0.76540	0.92646	1.15515	0.88416		
0.64	0.75913	0.92428	1.14515	0.86932		
0.65	0.75283	0.92208	1.13562	0.85493		
0.66	0.74650	0.91986	1.12654	0.84096		
0.67	0.74014	0.91762	1.11789	0.82739		
0.68	0.73376	0.91535	1.10965	0.81422		
0.69	0.72735	0.91306	1.10182	0.80141		

Table A–20

Isentropic Flow Parameters (Continued)

M	p/p_t	T/T_t	A/A^*	pA/p_tA^*	ν	μ
0.70	0.72093	0.91075	1.09437	0.78896		
0.71	0.71448	0.90841	1.08729	0.77685		
0.72	0.70803	0.90606	1.08057	0.76507		
0.73	0.70155	0.90369	1.07419	0.75360		
0.74	0.69507	0.90129	1.06814	0.74243		
0.75	0.68857	0.89888	1.06242	0.73155		
0.76	0.68207	0.89644	1.05700	0.72095		
0.77	0.67556	0.89399	1.05188	0.71061		
0.78	0.66905	0.89152	1.04705	0.70053		
0.79	0.66254	0.88903	1.04251	0.69070		
0.80	0.85602	0.88652	1.03823	0.68110		
0.81	0.64951	0.88400	1.03422	0.67173		
0.82	0.64300	0.88146	1.03046	0.66259		
0.83	0.63650	0.87890	1.02696	0.65366		
0.84	0.63000	0.87633	1.02370	0.64493		
0.85	0.62351	0.87374	1.02067	0.63640		
0.86	0.61703	0.87114	1.01787	0.62806		
0.87	0.61057	0.86852	1.01530	0.61991		
0.88	0.60412	0.86589	1.01294	0.61193		
0.89	0.59768	0.86324	1.01080	0.80413		
0.90	0.59126	0.86059	1.00886	0.59650		
0.91	0.58486	0.85791	1.00713	0.58903		
0.92	0.57848	0.85523	1.00560	0.58171		
0.93	0.57211	0.85253	1.00426	0.57455		
0.94	0.56578	0.84982	1.00311	0.56753		
0.95	0.55946	0.84710	1.00215	0.56066		
0.96	0.55317	0.84437	1.00136	0.55392		
0.97	0.54691	0.84162	1.00076	0.54732		
0.98	0.54067	0.83887	1.00034	0.54085		
0.99	0.53446	0.83611	1.00008	0.53451		
1.00	0.52828	0.83333	1.00000	0.52826	0.0	90.0000
1.01	0.52213	0.83055	1.00008	0.52218	3.04472	81.9307
1.02	0.51602	0.82776	1.00033	0.51619	0.12569	78.6351
1.03	0.50994	0.82496	1.00074	0.51031	0.22943	76.1376
1.04	0.50389	0.82215	1.00131	0.50454	0.35098	74.0576

(continued)

Table A–20

Isentropic Flow Parameters (Continued)

M	P/P_t	T/T_t	A/A^*	pA/p_tA^*	ν	μ
1.05	0.49787	0.81934	1.00203	0.49888	0.48741	72.2472
1.06	0.49189	0.81651	1.00291	0.49332	0.63669	70.6300
1.07	0.48595	0.81368	1.00394	0.48787	0.79729	69.1603
1.08	0.48005	0.81085	1.00512	0.48250	0.96804	67.8084
1.09	0.47418	0.80800	1.00645	0.47724	1.14795	66.5534
1.10	0.46835	0.80515	1.00793	0.47207	1.33620	65.3800
1.11	0.46257	0.80230	1.00955	0.46698	1.53210	64.2767
1.12	0.45682	0.79944	1.01131	0.46199	1.73504	63.2345
1.13	0.45111	0.79657	1.01322	0.45708	1.94448	62.2461
1.14	0.44545	0.79370	1.01527	0.45225	2.15996	61.3056
1.15	0.43983	0.79083	1.01745	0.44751	2.38104	60.4082
1.16	0.43425	0.78795	1.01978	0.44284	2.60735	59.5497
1.17	0.42872	0.78506	1.02224	0.43825	2.83852	58.7267
1.18	0.42322	0.78218	1.02484	0.43374	3.07426	57.9362
1.19	0.41778	0.77929	1.02757	0.42930	3.31425	57.1756
1.20	0.41238	0.77640	1.03044	0.42493	3.55823	56.4427
1.21	0.40702	0.77350	1.03344	0.42063	3.80596	55.7354
1.22	0.40171	0.77061	1.03657	0.41640	4.05720	55.0520
1.23	0.39645	0.76771	1.03983	0.41224	4.31173	54.3909
1.24	0.39123	0.76481	1.04323	0.40814	4.56936	53.7507
1.25	0.38606	0.76190	1.04675	0.40411	4.82989	53.1301
1.26	0.38093	0.75900	1.05041	0.40014	5.09315	52.5280
1.27	0.37586	0.75610	1.05419	0.39622	5.35897	51.9433
1.28	0.37083	0.75319	1.05810	0.39237	5.62720	51.3752
1.29	0.36585	0.75029	1.06214	0.38858	5.89768	50.8226
1.30	0.36091	0.74738	1.06630	0.38484	6.17029	50.2849
1.31	0.35603	0.74448	1.07060	0.38116	6.44488	49.7612
1.32	0.35119	0.74158	1.07502	0.37754	6.72133	49.2509
1.33	0.34640	0.73887	1.07957	0.37396	6.99953	48.7535
1.34	0.34166	0.73577	1.08424	0.37044	7.27937	48.2682
1.35	0.33697	0.73287	1.08904	0.36697	7.56072	47.7946
1.36	0.33233	0.72997	1.09396	0.36355	7.84351	47.3321
1.37	0.32773	0.72707	1.09902	0.36018	8.12762	46.8803
1.38	0.32319	0.72418	1.10419	0.35686	8.41297	46.4387
1.39	0.31869	0.72128	1.10950	0.35359	8.69946	46.0070

Table A–20

Isentropic Flow Parameters (Continued)

M	p/p_t	T/T_t	A/A^*	pA/p_tA^*	ν	μ
1.40	0.31424	0.71839	1.11493	0.35036	8.98702	45.5847
1.41	0.30984	0.71550	1.12048	0.34717	9.27556	45.1715
1.42	0.30549	0.71262	1.12616	0.34403	9.56502	44.7670
1.43	0.30118	0.70973	1.13197	0.34093	9.85531	44.3709
1.44	0.29693	0.70685	1.13790	0.33788	10.14636	43.9830
1.45	0.29272	0.70398	1.14396	0.33486	10.43811	43.6028
1.46	0.28856	0.70110	1.15015	0.33189	10.73050	43.2302
1.47	0.28445	0.69824	1.15646	0.32896	11.02346	42.8649
1.48	0.28039	0.69537	1.16290	0.32606	11.31694	42.5066
1.49	0.27637	0.69251	1.16947	0.32321	11.61087	42.1552
1.50	0.27240	0.68966	1.17617	0.32039	11.90521	41.8103
1.51	0.26848	0.68680	1.18299	0.31761	12.19990	41.4718
1.52	0.26461	0.68396	1.18994	0.31487	12.49489	41.1395
1.53	0.26078	0.68112	1.19702	0.31216	12.79014	40.8132
1.54	0.25700	0.57828	1.20423	0.30949	13.08559	40.4927
1.55	0.25326	0.67545	1.21157	0.30685	13.38121	40.1778
1.56	0.24957	0.67262	1.21904	0.30424	13.67696	39.8683
1.57	0.24593	0.66980	1.22664	0.30167	13.97278	39.5642
1.58	0.24233	0.66699	1.23438	0.29913	14.26865	39.2652
1.59	0.23878	0.66418	1.24224	0.29662	14.56452	38.9713
1.60	0.23527	0.66138	1.25023	0.29414	14.86035	38.6822
1.61	0.23181	0.65858	1.25836	0.29170	15.15612	38.3978
1.62	0.22839	0.65579	1.26663	0.28928	15.45180	38.1181
1.63	0.22501	0.65301	1.27502	0.28690	15.74733	37.8428
1.64	0.22168	0.65023	1.28355	0.28454	16.04271	37.5719
1.65	0.21839	0.64746	1.29222	0.28221	16.33789	37.3052
1.66	0.21515	0.64470	1.30102	0.27991	16.63284	37.0427
1.67	0.21195	0.64194	1.30996	0.27764	16.92755	36.7842
1.68	0.20879	0.63919	1.31904	0.27540	17.22198	36.5296
1.69	0.20567	0.63645	1.32825	0.27318	17.51611	36.2789
1.70	0.20259	0.63371	1.33761	0.27099	17.80991	36.0319
1.71	0.19956	0.63099	1.34710	0.26883	18.10336	35.7885
1.72	0.19656	0.62827	1.35674	0.26669	18.39643	35.5487
1.73	0.19361	0.62556	1.36651	0.26457	18.68911	35.3124
1.74	0.19070	0.62285	1.37643	0.26248	18.98137	35.0795

(*continued*)

Table A–20

Isentropic Flow Parameters (Continued)

M	P/P_t	T/T_t	A/A^*	pA/p_tA^*	ν	μ
1.75	0.18782	0.62016	1.38649	0.26042	19.27319	34.8499
1.76	0.18499	0.61747	1.39670	0.25837	19.56456	34.6235
1.77	0.18219	0.61479	1.40705	0.25636	19.85544	34.4003
1.78	0.17944	0.61211	1.41755	0.25436	20.14584	34.1802
1.79	0.17672	0.60945	1.42819	0.25239	20.43571	33.9631
1.80	0.17404	0.50680	1.43898	0.25044	20.72506	33.7490
1.81	0.17140	0.60415	1.44992	0.24851	21.01387	33.5377
1.82	0.16879	0.60151	1.46101	0.24661	21.30211	33.3293
1.83	0.16622	0.59886	1.47225	0.24472	21.58977	33.1237
1.84	0.16369	0.59626	1.48365	0.24286	21.87685	32.9207
1.85	0.16119	0.59365	1.49519	0.24102	22.16332	32.7204
1.86	0.15873	0.59104	1.50689	0.23920	22.44917	32.5227
1.87	0.15631	0.58845	1.51875	0.23739	22.73439	32.3276
1.88	0.15392	0.58586	1.53076	0.23561	23.01896	32.1349
1.89	0.15156	0.58329	1.54293	0.23385	23.30288	31.9447
1.90	0.14924	0.58072	1.55526	0.23211	23.58613	31.7569
1.91	0.14695	0.57816	1.56774	0.23038	23.86871	31.5714
1.92	0.14470	0.57561	1.58039	0.22868	24.15059	31.3882
1.93	0.14247	0.57307	1.59320	0.22699	24.43178	31.2072
1.94	0.14028	0.57054	1.60617	0.22532	24.71226	31.0285
1.95	0.13813	0.56802	1.61931	0.22367	24.99202	30.8519
1.96	0.13600	0.56551	1.63261	0.22203	25.27105	30.6774
1.97	0.13390	0.56301	1.64608	0.22042	25.54935	30.5050
1.98	0.13184	0.56051	1.65972	0.21882	25.82691	30.3347
1.99	0.12981	0.55803	1.67352	0.21724	26.10371	30.1664
2.00	0.12780	0.55556	1.68750	0.21567	26.37976	30.0000
2.01	0.12583	0.55309	1.70165	0.21412	26.65504	29.8356
2.02	0.12389	0.55064	1.71597	0.21259	26.92955	29.6730
2.03	0.12197	0.54819	1.73047	0.21107	27.20328	29.5123
2.04	0.12009	0.54576	1.74514	0.20957	27.47622	29.3535
2.05	0.11823	0.54333	1.75999	0.20808	27.74837	29.1964
2.06	0.11640	0.54091	1.77502	0.20661	28.01973	29.0411
2.07	0.11460	0.53851	1.79022	0.20516	28.29028	28.8875
2.08	0.11282	0.53611	1.80561	0.20371	28.56003	28.7357
2.09	0.11107	0.53373	1.82119	0.20229	28.82896	28.5855

Table A–20

Isentropic Flow Parameters (Continued)

M	P/P_t	T/T_t	A/A^*	pA/p_tA^*	ν	μ
2.10	0.10935	0.53135	1.83694	0.20088	29.09708	28.4369
2.11	0.10766	0.52898	1.85289	0.19948	29.36438	28.2899
2.12	0.10599	0.52663	1.86902	0.19809	29.63085	28.1446
2.13	0.10434	0.52428	1.88533	0.19672	29.89649	28.0008
2.14	0.10273	0.52194	1.90184	0.19537	30.16130	27.8585
2.15	0.10113	0.51962	1.91854	0.19403	30.42527	27.7177
2.16	0.09956	0.51730	1.93544	0.19270	30.68841	27.5785
2.17	0.09802	0.51499	1.95252	0.19138	30.95070	27.4406
2.18	0.09649	0.51269	1.96981	0.19008	31.21215	27.3043
2.19	0.09500	0.51041	1.98729	0.18879	31.47275	27.1693
2.20	0.09352	0.50813	2.00497	0.18751	31.73250	27.0357
2.21	0.09207	0.50586	2.02286	0.18624	31.99139	26.9035
2.22	0.09064	0.50361	2.04094	0.18499	32.24943	26.7726
2.23	0.08923	0.50136	2.05923	0.18375	32.50662	26.6430
2.24	0.08785	0.49912	2.07773	0.18252	32.76294	26.5148
2.25	0.08648	0.49689	2.09644	0.18130	33.01841	26.3878
2.26	0.08514	0.49468	2.11535	0.18010	33.27301	26.2621
2.27	0.06382	0.49247	2.13447	0.17890	33.52676	26.1376
2.28	0.08251	0.49027	2.15381	0.17772	33.77963	26.0144
2.29	0.08123	0.48809	2.17336	0.17655	34.03165	25.8923
2.30	0.07997	0.48591	2.19313	0.17539	34.28279	25.7715
2.31	0.07873	0.48374	2.21312	0.17424	34.53307	25.6518
2.32	0.07751	0.48158	2.23332	0.17310	34.78249	25.5332
2.33	0.07631	0.47944	2.25375	0.17198	35.03103	25.4158
2.34	0.07512	0.47730	2.27440	0.17086	35.27871	25.2995
2.35	0.07396	0.47517	2.29528	0.16975	35.52552	25.1843
2.36	0.07281	0.47305	2.31638	0.16866	35.77146	25.0702
2.37	0.07168	0.47095	2.33771	0.16757	36.01653	24.9572
2.38	0.07057	0.46885	2.35928	0.16649	36.26073	24.8452
2.39	0.06948	0.46676	2.38107	0.16543	36.50406	24.7342
2.40	0.06840	0.46468	2.40310	0.16437	36.74653	24.6243
2.41	0.06734	0.46262	2.42537	0.16332	36.98813	24.5154
2.42	0.06630	0.46056	2.44787	0.16229	37.22886	24.4075
2.43	0.06527	0.45851	2.47061	0.16126	37.46872	24.3005
2.44	0.06426	0.45647	2.49360	0.16024	37.70772	24.1945

(continued)

Isentropic Flow Parameters (Continued)

M	p/p_t	T/T_t	A/A^*	pA/p_tA^*	ν	μ
2.45	0.06327	0.45444	2.51683	0.15923	37.94585	24.0895
2.46	0.06229	0.45242	2.54031	0.15823	38.18312	23.9854
2.47	0.06133	0.45041	2.56403	0.15724	38.41952	23.8822
2.48	0.06038	0.44841	2.58801	0.15626	38.65507	23.7800
2.49	0.05945	0.44642	2.61224	0.15529	38.88974	23.6786
2.50	0.05853	0.44444	2.63672	0.15432	39.12356	23.5782
2.51	0.05762	0.44247	2.66146	0.15337	39.35652	23.4786
2.52	0.05674	0.44051	2.68645	0.15242	39.58862	23.3799
2.53	0.05586	0.43856	2.71171	0.15148	39.81987	23.2820
2.54	0.05500	0.43662	2.73723	0.15055	40.05026	23.1850
2.55	0.05415	0.43469	2.76301	0.14968	40.27979	23.0888
2.56	0.05332	0.43277	2.78906	0.14871	40.50847	22.9934
2.57	0.05250	0.43085	2.81538	0.14780	40.73630	22.8988
2.58	0.05169	0.42895	2.84197	0.14691	40.96329	22.8051
2.59	0.05090	0.42705	2.86884	0.14602	41.18942	22.7121
2.60	0.05012	0.42517	2.89598	0.14513	41.41471	22.6199
2.61	0.04935	0.42329	2.92339	0.14426	41.63915	22.5284
2.62	0.04859	0.42143	2.95109	0.14339	41.86275	22.4377
2.63	0.04784	0.41957	2.97907	0.14253	42.08551	22.3478
2.64	0.04711	0.41772	3.00733	0.14168	42.30744	22.2586
2.65	0.04639	0.41589	3.03588	0.14083	42.52852	22.1702
2.66	0.04568	0.41406	3.06472	0.13999	42.74877	22.0824
2.67	0.04498	0.41224	3.09385	0.13916	42.96819	21.9954
2.68	0.04429	0.41043	3.12327	0.13834	43.18678	21.9090
2.69	0.04362	0.40863	3.15299	0.13752	43.40454	21.8234
2.70	0.04295	0.40683	3.18301	0.13671	43.62148	21.7385
2.71	0.04229	0.40505	3.21333	0.13591	43.83759	21.6542
2.72	0.04165	0.40328	3.24395	0.13511	44.05288	21.5706
2.73	0.04102	0.40151	3.27488	0.13432	44.26735	21.4876
2.74	0.04039	0.39976	3.30611	0.13354	44.48100	21.4053
2.75	0.03978	0.39801	3.33766	0.13276	44.69384	21.3237
2.76	0.03917	0.39627	3.36952	0.13199	44.90586	21.2427
2.77	0.03858	0.39454	3.40169	0.13123	45.11708	21.1623
2.78	0.03799	0.39282	3.43418	0.13047	45.32749	21.0825
2.79	0.03742	0.39111	3.46699	0.12972	45.53709	21.0034

Table A–20

Isentropic Flow Parameters (Continued)

M	p/p_t	T/T_t	A/A^*	pA/p_tA^*	ν	μ
2.80	0.03685	0.38941	3.50012	0.12897	45.74589	20.9248
2.81	0.03629	0.38771	3.53358	0.12823	45.95389	20.8469
2.82	0.03574	0.38603	3.56737	0.12750	46.16109	20.7695
2.83	0.03520	0.38435	3.60148	0.12678	46.36750	20.6928
2.84	0.03467	0.38268	3.63593	0.12605	46.57312	20.6166
2.85	0.03415	0.38102	3.67072	0.12534	46.77794	20.5410
2.86	0.03363	0.37937	3.70584	0.12463	46.98198	20.4659
2.87	0.03312	0.37773	3.74131	0.12393	47.18523	20.3914
2.88	0.03263	0.37610	3.77711	0.12323	47.38770	20.3175
2.89	0.03213	0.37447	3.81327	0.12254	47.58940	20.2441
2.90	0.03165	0.37286	3.84977	0.12185	47.79031	20.1713
2.91	0.03118	0.37125	3.88662	0.12117	47.99045	20.0990
2.92	0.03071	0.36965	3.92383	0.12049	48.18982	20.0272
2.93	0.03025	0.36806	3.96139	0.11982	48.38842	19.9559
2.94	0.02980	0.36647	3.99932	0.11916	48.58626	19.8852
2.95	0.02935	0.36490	4.03760	0.11850	48.78333	19.8149
2.96	0.02891	0.36333	4.07625	0.11785	48.97965	19.7452
2.97	0.02848	0.36177	4.11527	0.11720	49.17520	19.6760
2.98	0.02805	0.36022	4.15466	0.11655	49.37000	19.6072
2.99	0.02764	0.35868	4.19443	0.11591	49.56405	19.5390
3.00	0.02722	0.35714	4.23457	0.11528	49.75735	19.4712
3.01	0.02682	0.35562	4.27509	0.11465	49.94990	19.4039
3.02	0.02642	0.35410	4.31599	0.11403	50.14171	19.3371
3.03	0.02603	0.35259	4.35726	0.11341	50.33277	19.2708
3.04	0.02564	0.35108	4.39895	0.11279	50.52310	19.2049
3.05	0.02526	0.34959	4.44102	0.11219	50.71270	19.1395
3.06	0.02439	0.34810	4.48347	0.11158	50.90156	19.0745
3.07	0.02452	0.34662	4.52633	0.11098	51.08969	19.0100
3.08	0.02416	0.34515	4.56959	0.11039	51.27710	18.9459
3.09	0.02380	0.34369	4.61325	0.10979	51.46378	18.8823
3.10	0.02345	0.34223	4.65731	0.10921	51.64974	18.8191
3.11	0.02310	0.34078	4.70178	0.10863	51.83499	18.7563
3.12	0.02276	0.33934	4.74667	0.10805	52.01952	18.6939
3.13	0.02243	0.33791	4.79197	0.10748	52.20333	18.6320
3.14	0.02210	0.33648	4.83769	0.10691	52.38644	18.5705

(*continued*)

Table A–20

Isentropic Flow Parameters (Continued)

M	P/P_t	T/T_t	A/A^*	pA/p_tA^*	ν	μ
3.15	0.02177	0.33506	4.88383	0.10634	52.56884	18.5094
3.16	0.02146	0.33365	4.93039	0.10578	52.75053	18.4487
3.17	0.02114	0.33225	4.97739	0.10523	52.93153	18.3884
3.18	0.02083	0.33085	5.02481	0.10468	53.11182	18.3285
3.19	0.02053	0.32947	5.07266	0.10413	53.29143	18.2691
3.20	0.02023	0.32808	5.12096	0.10359	53.47033	18.2100
3.21	0.01993	0.32671	5.16969	0.10305	53.64855	18.1512
3.22	0.01964	0.32534	5.21887	0.10251	53.82609	18.0929
3.23	0.01936	0.32398	5.26849	0.10198	54.00294	18.0350
3.24	0.01908	0.32263	5.31857	0.10145	54.17910	17.9774
3.25	0.01880	0.32129	5.36909	0.10093	54.35459	17.9202
3.26	0.01853	0.31995	5.42008	0.10041	54.52941	17.8634
3.27	0.01826	0.31862	5.47152	0.09989	54.70355	17.8069
3.28	0.01799	0.31729	5.52343	0.09938	54.87703	17.7508
3.29	0.01773	0.31597	5.57580	0.09887	55.04983	17.6951
3.30	0.01748	0.31466	5.62865	0.09837	55.22198	17.6397
3.31	0.01722	0.31336	5.68196	0.09787	55.39346	17.5847
3.32	0.01698	0.31206	5.73576	0.09737	55.56428	17.5300
3.33	0.01673	0.31077	5.79008	0.09688	55.73445	17.4756
3.34	0.01649	0.30949	5.84479	0.09639	55.90396	17.4216
3.35	0.01625	0.30821	5.90004	0.09590	56.07283	17.3680
3.36	0.01602	0.30694	5.95577	0.09542	56.24105	17.3147
3.37	0.01579	0.30568	6.01201	0.09494	56.40862	17.2617
3.38	0.01557	0.30443	6.06873	0.09447	56.57556	17.2090
3.39	0.01534	0.30318	6.12596	0.09399	56.74185	17.1567
3.40	0.01512	0.30193	6.18370	0.09353	56.90751	17.1046
3.41	0.01491	0.30070	6.24194	0.09306	57.07254	17.0529
3.42	0.01470	0.29947	6.30070	0.09260	57.23694	17.0016
3.43	0.01449	0.29824	6.35997	0.09214	57.40071	16.9505
3.44	0.01428	0.29702	6.41976	0.09168	57.56385	16.8997
3.45	0.01408	0.29581	6.48007	0.09123	57.72637	16.8493
3.46	0.01388	0.29461	6.54092	0.09078	57.88828	16.7991
3.47	0.01368	0.29341	6.60229	0.09034	58.04957	16.7493
3.48	0.01349	0.29222	6.66419	0.08989	58.21024	16.6997
3.49	0.01330	0.29103	6.72664	0.08945	53.37030	16.6505

Table A–20

Isentropic Flow Parameters (Continued)

M	P/P_t	T/T_t	A/A^*	pA/p_tA^*	ν	μ
3.50	0.01311	0.28986	6.78962	0.08902	58.52976	16.6015
3.51	0.01293	0.28868	6.85315	0.08858	58.68861	16.5529
3.52	0.01274	0.28751	6.91723	0.08815	58.84685	16.5045
3.53	0.01256	0.28635	6.98186	0.08773	59.00450	16.4564
3.54	0.01239	0.28520	7.04705	0.08730	59.16155	16.4086
3.55	0.01221	0.28405	7.11281	0.08688	59.31801	16.3611
3.56	0.01204	0.28291	7.17912	0.08646	59.47387	16.3139
3.57	0.01188	0.28177	7.24601	0.08605	59.62914	16.2669
3.58	0.01171	0.28064	7.31346	0.08563	59.78383	16.2202
3.59	0.01155	0.27952	7.38150	0.08522	59.93793	16.1738
3.60	0.01138	0.27840	7.45011	0.08482	60.09146	16.1276
3.61	0.01123	0.27728	7.51931	0.08441	60.24440	16.0817
3.62	0.01107	0.27618	7.58910	0.08401	60.39677	16.0361
3.63	0.01092	0.27507	7.65948	0.08361	60.54856	15.9907
3.64	0.01076	0.27398	7.73045	0.08322	60.69978	15.9456
3.65	0.01062	0.27289	7.80203	0.08282	60.85044	15.9008
3.66	0.01047	0.27180	7.87421	0.08243	61.00052	15.8562
3.67	0.01032	0.27073	7.94700	0.08205	61.15005	15.8119
3.68	0.01018	0.26965	8.02040	0.08166	61.29902	15.7678
3.69	0.01004	0.26858	8.09442	0.08128	61.44742	15.7239
3.70	0.00990	0.26752	8.16907	0.08090	61.59527	15.6803
3.71	0.00977	0.26647	8.24433	0.08052	61.74257	15.6370
3.72	0.00963	0.26542	8.32023	0.08014	61.88932	15.5939
3.73	0.00950	0.26437	8.39676	0.07977	62.03552	15.5510
3.74	0.00937	0.26333	8.47393	0.07940	62.18118	15.5084
3.75	0.00924	0.26230	8.55174	0.07904	62.32629	15.4660
3.76	0.00912	0.26127	8.63020	0.07867	62.47086	15.4239
3.77	0.00899	0.26024	8.70931	0.07831	62.61490	15.3819
3.78	0.00887	0.25922	8.78907	0.07795	62.75840	15.3402
3.79	0.00875	0.25821	8.86950	0.07759	62.90136	15.2988
3.80	0.00863	0.25720	8.95059	0.07723	63.04380	15.2575
3.81	0.00851	0.25620	9.03234	0.07688	63.18571	15.2165
3.82	0.00840	0.25520	9.11477	0.07653	63.32709	15.1757
3.83	0.00828	0.25421	9.19788	0.07618	63.46795	15.1351
3.84	0.00817	0.25322	9.28167	0.07584	63.60829	15.0948

(continued)

Table A–20

Isentropic Flow Parameters (Continued)

M	p/p_t	T/T_t	A/A^*	pA/p_tA^*	ν	μ
3.85	0.00806	0.25224	9.36614	0.07549	63.74811	15.0547
3.86	0.00795	0.25126	9.45131	0.07515	63.88741	15.0147
3.87	0.00784	0.25029	9.53717	0.07481	64.02620	14.9750
3.88	0.00774	0.24932	9.62373	0.07447	64.16448	14.9355
3.89	0.00763	0.24836	9.71100	0.07414	64.30225	14.8962
3.90	0.00753	0.24740	9.79897	0.07381	64.43952	14.8572
3.91	0.00743	0.24645	9.88766	0.07348	64.57628	14.8183
3.92	0.00733	0.24550	9.97707	0.07315	64.71254	14.7796
3.93	0.00723	0.24456	10.06720	0.07282	64.84829	14.7412
3.94	0.00714	0.24362	10.15806	0.07250	64.98356	14.7029
3.95	0.00704	0.24269	10.24965	0.07217	65.11832	14.6649
3.96	0.00695	0.24176	10.34197	0.07185	65.25260	14.6270
3.97	0.00686	0.24084	10.43504	0.07154	65.33638	14.5893
3.98	0.00676	0.23992	10.52886	0.07122	65.51968	14.5519
3.99	0.00667	0.23900	10.62343	0.07091	65.65249	14.5146
4.00	0.00659	0.23810	10.71875	0.07059	65.78482	14.4775
4.10	0.00577	0.22925	11.71465	0.06758	67.08200	14.1170
4.20	0.00506	0.22085	12.79164	0.06475	68.33324	13.7741
4.30	0.00445	0.21286	13.95490	0.06209	69.54063	13.4477
4.40	0.00392	0.20525	15.20987	0.05959	70.70616	13.1366
4.50	0.00346	0.19802	16.56219	0.05723	71.83174	12.8396
4.60	0.00305	0.19113	18.01779	0.05500	72.91915	12.5559
4.70	0.00270	0.18457	19.58283	0.05290	73.97012	12.2845
4.80	0.00239	0.17832	21.26371	0.05091	74.98627	12.0247
4.90	0.00213	0.17235	23.06712	0.04903	75.96915	11.7757
5.00	0.00189	0.16667	25.00000	0.04725	76.92021	11.5370
5.10	0.00168	0.16124	27.06957	0.04556	77.84087	11.3077
5.20	0.00150	0.15605	29.28333	0.04396	78.73243	11.0875
5.30	0.00134	0.15110	31.64905	0.04244	79.59616	10.8757
5.40	0.00120	0.14637	34.17481	0.04100	80.43323	10.6719
5.50	0.00107	0.14184	36.86896	0.03963	81.24479	10.4757
5.60	0.000964	0.13751	39.74018	0.03832	82.03190	10.2866
5.70	0.000866	0.13337	42.79743	0.03708	82.79558	10.1042
5.80	0.000779	0.12940	46.05000	0.03589	83.53681	9.9282
5.90	0.000702	0.12560	49.50747	0.03476	84.25649	9.7583

Table A–20

Isentropic Flow Parameters (Continued)

M	p/p_t	T/T_t	A/A^*	pA/p_tA^*	ν	μ
6.00	0.000633	0.12195	53.17978	0.03368	84.95550	9.5941
6.10	0.000572	0.11846	57.07718	0.03265	85.63467	9.4353
6.20	0.000517	0.11510	61.21023	0.03167	86.29479	9.2818
6.30	0.000468	0.11188	65.58987	0.03073	86.93661	9.1332
6.40	0.000425	0.10879	70.22736	0.02982	87.56084	8.9893
6.50	0.000385	0.10582	75.13431	0.02896	88.16816	8.8499
6.60	0.000350	0.10297	80.32271	0.02814	88.75922	8.7147
6.70	0.000319	0.10022	85.80487	0.02734	89.33463	8.5837
6.80	0.000290	0.09758	91.59351	0.02658	89.89499	8.4565
6.90	0.000265	0.09504	97.70169	0.02586	90.44084	8.3331
7.00	0.000242	0.09259	104.14280	0.02516	90.97273	8.2132
7.50	0.000155	0.08163	141.84148	0.02205	93.43967	7.6623
8.00	0.000102	0.07246	190.10937	0.01947	95.62467	7.1808
8.50	0.0000690	0.06472	251.086167	0.01732	97.57220	6.7563
9.00	0.0000474	0.05814	327.189300	0.01550	99.31810	6.3794
9.50	0.0000331	0.05249	421.131373	0.01396	100.89148	6.0423
10.00	0.0000236	0.04762	535.937500	0.01263	102.31625	5.7392
∞	0.0	0.0	∞	0.0	130.4541	0.0

Table A–21

Normal Shock Parameters

M_1	M_2	P_2/P_1	T_2/T_1	$\Delta V/a_1$	$P_{t,2}/P_{t,1}$	$P_{t,2}/P_1$
1.00	1.00000	1.00000	1.00000	0.0	1.00000	1.89293
1.01	0.99013	1.02345	1.00664	0.01658	1.00000	1.91521
1.02	0.98052	1.04713	1.01325	0.03301	0.99999	1.93790
1.03	0.97115	1.07105	1.01981	0.04927	0.99997	1.96097
1.04	0.96203	1.09520	1.02634	0.06538	0.99992	1.98442
1.05	0.95313	1.11958	1.03284	0.08135	0.99985	2.00825
1.06	0.94445	1.14420	1.03931	0.09717	0.99975	2.03245
1.07	0.93598	1.16905	1.04575	0.11285	0.99961	2.05702
1.08	0.92771	1.19413	1.05217	0.12840	0.99943	2.08194
1.09	0.91965	1.21945	1.05856	0.14381	0.99920	2.10722
1.10	0.91177	1.24500	1.06494	0.15909	0.99893	2.13285
1.11	0.90408	1.27078	1.07129	0.17425	0.99860	2.15882
1.12	0.89656	1.29680	1.07763	0.18929	0.99821	2.18513
1.13	0.88922	1.32305	1.08396	0.20420	0.99777	2.21178
1.14	0.88204	1.34953	1.09027	0.21901	0.99726	2.23877
1.15	0.87502	1.37625	1.09658	0.23370	0.99669	2.26608
1.16	0.86816	1.40320	1.10287	0.24828	0.99605	2.29372
1.17	0.86145	1.43038	1.10916	0.26275	0.99535	2.32169
1.18	0.85488	1.45780	1.11544	0.27712	0.99457	2.34998
1.19	0.84846	1.48545	1.12172	0.29139	0.99372	2.37658
1.20	0.84217	1.51333	1.12799	0.30556	0.99280	2.40750
1.21	0.83601	1.54145	1.13427	0.31963	0.99180	2.43674
1.22	0.62999	1.56980	1.14054	0.33361	0.99073	2.46628
1.23	0.82408	1.59838	1.14682	0.34749	0.98958	2.49613
1.24	0.81830	1.62720	1.15309	0.36129	0.98836	2.52629
1.25	0.81264	1.65625	1.15937	0.37500	0.98706	2.55676
1.26	0.80709	1.68553	1.16566	0.38862	0.98568	2.58753
1.27	0.80164	1.71505	1.17195	0.40217	0.98422	2.61860
1.28	0.79631	1.74480	1.17825	0.41562	0.98268	2.64996
1.29	0.79108	1.77478	1.18456	0.42901	0.98107	2.68163
1.30	0.78596	1.80500	1.19087	0.44231	0.97937	2.71359
1.31	0.78093	1.83545	1.19720	0.45553	0.97760	2.74585
1.32	0.77600	1.86613	1.20353	0.46869	0.97575	2.77840
1.33	0.77116	1.89705	1.20988	0.48177	0.97382	2.81125
1.34	0.78641	1.92820	1.21624	0.49478	0.97132	2.84438

Table A–21

Normal Shock Parameters (Continued)

M_1	M_2	P_2/P_1	T_2/T_1	$\Delta V/a_1$	$P_{t,2}/P_{t,1}$	$P_{t,2}/P_1$
1.35	0.76175	1.95958	1.22261	0.50772	0.96974	2.87781
1.36	0.75718	1.99120	1.22900	0.52059	0.96758	2.91152
1.37	0.75269	2.02305	1.23540	0.53339	0.96534	2.94552
1.38	0.74829	2.05513	1.24181	0.54614	0.96304	2.97981
1.39	0.74396	2.08745	1.24825	0.55881	0.96065	3.01438
1.40	0.73971	2.12000	1.25469	0.57143	0.95819	3.04924
1.41	0.73554	2.15278	1.26116	0.58398	0.95566	3.08438
1.42	0.73144	2.18580	1.26764	0.59648	0.95306	3.11980
1.43	0.72741	2.21905	1.27414	0.60892	0.95039	3.15551
1.44	0.72345	2.25253	1.28066	0.62130	0.94765	3.19149
1.45	0.71956	2.28625	1.28720	0.63362	0.94484	3.22776
1.46	0.71574	2.32020	1.29377	0.64589	0.94196	3.26431
1.47	0.71198	2.35438	1.30035	0.65811	0.93901	3.30113
1.48	0.70829	2.38880	1.30695	0.67027	0.93600	3.33823
1.49	0.70466	2.42346	1.31357	0.68238	0.93293	3.37562
1.50	0.70109	2.45833	1.32022	0.69444	0.92979	3.41327
1.51	0.69758	2.49345	1.32688	0.07046	0.92659	3.45121
1.52	0.69413	2.52880	1.33357	0.71842	0.92332	3.48942
1.53	0.69073	2.56438	1.34029	0.73034	0.92000	3.52791
1.54	0.68739	2.60020	1.34703	0.74221	0.91662	3.56667
1.55	0.68410	2.63625	1.35379	0.75403	0.91319	3.60570
1.56	0.68087	2.67253	1.36057	0.76581	0.90970	3.64501
1.57	0.67768	2.70906	1.36738	0.77755	0.90615	3.68459
1.58	0.67455	2.74580	1.37422	0.78924	0.90255	3.72445
1.59	0.67147	2.78278	1.38108	0.80089	0.89890	3.76457
1.60	0.66844	2.82000	1.38797	0.81250	0.89520	3.80497
1.61	0.66545	2.85745	1.39488	0.82407	0.89145	3.84564
1.62	0.66251	2.89513	1.40182	0.83560	0.88765	3.88658
1.63	0.65962	2.93305	1.40879	0.84709	0.88381	3.92780
1.64	0.65677	2.97120	1.41578	0.85854	0.87992	3.96928
1.65	0.65396	3.00958	1.42280	0.86995	0.87599	4.01103
1.66	0.65119	3.04820	1.42985	0.88133	0.87201	4.05305
1.67	0.64847	3.08705	1.43693	0.89266	0.86800	4.09535
1.68	0.64579	3.12613	1.44403	0.90397	0.86394	4.13791
1.69	0.64315	3.18545	1.45117	0.91524	0.85985	4.18074

(*continued*)

Table A–21

Normal Shock Parameters (Continued)

M_1	M_2	P_2/P_1	T_2/T_1	$\Delta V/a_1$	$P_{t,2}/P_{t,1}$	$P_{t,2}/P_1$
1.70	0.64054	3.20500	1.45833	0.92647	0.85572	4.22383
1.71	0.63798	3.24478	1.46552	0.93767	0.85156	4.26720
1.72	0.63545	3.28480	1.47274	0.94884	0.84736	4.31083
1.73	0.63296	3.32505	1.47999	0.95997	0.84312	4.35473
1.74	0.63051	3.38553	1.48727	0.97107	0.83886	4.39890
1.75	0.62809	3.40625	1.49458	0.98214	0.83457	4.44334
1.76	0.62570	3.44720	1.50192	0.99318	0.83024	4.48804
1.77	0.62335	3.48838	1.50929	1.00419	0.82589	4.53301
1.78	0.62104	3.52980	1.51669	1.01517	0.82151	4.57825
1.79	0.61875	3.57145	1.52412	1.02612	0.81711	4.62375
1.80	0.61650	3.61333	1.53158	1.03704	0.81268	4.66952
1.81	6.61428	3.65545	1.53907	1.04793	0.80823	4.71555
1.82	0.61209	3.69780	1.54659	1.05879	0.80376	4.76185
1.83	0.60993	3.74038	1.55415	1.06963	0.79927	4.80841
1.84	0.60780	3.78320	1.56173	1.08043	0.79476	4.85524
1.85	0.60570	3.82625	1.56935	1.09122	0.79023	4.90234
1.86	0.60363	3.86953	1.57700	1.10197	0.78569	4.94970
1.87	0.60158	3.91305	1.58468	1.11270	0.78112	4.99732
1.88	0.59957	3.95680	1.59239	1.12340	0.77655	5.04521
1.89	0.59758	4.00078	1.60014	1.13408	0.77196	5.09336
1.90	0.59562	4.04500	1.60792	1.14474	0.76736	5.14178
1.91	0.59368	4.08945	1.61573	1.15537	0.76274	5.19046
1.92	0.59177	4.13413	1.62357	1.16597	0.75812	5.23940
1.93	0.58988	4.17905	1.63144	1.17655	0.75349	5.28861
1.94	0.58802	4.22420	1.63935	1.18711	0.74884	5.33808
1.95	0.58618	4.26958	1.64729	1.19765	0.74420	5.38782
1.96	0.58437	4.31520	1.65527	1.20816	0.73954	5.43782
1.97	0.58258	4.36105	1.66328	1.21865	0.73488	5.48808
1.98	0.58082	4.40713	1.67132	1.22912	0.73021	5.53860
1.99	0.57907	4.45345	1.67939	1.23957	0.72555	5.58939
2.00	0.57735	4.50000	1.68750	1.25000	0.72087	5.64044
2.01	0.57565	4.54678	1.69564	1.26041	0.71620	5.69175
2.02	0.57397	4.59380	1.70382	1.27079	0.71153	5.74333
2.03	0.57231	4.64105	1.71203	1.28116	0.70685	5.79517
2.04	0.57068	4.63853	1.72027	1.29150	0.70218	5.84727

Table A–21

Normal Shock Parameters (Continued)

M_1	M_2	P_2/P_1	T_2/T_1	$\Delta V/a_1$	$P_{t,2}/P_{t,1}$	$P_{t,2}/P_1$
2.05	0.56906	4.73625	1.72855	1.30183	0.69751	5.89963
2.06	0.56747	4.78420	1.73686	1.31214	0.69284	5.95226
2.07	0.56589	4.83238	1.74521	1.32242	0.68817	6.00514
2.08	0.56433	4.88080	1.75359	1.33269	0.68351	6.05829
2.09	0.56280	4.92945	1.76200	1.34294	0.67885	6.11170
2.10	0.56128	4.97833	1.77045	1.35317	0.67420	6.16537
2.11	0.55978	5.02745	1.77893	1.36339	0.66956	6.21931
2.12	0.55829	5.07680	1.78745	1.37358	0.66492	6.27351
2.13	0.55683	5.12638	1.79601	1.38376	0.66029	6.32796
2.14	0.55538	5.17620	1.80459	1.39393	0.65567	6.38268
2.15	0.55395	5.22625	1.81322	1.40407	0.65105	6.43766
2.16	0.55254	5.27653	1.82188	1.41420	0.64645	6.49290
2.17	0.55115	5.32705	1.83057	1.42431	0.64185	6.54841
2.18	0.54977	5.37780	1.83930	1.43440	0.63727	6.60417
2.19	0.54840	5.42878	1.84806	1.44448	0.63270	6.66019
2.20	0.54706	5.48000	1.85686	1.45455	0.62814	6.71648
2.21	0.54572	5.53145	1.86569	1.46459	0.62359	6.77303
2.22	0.54441	5.58313	1.87456	1.47462	0.61905	6.82983
2.23	0.54311	5.63505	1.88347	1.48464	0.61453	6.88690
2.24	0.54182	5.68720	1.89241	1.49464	0.61002	6.94423
2.25	0.54055	5.73958	1.90138	1.50463	0.60553	7.00182
2.26	0.53930	5.79220	1.91040	1.51460	0.60105	7.05967
2.27	0.53805	5.84505	1.91944	1.52456	0.59659	7.11778
2.28	0.53683	5.89813	1.92853	1.53450	0.59214	7.17616
2.29	0.53561	5.95145	1.93765	1.54443	0.58771	7.23479
2.30	0.53441	6.00500	1.94680	1.55435	0.58329	7.29368
2.31	0.53322	6.05878	1.95599	1.56425	0.57890	7.35283
2.32	0.53205	6.11280	1.96522	1.57414	0.57452	7.41225
2.33	0.53089	6.16705	1.97448	1.58401	0.57015	7.47192
2.34	0.52974	6.22153	1.98378	1.59387	0.58581	7.53185
2.35	0.52861	6.27625	1.99311	1.60372	0.56148	7.59205
2.36	0.52749	6.33120	2.00249	1.61356	0.55718	7.65250
2.37	0.52638	6.38638	2.01189	1.62338	0.55289	7.71321
2.38	0.52528	6.44180	2.02134	1.63319	0.54862	7.77419
2.39	0.52419	6.49745	2.03082	1.64299	0.54437	7.83542

(continued)

Table A–21

Normal Shock Parameters (Continued)

M_1	M_2	P_2/P_1	T_2/T_1	$\Delta V/a_1$	$P_{t,2}/P_{t,1}$	$P_{t,2}/P_1$
2.40	0.52312	6.55333	2.04033	1.65278	0.54014	7.89691
2.41	0.52206	6.60945	2.04988	1.66255	0.53594	7.95867
2.42	0.52100	6.66560	2.05947	1.67231	0.53175	8.02068
2.43	0.51996	6.72238	2.06910	1.68206	0.52758	8.08295
2.44	0.51894	6.77920	2.07876	1.69180	0.52344	8.14549
2.45	0.51792	6.83625	2.08846	1.70153	0.51931	8.20828
2.46	0.51691	6.89353	2.09819	1.71125	0.51521	8.27133
2.47	0.51592	6.95105	2.10797	1.72095	0.51113	8.33464
2.48	0.51493	7.00880	2.11777	1.73065	0.50707	8.39821
2.49	0.51395	7.06678	2.12762	1.74033	0.503603	8.46205
2.50	0.51299	7.12500	2.13750	1.75000	0.49901	8.52614
2.51	0.51203	7.18345	2.14742	1.75966	0.49502	8.59049
2.52	0.51109	7.24213	2.15737	1.76931	0.49105	8.65510
2.53	0.51015	7.30105	2.16737	1.77895	0.48711	8.71996
2.54	0.50923	7.36020	2.17739	1.78858	0.48318	8.78509
2.55	0.50831	7.41958	2.18746	1.79820	0.47928	8.85048
2.56	0.50741	7.47920	2.19756	1.80781	0.47540	8.91613
2.57	0.50651	7.53905	2.20770	1.81741	0.47155	8.98203
2.58	0.50562	7.59913	2.21788	1.82700	0.46772	9.04820
2.59	0.50474	7.65945	2.22809	1.83658	0.46391	9.11462
2.60	0.50387	7.72000	2.23834	1.84615	0.46012	9.18131
2.61	0.50301	7.78078	2.24863	1.85572	0.45636	9.24825
2.62	0.50216	7.84180	2.25396	1.86527	0.45263	9.31545
2.63	0.50131	7.90305	2.26932	1.87481	0.44891	9.38291
2.64	0.50048	7.96453	2.27972	1.88434	0.44522	9.45064
2.65	0.49965	8.02625	2.29015	1.89387	0.44156	9.51862
2.66	0.49883	8.08820	2.30063	1.90338	0.43792	9.58685
2.67	0.49802	8.15038	2.31114	1.91289	0.43430	9.65535
2.68	0.49722	8.21280	2.32168	1.92239	0.43070	9.72411
2.69	0.49642	8.27545	2.33227	1.93188	0.42714	9.79312
2.70	0.49563	8.33833	2.34289	1.94136	0.42359	9.86240
2.71	0.49485	8.40145	2.35355	1.95083	0.42007	9.93193
2.72	0.49408	8.46480	2.36425	1.96029	0.41657	10.00173
2.73	0.49332	8.52838	2.37498	1.96975	0.41310	10.07178
2.74	0.49256	8.59220	2.38576	1.97920	0.40965	10.14209

Table A–21

Normal Shock Parameters (Continued)

M_1	M_2	P_2/P_1	T_2/T_1	$\Delta V/a_1$	$P_{t,2}/P_{t,1}$	$P_{t,2}/P_1$
2.75	0.49181	8.65625	2.39657	1.98864	0.40623	10.21266
2.76	0.49107	8.72053	2.40741	1.99807	0.40283	10.28349
2.77	0.49033	8.78505	2.41830	2.00749	0.39945	10.35457
2.78	0.48960	8.84980	2.42922	2.01691	0.39610	10.42592
2.79	0.48888	8.91478	2.44018	2.02631	0.39277	10.49752
2.80	0.48817	8.98000	2.45117	2.03571	0.38946	10.56939
2.81	0.48746	9.04545	2.46221	2.04511	0.38618	10.64151
2.82	0.48676	9.11113	2.47328	2.05449	0.38293	10.71389
2.83	0.48606	9.17705	2.48439	2.06387	0.37969	10.78653
2.84	0.48538	9.24320	2.49554	2.07324	0.37849	10.85943
2.85	0.48469	9.30956	2.50672	2.08260	0.37330	10.93258
2.86	0.48402	9.37620	2.51794	2.09196	0.37014	11.00600
2.87	0.48335	9.44305	2.52920	2.10131	0.36700	11.07967
2.88	0.48269	9.51013	2.54050	2.11065	0.36389	11.15361
2.89	0.48203	9.57745	2.55183	2.11998	0.36080	11.22780
2.90	0.48138	9.64500	2.56321	2.12931	0.35778	11.30225
2.91	0.48073	9.71278	2.57462	2.13863	0.35469	11.37695
2.92	0.48010	9.78080	2.58607	2.14795	0.35167	11.45192
2.93	0.47946	9.84905	2.59755	2.15725	0.34867	11.52715
2.94	0.47884	9.91753	2.60908	2.16655	0.34570	11.60263
2.95	0.47821	9.98625	2.62064	2.17585	0.34275	11.67837
2.96	0.47760	10.05520	2.63224	2.18514	0.33982	11.75438
2.97	0.47699	10.12438	2.64387	2.19442	0.33692	11.83064
2.98	0.47638	10.19380	2.65555	2.20369	0.33404	11.90715
2.99	0.47578	10.26345	2.66726	2.21296	0.33118	11.98393
3.00	0.47519	10.33333	2.67901	2.22222	0.32834	12.06096
3.01	0.47460	10.40345	2.69080	2.23148	0.32553	12.13826
3.02	0.47402	10.47380	2.70263	2.24073	0.32274	12.21581
3.03	0.47344	10.54438	2.71449	2.24997	0.31997	12.29362
3.04	0.47287	10.61520	2.72639	2.25921	0.31723	12.37169
3.05	0.47230	10.68625	2.73833	2.26844	0.31450	12.45002
3.06	0.47174	10.75753	2.75031	2.27767	0.31180	12.52860
3.07	0.47118	10.82905	2.76233	2.28689	0.30912	12.60745
3.08	0.47063	10.90080	2.77438	2.29610	0.30646	12.68655
3.09	0.47008	10.97278	2.78647	2.30531	0.30383	12.76591

(*continued*)

Table A–21

Normal Shock Parameters (Continued)

M_1	M_2	P_2/P_1	T_2/T_1	$\Delta V/a_1$	$P_{t.2}/P_{t.1}$	$P_{t.2}/P_1$
3.10	0.46953	11.04500	2.79860	2.31452	0.30121	12.84553
3.11	0.46899	11.11745	2.81077	2.32371	0.29862	12.92540
3.12	0.46846	11.19013	2.82298	2.33291	0.29605	13.00554
3.13	0.46793	11.26305	2.83522	2.34209	0.29350	13.08593
3.14	0.45741	11.33620	2.84750	2.35127	0.29097	13.16659
3.15	0.46689	11.40958	2.85982	2.36045	0.28846	13.24750
3.16	0.46637	11.48320	2.87218	2.36962	0.28597	13.32866
3.17	0.46586	11.55705	2.88458	2.37879	0.28350	13.41009
3.18	0.46535	11.63113	2.89701	2.38795	0.28106	13.49178
3.19	0.46485	11.70545	2.90948	2.39710	0.27863	13.57372
3.20	0.46435	11.78000	2.92199	2.40625	0.27623	13.65592
3.21	0.46385	11.85478	2.93454	2.41539	0.27384	13.73838
3.22	0.46336	11.92980	2.94713	2.42453	0.27148	13.82110
3.23	0.46288	12.00505	2.95975	2.43367	0.26914	13.90407
3.24	0.45240	12.08053	2.97241	2.44280	0.26681	13.98731
3.25	0.46192	12.15625	2.98511	2.45192	0.26451	14.07080
3.26	0.46144	12.23220	2.99785	2.46104	0.26222	14.15455
3.27	0.46097	12.30838	3.01063	2.47016	0.25996	14.23856
3.28	0.46051	12.38480	3.02345	2.47927	0.25771	14.32283
3.29	0.46004	12.46145	3.03630	2.48837	0.25548	14.40735
3.30	0.45959	12.53833	3.04919	2.49747	0.25328	14.49214
3.31	0.45913	12.61545	3.06212	2.50657	0.25109	14.57718
3.32	0.45868	12.69280	3.07509	2.51566	0.24892	14.66248
3.33	0.45823	12.77038	3.08809	2.52475	0.24677	14.74804
3.34	0.45779	12.84820	3.10114	2.53383	0.24463	14.83385
3.35	0.45735	12.92625	3.11422	2.54291	0.24252	14.91992
3.36	0.45691	13.00453	3.12734	2.55198	0.24043	15.00626
3.37	0.45648	13.08305	3.14050	2.56105	0.23835	15.09285
3.38	0.45605	13.16180	3.15370	2.57012	0.23629	15.17969
3.39	0.45562	13.24078	3.16693	2.57918	0.23425	15.26680
3.40	0.45520	13.32000	3.18021	2.58824	0.23223	15.35417
3.41	0.45478	13.39945	3.19352	2.59729	0.23022	15.44179
3.42	0.45436	13.47913	3.20687	2.60634	0.22823	15.52967
3.43	0.45395	13.55905	3.22026	2.61538	0.22626	15.61781
3.44	0.45354	12.63920	3.23369	2.62442	0.22431	15.70620

Table A–21

Normal Shock Parameters (Continued)

M_1	M_2	P_2/P_1	T_2/T_1	$\Delta V/a_1$	$P_{t,2}/P_{t,1}$	$P_{t,2}/P_1$
3.45	0.45314	13.71958	3.24715	2.63345	0.22237	15.79486
3.46	0.45273	13.80020	3.26065	2.64249	0.22045	15.88377
3.47	0.45233	13.88105	3.27420	2.65151	0.21855	15.97294
3.48	0.45194	13.96213	3.28778	2.66054	0.21667	16.06237
3.49	0.45154	14.04345	3.30139	2.66956	0.21480	16.15206
3.50	0.45115	14.12500	3.31505	2.67857	0.21295	16.24200
3.51	0.45077	14.20678	3.32875	2.68758	0.21111	16.33220
3.52	0.45038	14.28880	3.34248	2.69659	0.20929	16.42266
3.53	0.45000	14.37105	3.35625	2.70559	0.20749	16.51338
3.54	0.44962	14.45353	3.37006	2.71460	0.20570	16.60436
3.55	0.44925	14.53625	3.38391	2.72359	0.20393	16.69559
3.56	0.44887	14.61920	3.39780	2.73258	0.20218	16.78709
3.57	0.44850	14.70238	3.41172	2.74157	0.20044	16.87884
3.58	0.44814	14.78580	3.42569	2.75056	0.19871	16.97065
3.59	0.44777	14.86945	3.43969	2.75954	0.19701	17.06311
3.60	0.44741	14.95333	3.45373	2.76852	0.19531	17.15564
3.61	0.44705	15.03745	3.46781	2.77749	0.19363	17.24842
3.62	0.44670	15.12180	3.48192	2.78646	0.19197	17.34146
3.63	0.44635	15.20638	3.49608	2.79543	0.19032	17.43476
3.64	0.44600	15.29120	3.51027	2.80440	0.18869	17.52831
3.65	0.44565	15.37625	3.52451	2.81336	0.18707	17.62213
3.66	3.44530	15.46153	3.53878	2.82231	0.18547	17.71620
3.67	0.44496	15.54705	3.55309	2.83127	0.18388	17.81053
3.68	0.44462	15.63280	3.56743	2.84022	0.18230	17.90512
3.69	0.44428	15.71878	3.58182	2.84916	0.18074	17.99996
3.70	0.44395	15.80500	3.59624	2.85811	0.17919	18.09507
3.71	0.44362	15.89145	3.61071	2.86705	0.17766	18.19043
3.72	0.44329	15.97813	3.62521	2.87599	0.17614	18.28605
3.73	0.44296	16.08505	3.63975	2.88492	0.17464	18.38192
3.74	0.44263	16.15220	3.65433	2.89385	0.17314	18.47806
3.75	0.44231	16.23958	3.66894	2.90278	0.17166	18.57445
3.76	0.44199	16.32720	3.68360	2.91170	0.17020	18.67110
3.77	0.44167	16.41505	3.69829	2.92062	0.16875	18.76801
3.78	0.44136	16.50313	3.71302	2.92954	0.16781	18.86518
3.79	0.44104	16.59145	3.72779	2.93846	0.16588	18.96260

(continued)

Normal Shock Parameters (Continued)

M_1	M_2	p_2/p_1	T_2/T_1	$\Delta V/a_1$	$p_{t,2}/p_{t,1}$	$p_{t,2}/p_1$
3.80	0.44073	16.68000	3.74260	2.94737	0.16447	19.06029
3.81	0.44042	16.76878	3.75745	2.95628	0.16307	19.15823
3.82	0.44012	16.85780	3.77234	2.96518	0.16168	19.25642
3.83	0.43981	16.94705	3.78726	2.97409	0.16031	19.35488
3.84	0.43951	17.03653	3.80223	2.98299	0.15895	19.45359
3.85	0.43921	17.12623	3.81723	2.99188	0.15760	19.55257
3.86	0.43891	17.21620	3.83227	3.00076	0.15626	19.65180
3.87	0.43862	17.30638	3.84735	3.00967	0.15493	19.75128
3.88	0.43832	17.39680	3.86246	3.01856	0.15362	19.85103
3.89	0.43803	17.48745	3.87762	3.02744	0.15232	19.95103
3.90	0.43774	17.57833	3.89281	3.03632	0.15103	20.05129
3.91	0.43746	17.66945	3.90805	3.04520	0.14975	20.15181
3.92	0.43717	17.76080	3.92332	3.05408	0.14848	20.25259
3.93	0.43689	17.85238	3.93863	3.06296	0.14723	20.35362
3.94	0.43661	17.94420	3.95398	3.07183	0.14598	20.45491
3.95	0.43633	18.03625	3.96936	3.08070	0.14475	20.55646
3.96	0.43605	18.12853	3.98479	3.08956	0.14353	20.65827
3.97	0.43577	18.22105	4.00025	3.09843	0.14232	20.76034
3.98	0.43550	18.31380	4.01575	3.10729	0.14112	20.86266
3.99	0.43523	18.40678	4.03130	3.11614	0.13993	20.96524
4.00	0.43496	18.50000	4.04687	3.12500	0.13876	21.06808
4.10	0.43236	19.44500	4.20479	3.21341	0.12756	22.11065
4.20	0.42994	20.41333	4.36657	3.30159	0.11733	23.17899
4.30	0.42767	21.40500	4.53221	3.38953	0.10800	24.27311
4.40	0.42554	22.42000	4.70171	3.47727	0.09948	25.39300
4.50	0.42355	23.45833	4.87509	3.56481	0.09170	26.53867
4.60	0.42168	24.52000	5.05233	3.65217	0.08459	27.71010
4.70	0.41992	25.60500	5.23343	3.73936	0.07809	28.90729
4.80	0.41826	26.71333	5.41842	3.82639	0.07214	30.13026
4.90	0.41670	27.84500	5.60727	3.91327	0.06670	31.37898
5.00	0.41523	29.00000	5.80000	4.00000	0.06172	32.65347
5.10	0.41384	30.17833	5.99660	4.08660	0.05715	33.95373
5.20	0.41252	31.38000	6.19709	4.17308	0.05297	35.27974
5.30	0.41127	32.60500	6.40144	4.25943	0.04913	36.63152
5.40	0.41009	33.85333	6.60968	4.34568	0.04560	38.00906

Table A–21

Normal Shock Parameters (Continued)

M_1	M_2	P_2/P_1	T_2/T_1	$\Delta V/a_1$	$P_{t,2}/P_{t,1}$	$P_{t,2}/P_1$
5.50	0.40897	35.12500	6.82180	4.43182	0.04236	39.41235
5.60	0.40791	36.42000	7.03779	4.51786	0.03938	40.84141
5.70	0.40690	37.73833	7.25767	4.60380	0.03664	42.29622
5.80	0.40594	39.08000	7.48143	4.68966	0.03412	43.77679
5.90	0.40503	40.44500	7.70907	4.77542	0.03179	45.28312
6.00	0.40416	41.83333	7.94059	4.86111	0.02965	46.81521
6.10	0.40333	43.24500	9.17599	4.74672	0.02767	48.37305
6.20	0.40254	44.68000	8.41528	5.03226	0.02584	49.95665
6.30	0.40179	46.13833	8.65845	5.11772	0.02416	51.56600
6.40	0.40107	47.62000	8.90550	5.20312	0.02259	53.20111
6.50	0.40038	49.12500	9.15643	5.28846	0.02115	54.86198
6.60	0.39972	50.65333	9.41126	5.37374	0.01981	56.54860
6.70	3.39909	52.20500	9.66996	5.45896	0.01857	58.26097
6.80	0.39849	53.78000	9.93255	5.54412	0.01741	59.99910
6.90	0.39791	55.37833	10.19903	5.62923	0.01634	61.76299
7.00	0.39736	57.00000	10.46939	5.71429	0.01535	63.55263
7.50	0.39491	65.45833	11.87948	6.13889	0.01133	72.88713
8.00	0.39289	74.50000	13.38672	6.56250	0.00849	82.86547
8.50	0.39121	84.12500	14.99113	6.98529	0.00645	93.48763
9.00	0.38980	94.33333	16.69273	7.40741	0.00496	104.75360
9.50	0.38860	105.12500	18.49152	7.82895	0.00387	116.66339
10.00	0.38758	116.50000	20.38750	8.25000	0.00304	129.21697
∞	0.37796	∞	∞	∞	0.0	∞

Table A–22

Fanno Flow Parameters

M	T/T*	p/p*	P_t/P_t^*	V/V*	fL_{max}/D	S_{max}/R
0.0	1.20000	∞	∞	0.0	∞	∞
0.01	1.19998	109.54342	57.87384	0.01095	7134.40454	4.05827
0.02	1.19990	54.77006	28.94213	0.02191	1778.44988	3.36530
0.03	1.19978	36.31155	19.30054	0.03286	787.08139	2.96013
0.04	1.19962	27.38175	14.48149	0.04381	440.35221	2.67287
0.05	1.19940	21.90343	11.59144	0.05476	280.02031	2.45027
0.06	1.19914	18.25085	9.66591	0.06570	193.03108	2.26861
0.07	1.19883	15.64155	8.29153	0.07664	140.65501	2.11523
0.08	1.19847	13.68431	7.26161	0.08758	106.71822	1.98260
0.09	1.19806	12.16177	4.46134	0.09851	83.49612	1.86584
0.10	1.19760	10.94351	5.62183	0.10944	66.92156	1.76161
0.11	1.19710	9.94656	5.29923	0.12035	54.68790	1.66756
0.12	1.19655	9.11559	4.86432	0.13126	45.40796	1.58193
0.13	1.19596	8.41230	4.49686	0.14217	38.20700	1.50338
0.14	1.19531	7.80932	4.18240	0.15306	32.51131	1.43089
0.15	1.19462	7.28659	3.91034	0.16395	27.93197	1.36363
0.16	1.19389	6.82907	3.67274	0.17482	24.19783	1.30094
0.17	1.19310	6.42525	3.46351	0.18569	21.11518	1.24228
0.18	1.19227	6.06618	3.27793	0.19654	18.54265	1.18721
0.19	1.19140	5.74480	3.11226	0.20739	16.37516	1.13535
0.20	1.19048	5.45545	2.96352	0.21822	14.53327	1.08638
0.21	1.18951	5.19355	2.82929	0.22904	12.95602	1.04003
0.22	1.18850	4.95537	2.70760	0.23984	11.59605	0.99606
0.23	1.18744	4.73781	2.59681	0.25063	10.41609	0.95428
0.24	1.18633	4.53829	2.49556	0.26141	9.38648	0.91451
0.25	1.18519	4.35465	2.40271	0.27217	8.48341	0.87660
0.26	1.18399	4.18505	1.31729	0.28291	7.68757	0.84040
0.27	1.18276	4.02795	2.23847	0.29364	6.98317	0.80579
0.28	1.18147	3.88199	2.16555	0.30435	6.35721	0.77268
0.29	1.18015	3.74602	2.09793	0.31504	5.79891	0.74095
0.30	1.17878	3.61906	2.03507	0.32572	5.29925	0.71053
0.31	1.17737	3.50022	1.97651	0.33637	4.85066	0.68133
0.32	1.17592	3.38874	1.92185	0.34701	4.44674	0.65329
0.33	1.17442	3.28396	1.87074	0.35762	4.08205	0.62634
0.34	1.17288	3.18529	1.82288	0.36822	3.75195	0.60042

Table A–22

Fanno Flow Parameters (Continued)

M	T/T^*	p/p^*	P_t/P_t^*	V/V^*	fL_{max}/D	S_{max}/R
0.35	1.17130	3.09219	1.77797	0.37879	3.45245	0.57547
0.36	1.16968	3.00422	1.73578	0.38935	3.18012	0.55146
0.37	1.16802	2.92094	1.69609	0.39988	2.93198	0.52832
0.38	1.16632	2.84200	1.65870	0.41039	2.70545	0.50603
0.39	1.16457	2.76706	1.62343	0.42087	2.49828	0.48454
0.40	1.16279	2.69582	1.59014	0.43133	2.30849	0.46382
0.41	1.16097	2.62801	1.55867	0.44177	2.13436	0.44384
0.42	1.15911	2.56338	1.52890	0.45218	1.97437	0.42455
0.43	1.15721	2.50171	1.50072	0.46257	1.82715	0.40594
0.44	1.15527	2.44280	1.47401	0.47293	1.69152	0.38798
0.45	1.15329	2.38648	1.44367	0.48326	1.56643	0.37065
0.46	1.15128	2.33256	1.42463	0.49357	1.45091	0.35391
0.47	1.14923	2.28089	1.40180	0.50385	1.34413	0.33775
0.48	1.14714	2.23135	1.38010	0.51410	1.24534	0.32215
0.49	1.14502	2.18378	1.35947	0.52433	1.15385	0.30709
0.50	1.14286	2.13809	1.33984	0.53452	1.06906	0.29255
0.51	1.14066	2.09415	1.32117	0.54469	0.99041	0.27852
0.52	1.13843	2.05187	1.30339	0.55483	0.91742	0.26497
0.53	1.13617	2.01116	1.28645	0.56493	0.84962	0.25189
0.54	1.13387	1.97192	1.27032	0.57501	0.78663	0.23927
0.55	1.13154	1.93407	1.25495	0.58506	0.72805	0.22709
0.56	1.12918	1.89755	1.24029	0.59507	0.67357	0.21535
0.57	1.12678	1.86228	1.22633	0.60505	0.62287	0.20402
0.58	1.12435	1.82820	1.21301	0.61501	0.57568	0.19310
0.59	1.12189	1.79525	1.20031	0.62492	0.53174	0.18258
0.60	1.11940	1.76336	1.18820	0.63481	0.49082	0.17244
0.61	1.11688	1.73250	1.17665	0.64466	0.45271	0.16267
0.62	1.11433	1.70261	1.16565	0.65448	0.41720	0.15328
0.63	1.11175	1.67364	1.15515	0.66427	0.38412	0.14423
0.64	1.10914	1.64556	1.14515	0.67402	0.35330	0.13553
0.65	1.10650	1.61831	1.13562	0.68374	0.32459	0.12718
0.66	1.10383	1.59187	1.12654	0.69342	0.29785	0.11915
0.67	1.10114	1.56620	1.11789	0.70307	0.27295	0.11144
0.68	1.09842	1.54126	1.10965	0.71268	0.24978	0.10405
0.69	1.09567	1.51702	1.10182	0.72225	0.22820	0.09696

(*continued*)

Table A–22

Fanno Flow Parameters (Continued)

M	T/T^*	p/p^*	p_t/p_t^*	V/V^*	fL_{max}/D	S_{max}/R
0.70	1.09290	1.49345	1.09437	0.73179	0.20814	0.09018
0.71	1.09010	1.47053	1.08729	0.74129	0.18948	0.08369
0.72	1.08727	1.44823	1.08057	0.75076	0.17215	0.07749
0.73	1.08442	1.42652	1.07419	0.76019	0.15605	0.07157
0.74	1.08155	1.40537	1.06814	0.76958	0.14112	0.06592
0.75	1.07865	1.38478	1.06242	0.77894	0.12728	0.06055
0.76	1.07573	1.36470	1.05700	0.78825	0.11447	0.05543
0.77	1.07279	1.34514	1.05188	0.79753	0.10262	0.05058
0.78	1.06982	1.32605	1.04705	0.80677	0.09167	0.04598
0.79	1.06684	1.30744	1.04251	0.81597	0.08158	0.04163
0.80	1.06383	1.28928	1.03823	0.82514	0.07229	0.03752
0.81	1.06080	1.27155	1.03422	0.83426	0.06376	0.03365
0.82	1.05775	1.25423	1.03046	0.84335	0.05593	0.03001
0.83	1.05469	1.23732	1.02696	0.85239	0.04878	0.02660
0.84	1.05160	1.22080	1.02370	0.86140	0.04226	0.02342
0.85	1.04849	1.20466	1.02067	0.87037	0.03633	0.02046
0.86	1.04537	1.18888	1.01787	0.87929	0.03097	0.01771
0.87	1.04223	1.17344	1.01530	0.88818	0.02613	0.01518
0.88	1.03907	1.15835	1.01294	0.89703	0.02179	0.01286
0.89	1.03589	1.14358	1.01080	0.90583	0.01793	0.01074
0.90	1.03270	1.12913	1.00886	0.91460	0.01451	0.00882
0.91	1.02950	1.11499	1.00713	0.92332	0.01151	0.00711
0.92	1.02627	1.10114	1.00560	0.93201	0.00891	0.00558
0.93	1.02304	1.08759	1.00426	0.94065	0.00669	0.00425
0.94	1.01978	1.07430	1.00311	0.94925	0.00482	0.00310
0.95	1.01652	1.06129	1.00215	0.95781	0.00328	0.00214
0.96	1.01324	1.04854	1.00136	0.96633	0.00206	0.00136
0.97	1.00995	1.03604	1.00076	0.97481	0.00113	0.00076
0.98	1.00664	1.02379	1.00034	0.98325	0.00049	0.00034
0.99	1.00333	1.01178	1.00008	0.99165	0.00012	0.00008
1.00	1.00000	1.00000	1.00000	1.00000	0.00000	0.00000
1.01	0.99666	0.98844	1.00008	1.00831	0.00012	0.00008
1.02	0.99331	0.97711	1.00033	1.01658	0.00046	0.00033
1.03	0.98995	0.96598	1.00074	1.02481	0.00101	0.00074
1.04	0.98658	0.95507	1.00131	1.03300	0.00177	0.00130

Table A–22

Fanno Flow Parameters (Continued)

M	T/T^*	P/P^*	P_t/P_t^*	V/V^*	fL_{max}/D	S_{max}/R
1.05	0.98320	0.94435	1.00203	1.04114	0.00271	0.00203
1.06	0.97982	0.93383	1.00291	1.04925	0.00384	0.00290
1.07	0.97542	0.92349	1.00394	1.05731	0.00513	0.00393
1.08	0.97302	0.91335	1.00512	1.06533	0.00658	0.00511
1.09	0.96960	0.90338	1.00645	1.07331	0.00819	0.00643
1.10	0.96618	0.89359	1.00793	1.08124	0.00994	0.00789
1.11	0.96276	0.88397	1.00955	1.08913	0.01182	0.00950
1.12	0.95932	0.87451	1.01131	1.09699	0.01382	0.01125
1.13	0.95589	0.86522	1.01322	1.10479	0.01595	0.01313
1.14	0.95244	0.85608	1.01527	1.11256	0.01819	0.01515
1.15	0.94899	0.84710	1.01745	1.12029	0.02053	0.01730
1.16	0.94554	0.83826	1.01978	1.12797	0.02293	0.01959
1.17	0.94208	0.82958	1.02224	1.13561	0.02552	0.02200
1.18	0.93861	0.82103	1.02484	1.14321	0.02814	0.02454
1.19	0.93515	0.81263	1.02757	1.15077	0.03085	0.02720
1.20	0.93168	0.80436	1.03044	1.15828	0.03364	0.02999
1.21	0.92820	0.79623	1.03344	1.16575	0.03650	0.03289
1.22	0.92473	0.78822	1.03657	1.17319	0.03943	0.03592
1.23	0.92125	0.78034	1.03983	1.18057	0.04242	0.03906
1.24	0.91777	0.77258	1.04323	1.18792	0.04547	0.04232
1.25	0.91429	0.76495	1.04675	1.19523	0.04858	0.04569
1.26	0.91080	0.75743	1.05041	1.20249	0.05174	0.04918
1.27	0.90732	0.75003	1.05419	1.20972	0.05495	0.05277
1.28	0.90383	0.74274	1.05810	1.21690	0.05820	0.05647
1.29	0.90035	0.73556	1.06214	1.22404	0.06150	0.06028
1.30	0.89686	0.72848	1.06630	1.23114	0.06483	0.06420
1.31	0.89338	0.72152	1.07060	1.23819	0.06820	0.06822
1.32	0.88989	0.71465	1.07502	1.24521	0.07161	0.07234
1.33	0.88641	0.70789	1.07957	1.25218	0.07504	0.07656
1.34	0.88292	0.70122	1.08424	1.25912	0.07850	0.08088
1.35	0.87944	0.69466	1.08904	1.26601	0.08199	0.08529
1.36	0.87596	0.68818	1.09396	1.27286	0.08550	0.08981
1.37	0.87249	0.68180	1.09902	1.27968	0.08904	0.09441
1.38	0.86901	0.67551	1.10419	1.28645	0.09259	0.09911
1.39	0.86554	0.66931	1.10950	1.29318	0.09615	0.10391

(continued)

Table A–22

Fanno Flow Parameters (Continued)

M	T/T^*	p/p^*	p_t/p_t^*	V/V^*	fL_{max}/D	S_{max}/R
1.40	0.86207	0.66320	1.11493	1.29987	0.09974	0.10879
1.41	0.85860	0.65717	1.12048	1.30652	0.10334	0.11376
1.42	0.85514	0.65122	1.12616	1.31313	0.10694	0.11882
1.43	0.85168	0.64536	1.13197	1.31970	0.11056	0.12396
1.44	0.84822	0.63958	1.13790	1.32623	0.11419	0.12919
1.45	0.84477	0.63387	1.14396	1.33272	0.11782	0.13450
1.46	0.84133	0.62825	1.15015	1.33917	0.12146	0.13989
1.47	0.83788	0.62269	1.15646	1.34558	0.12511	0.14537
1.48	0.83445	0.61722	1.16290	1.35195	0.12875	0.15092
1.49	0.83101	0.61181	1.16947	1.35828	0.13240	0.15655
1.50	0.82759	0.60648	1.17617	1.36458	0.13605	0.16226
1.51	0.82416	0.60122	1.18299	1.37083	0.13970	0.16805
1.52	0.82075	0.59602	1.18994	1.37705	0.14335	0.17391
1.53	0.81734	0.59089	1.19702	1.38322	0.14699	0.17984
1.54	0.81393	0.58583	1.20423	1.38936	0.15063	0.18584
1.55	0.81054	0.58084	1.21157	1.39546	0.15427	0.19192
1.56	0.80715	0.57591	1.21904	1.40152	0.15790	0.19807
1.57	0.80376	0.57104	1.22664	1.40755	0.16152	0.20428
1.58	0.80038	0.56623	1.23438	1.41353	0.16514	0.21057
1.59	0.79701	0.56148	1.24224	1.41948	0.16875	0.21692
1.60	0.79365	0.55679	1.25023	1.42539	0.17236	0.22333
1.61	0.79080	0.55216	1.25386	1.43127	0.17595	0.22981
1.62	0.78695	0.54759	1.26663	1.43710	0.17954	0.23636
1.63	0.78361	0.54308	1.27502	1.44290	0.18311	0.24296
1.64	0.78027	0.53862	1.28355	1.44866	0.18667	0.24963
1.65	0.77695	0.53421	1.29222	1.45439	0.19023	0.25636
1.66	0.77663	0.52986	1.30102	1.46008	0.19377	0.26315
1.67	0.77033	0.52556	1.30996	1.46573	0.19729	0.27000
1.68	0.76703	0.52131	1.31904	1.47135	0.20081	0.27690
1.69	0.76374	0.51711	1.32825	1.47693	0.20431	0.28386
1.70	0.76046	0.51297	1.33761	1.48247	0.20780	0.29088
1.71	0.75718	0.50887	1.34710	1.48798	0.21128	0.29795
1.72	0.75392	0.50482	1.35674	1.49345	0.21474	0.30508
1.73	0.75067	0.50082	1.36651	1.49889	0.21819	0.31226
1.74	0.74742	0.49686	1.37643	1.50429	0.22162	0.31949

Table A–22

Fanno Flow Parameters (Continued)

M	T/T^*	p/p^*	P_t/P_t^*	V/V^*	fL_{max}/D	S_{max}/R
1.75	0.74419	0.49295	1.38649	1.50966	0.22504	0.32678
1.76	0.74096	0.48909	1.39670	1.51499	0.22844	0.33411
1.77	0.73774	0.48527	1.40705	1.52029	0.23182	0.34149
1.78	0.73454	0.48149	1.41755	1.52555	0.23519	0.34893
1.79	0.73134	0.47776	1.42819	1.53078	0.23855	0.35641
1.80	0.72816	0.47407	1.43898	1.53598	0.24189	0.36394
1.81	0.72498	0.47042	1.44992	1.54114	0.24521	0.37151
1.82	0.72181	0.46681	1.46101	1.54626	0.24851	0.37913
1.83	0.71866	0.46324	1.47225	1.55136	0.25180	0.38680
1.84	0.71551	0.45972	1.48365	1.55642	0.25507	0.39450
1.85	0.71238	0.45623	1.49519	1.56145	0.25832	0.40226
1.86	0.70925	0.45278	1.50689	1.56644	0.26156	0.41005
1.87	0.70614	0.44937	1.51875	1.57140	0.26478	0.41789
1.88	0.70304	0.44600	1.53076	1.57633	0.26798	0.42576
1.89	0.69995	0.44266	1.54293	1.58123	0.27116	0.43368
1.90	0.69686	0.43936	1.55326	1.58609	0.27433	0.44164
1.91	0.69379	0.43610	1.56774	1.59092	0.27748	0.44964
1.92	0.69073	0.43287	1.58039	1.59572	0.28061	0.45767
1.93	0.68769	0.42967	1.59320	1.60049	0.28372	0.46574
1.94	0.68465	0.42651	1.60617	1.60523	0.28681	0.47385
1.95	0.68162	0.42339	1.61931	1.60993	0.28989	0.48200
1.96	0.67861	0.42029	1.63261	1.61460	0.29295	0.49018
1.97	0.67561	0.41724	1.64608	1.61925	0.29599	0.49840
1.98	0.67262	0.41421	1.65972	1.62386	0.29901	0.50665
1.99	0.66964	0.41121	1.67352	1.62844	0.30201	0.51493
2.00	0.66667	0.40825	1.68750	1.63299	0.30500	0.52325
2.01	0.66371	0.40532	1.70165	1.63751	0.30796	0.53160
2.02	0.66076	0.40241	1.71597	1.64201	0.31091	0.53998
2.03	0.65783	0.39954	1.73047	1.64647	0.31384	0.54839
2.04	0.65491	0.39670	1.74514	1.65090	0.31676	0.55683
2.05	0.65200	0.39388	1.75999	1.65530	0.31965	0.56531
2.06	0.64910	0.39110	1.77502	1.65967	0.32253	0.57381
2.07	0.64621	0.38834	1.79022	1.66402	0.32538	0.58234
2.08	0.64334	0.38562	1.80561	1.66833	0.32822	0.59090
2.09	0.64047	0.38292	1.82119	1.67262	0.33105	0.59949

(*continued*)

Table A–22

Fanno Flow Parameters (Continued)

M	T/T^*	p/p^*	P_t/P_t^*	V/V^*	fL_{max}/D	S_{max}/R
2.10	0.63762	0.38024	1.83694	1.67687	0.33385	0.60810
2.11	0.63478	0.37760	1.85289	1.68110	0.33664	0.61674
2.12	0.63195	0.37498	1.86902	1.68530	0.33940	0.62541
2.13	0.62914	0.37239	1.88533	1.68947	0.34215	0.63411
2.14	0.62633	0.36982	1.90184	1.69362	0.34489	0.64282
2.15	0.62354	0.36728	1.91854	1.69774	0.34760	0.65157
2.16	0.62076	0.36476	1.98544	1.70183	0.35030	0.66033
2.17	0.61799	0.36227	1.95252	1.70589	0.35298	0.66912
2.18	0.61523	0.35980	1.96981	1.70992	0.35554	0.67794
2.19	0.61249	0.35736	1.98729	1.71393	0.35828	0.68677
2.20	0.60976	0.35494	2.00497	1.71791	0.36091	0.69563
2.21	0.60704	0.35255	2.02286	1.72187	0.36352	0.70451
2.22	0.60433	0.35017	2.04094	1.72579	0.36611	0.71341
2.23	0.60163	0.34782	2.05923	1.72970	0.36869	0.72233
2.24	0.59895	0.34550	2.07773	1.73357	0.37124	0.73128
2.25	0.59627	3.34319	2.09644	1.73742	0.37378	0.74024
2.26	0.59361	0.34091	2.11535	1.74125	0.37631	0.74922
2.27	0.59096	0.33865	2.13447	1.74504	0.37881	0.75822
2.28	0.58833	0.33641	2.15381	1.74882	0.38130	0.76724
2.29	0.58570	0.33420	2.17336	1.75257	0.38377	0.77628
2.30	0.58309	0.33200	2.19313	1.75629	0.38623	0.78533
2.31	0.58049	0.32983	2.21312	1.75999	0.38867	0.79440
2.32	0.57790	0.32767	2.23332	1.76366	0.39109	0.80349
2.33	0.57532	0.32554	2.25375	1.76731	0.39350	0.81260
2.34	0.57276	0.32342	2.27440	1.77093	0.39589	0.82172
2.35	0.57021	0.32133	2.29528	1.77453	0.39826	0.83085
2.36	0.56767	0.31925	2.31638	1.77811	0.40062	0.84001
2.37	0.56514	0.31720	2.33771	1.78166	0.40296	0.84917
2.38	0.56262	0.31516	2.35928	1.78519	0.40529	0.85835
2.39	0.56011	0.31314	2.38107	1.78869	0.40760	0.86755
2.40	0.55762	0.31114	2.40310	1.79218	0.40989	0.87676
2.41	0.55514	0.30916	2.42537	1.79563	0.41217	0.88598
2.42	0.55267	0.30720	2.44787	1.79907	0.41443	0.89522
2.43	0.55021	0.30525	2.47061	1.80248	0.41668	0.90447
2.44	0.54777	0.30332	2.49360	1.80587	0.41891	0.91373

Table A–22

Fanno Flow Parameters (Continued)

M	T/T^*	p/p^*	p_t/p_t^*	V/V^*	fL_{max}/D	S_{max}/R
2.45	0.54533	0.30141	2.51683	1.80924	0.42112	0.92300
2.46	0.54291	0.29952	2.54031	1.81258	0.42332	0.93229
2.47	0.54050	0.29765	2.56403	1.81591	0.42551	0.94158
2.48	0.53810	0.29579	2.58801	1.81921	0.42768	0.95089
2.49	0.53571	0.29394	2.61224	1.82249	0.42984	0.96021
2.50	0.53333	0.29212	2.63672	1.82574	0.43198	0.96954
2.51	0.53097	0.29031	2.66146	1.82898	0.43410	0.97887
2.52	0.52862	0.28852	2.68645	1.83219	0.43621	0.98822
2.53	0.52627	0.28674	2.71171	1.83536	0.43831	0.99758
2.54	0.52394	0.28498	2.73723	1.83855	0.44039	1.00695
2.55	0.52163	0.28323	2.76301	1.84170	0.44246	1.01632
2.56	0.51932	0.28150	2.78906	1.84483	0.44451	1.02571
2.57	0.51702	0.27978	2.81538	1.84794	0.44655	1.03510
2.58	0.51474	0.27808	2.84197	1.85103	0.44858	1.04450
2.59	0.51247	0.27640	2.86884	1.85410	0.45059	1.05391
2.60	0.51020	0.27473	2.89598	1.85714	0.45259	1.06332
2.61	0.50795	0.27307	2.92339	1.86017	0.45457	1.07274
2.62	0.50571	0.27143	2.95109	1.86318	0.45654	1.08217
2.63	0.50349	0.26980	2.97907	1.86616	0.45850	1.09161
2.64	0.50127	0.26818	3.00733	1.86913	0.46044	1.10105
2.65	0.49906	0.26658	3.03588	1.87208	0.46287	1.11050
2.66	0.49687	0.26500	3.06472	1.87501	0.46429	1.11996
2.67	0.49469	0.26342	3.09385	1.87792	0.46619	1.12942
2.68	0.49251	0.26186	3.12327	1.88081	0.46808	1.13888
2.69	0.49035	0.26032	3.15299	1.88368	0.46996	1.14835
2.70	0.48820	0.25878	3.18301	1.88653	0.47182	1.15783
2.71	0.48606	0.25726	3.21333	1.88936	0.47367	1.16731
2.72	0.48393	0.25575	3.24395	1.89218	0.47551	1.17679
2.73	0.48182	0.25426	3.27488	1.89497	0.47733	1.18628
2.74	0.47971	0.25278	3.30611	1.89775	0.47915	1.19577
2.75	0.47761	0.25131	3.33766	1.90051	0.48095	1.20527
2.76	0.47553	0.24985	3.36952	1.90325	0.48273	1.21477
2.77	0.47345	0.24840	3.40169	1.90598	0.48451	1.22427
2.78	0.47139	0.24697	3.43418	1.90868	0.48627	1.23378
2.79	0.46933	0.24555	3.46699	1.91137	0.48803	1.24329

(continued)

Table A-22

Fanno Flow Parameters (Continued)

M	T/T^*	p/p^*	p_t/p_t^*	V/V^*	fL_{max}/D	S_{max}/R
2.80	0.46729	0.24414	3.50012	1.91404	0.48976	1.25280
2.81	0.46526	0.24274	3.53358	1.91669	0.49149	1.26231
2.82	0.46323	0.24135	3.56737	1.91933	0.49321	1.27183
2.83	0.46122	0.23998	3.60148	1.92195	0.49491	1.28135
2.84	0.45922	0.23861	3.63593	1.92455	0.49660	1.29087
2.85	0.45723	0.23726	3.67072	1.92714	0.49828	1.30039
2.86	0.45525	0.23592	3.70584	1.92970	0.49995	1.30991
2.87	0.45328	0.23459	3.74131	1.93225	0.50161	1.31943
2.88	0.45132	0.23326	3.77711	1.93479	0.50326	1.32896
2.89	0.44937	0.23195	3.81327	1.93731	0.50489	1.33849
2.90	0.44743	0.23066	3.84977	1.93981	0.50652	1.34801
2.91	0.44550	0.22937	3.88662	1.94230	0.50813	1.35754
2.92	0.44358	0.22809	3.92383	1.94477	0.50973	1.36707
2.93	0.44167	0.22682	3.96139	1.94722	0.51132	1.37660
2.94	0.43977	0.22556	3.99932	1.94966	0.51290	1.38612
2.95	0.43788	0.22431	4.03760	1.95208	0.51447	1.39565
2.96	0.43600	0.22307	4.07625	1.95449	0.51603	1.40518
2.97	0.43413	0.22185	4.11527	1.95688	0.51758	1.41471
2.98	0.43226	0.22063	4.15466	1.95925	0.51912	1.42423
2.99	0.43041	0.21942	4.19443	1.96162	0.52064	1.43376
3.00	0.42857	0.26822	4.23457	1.96396	0.52216	1.44328
3.01	0.42674	0.21703	4.27509	1.96629	0.52367	1.45280
3.02	0.42492	0.21585	4.31599	1.96861	0.52516	1.46233
3.03	0.42310	0.21467	4.35728	1.97091	0.52665	1.47185
3.04	0.42130	0.21351	4.39895	1.97319	0.52813	1.48137
3.05	0.41951	0.21236	4.44102	1.97647	0.52959	1.49088
3.06	0.41772	0.21121	4.48347	1.97772	0.53105	1.50040
3.07	0.41595	0.21008	4.52633	1.97997	0.53249	1.50991
3.08	0.41418	0.20895	4.56959	1.98219	0.53393	1.51942
3.09	0.41242	0.20783	4.61325	1.96441	0.53536	1.52893
3.10	0.41068	0.20672	4.65731	1.98661	0.53678	1.53844
3.11	0.40894	0.20562	4.70178	1.98879	0.53818	1.54794
3.12	0.40721	0.20453	4.74667	1.99097	0.53958	1.55744
3.13	0.40549	0.20344	4.79197	1.99313	0.54097	1.56694
3.14	0.40378	0.20237	4.83769	1.99527	0.54235	1.57644

Table A–22

Fanno Flow Parameters (Continued)

M	T/T*	p/p*	p_t/p_t^*	V/V*	fL_{max}/D	S_{max}/R
3.15	0.40208	0.20130	4.88383	1.99740	0.54372	1.58593
3.16	0.40038	0.20024	4.93039	1.99952	0.54509	1.59542
3.17	0.39870	0.19919	4.97739	2.00162	0.54644	1.60490
3.18	0.39702	0.19814	5.02481	2.00372	0.54778	1.61439
3.19	0.39536	0.19711	5.07266	2.00579	0.54912	1.62387
3.20	0.39370	0.19608	5.12096	2.00786	0.55044	1.63334
3.21	0.39205	0.19506	5.16969	2.00991	0.55176	1.64281
3.22	0.39041	0.19405	5.21887	2.01195	0.55307	1.65228
3.23	0.38878	0.19304	5.25849	2.01398	0.55437	1.66174
3.24	0.38716	0.19204	5.31857	2.01599	0.55566	1.67123
3.25	0.38554	0.19105	5.36909	2.01799	0.55694	1.68066
3.26	0.38394	0.19007	5.42008	2.01998	0.55622	1.69011
3.27	0.38234	0.18909	5.47152	2.02196	0.55948	1.69956
3.28	0.38075	0.18812	5.52343	2.02392	0.56074	1.70900
3.29	0.37917	0.18716	5.57580	2.02587	0.56199	1.71844
3.30	0.37760	0.18621	5.62865	2.02781	0.56323	1.72787
3.31	0.37603	0.18526	5.68196	2.02974	0.56446	1.73730
3.32	0.37448	0.18432	5.73576	2.03165	0.56569	1.74672
3.33	0.37293	0.18339	5.79003	2.03356	0.56691	1.75614
3.34	0.37139	0.18246	5.84479	2.03545	0.56812	1.76555
3.35	0.36986	0.18154	3.90004	2.03733	0.56932	1.77496
3.36	0.36833	0.18063	5.95577	2.03920	0.57051	1.78436
3.37	0.36682	0.17972	6.01201	2.04106	0.57170	1.79376
3.38	0.36531	0.17882	6.06873	2.04290	0.57287	1.80315
3.39	0.36381	0.17793	6.12596	2.04474	0.57404	1.81254
3.40	0.36232	0.17704	6.18370	2.04656	0.57521	1.82192
3.41	0.36083	0.17616	6.24194	2.04837	0.57636	1.83129
3.42	0.35936	0.17528	6.30070	2.05017	0.57751	1.84066
3.43	0.35789	0.17441	6.35997	2.05196	0.57865	1.85002
3.44	0.35643	0.17355	6.41976	2.05374	0.57978	1.85638
3.45	0.35498	0.17270	6.48007	2.05551	0.58091	1.86873
3.46	0.35353	0.17185	6.54092	2.05727	0.58203	1.87808
3.47	0.35209	0.17100	6.60229	2.05901	0.58314	1.88742
3.48	0.35066	0.17016	6.66419	2.06075	0.58424	1.89675
3.49	0.34924	0.16933	6.72664	2.06247	0.58534	1.90608

(continued)

Table A–22

Fanno Flow Parameters (Continued)

M	T/T^*	p/p^*	p_t/p_t^*	V/V^*	fL_{max}/D	S_{max}/R
3.50	0.34783	0.16851	6.78962	2.06419	0.58643	1.91540
3.51	0.34642	0.16768	6.85315	2.06589	0.58751	1.92471
3.52	0.34502	0.16687	6.91723	2.06759	0.58859	1.93402
3.53	0.34362	0.16606	6.98186	2.06927	0.58966	1.94332
3.54	0.34224	0.16526	7.04705	2.07094	0.59072	1.95261
3.55	0.34086	0.16446	7.11281	2.07261	0.59178	1.96190
3.56	0.33949	0.16367	7.17912	2.07426	0.59282	1.97118
3.57	0.33813	0.16288	7.24601	2.07590	0.59387	1.98045
3.58	0.33677	0.16210	7.31346	2.07754	0.59490	1.98972
3.59	0.33542	0.16132	7.38150	2.07916	0.59593	1.99898
3.60	0.33408	0.16055	7.45011	2.08077	0.59695	2.00826
3.61	0.33274	0.15979	7.51931	2.08238	0.59797	2.01747
3.62	0.33141	0.15906	7.58910	2.08397	0.59898	2.02671
3.63	0.33009	0.15827	7.65948	2.08556	0.59998	2.03596
3.64	0.32877	0.15752	7.73045	2.08713	0.60098	2.04566
3.65	0.32747	0.15678	7.80203	2.08870	0.60197	2.05436
3.66	0.32616	0.15604	7.87421	2.09026	0.60296	2.06356
3.67	0.32487	0.15531	7.94700	2.09180	0.60394	2.07276
3.68	0.32358	0.15458	8.02040	2.09334	0.60491	2.08199
3.69	0.32260	0.15385	8.09442	2.09487	0.60588	2.09116
3.70	0.32103	0.15313	8.16907	2.09639	0.60684	2.10035
3.71	0.31976	0.15242	8.24433	2.09790	0.60779	2.10953
3.72	0.31850	0.15171	8.32023	2.09941	0.60874	2.11666
3.73	0.31724	0.15100	8.39676	2.10090	0.60968	2.12785
3.74	0.31600	0.15030	8.47393	2.10238	0.61062	2.13699
3.75	0.31475	0.14961	8.55174	2.10386	0.61155	2.14616
3.76	0.31352	0.14892	8.63020	2.10533	0.61247	2.15527
3.77	0.31229	0.14823	8.70931	2.10679	0.61339	2.16439
3.78	0.31107	0.14755	8.78907	2.10824	0.61431	2.17351
3.79	0.30985	0.14687	8.86950	2.10968	0.61522	2.18262
3.80	0.30864	3.14620	8.95059	2.11111	0.61612	2.19172
3.81	0.30744	0.14553	9.03234	2.11254	0.61702	2.20081
3.82	0.30624	0.14487	9.11477	2.11396	0.61791	2.20990
3.83	0.30505	0.14421	9.19788	2.11536	0.61879	2.21897
3.84	0.30387	0.14355	9.28167	2.11676	0.61968	2.22804

Table A–22

Fanno Flow Parameters (Continued)

M	T/T^*	p/p^*	P_t/P_t^*	V/V^*	fL_{max}/D	S_{max}/R
3.85	0.30269	0.14290	9.36614	2.11815	0.62055	2.23710
3.86	0.30151	0.14225	9.45131	2.11954	0.62142	2.24615
3.87	0.30035	0.14161	9.53717	2.12091	0.62229	2.25520
3.88	0.29919	0.14097	9.62373	2.12228	0.62315	2.26423
3.89	0.29803	0.14034	9.71100	2.12364	0.62400	2.27326
3.90	0.29688	0.13971	9.79897	2.12499	0.62485	2.28228
3.91	0.29574	0.13908	9.88766	2.12634	0.62569	2.29129
3.92	0.29460	0.13846	9.97707	2.12767	0.62653	2.30029
3.93	0.29347	0.13784	10.06720	2.12900	0.62737	2.30928
3.94	0.29235	0.13723	10.15806	2.13032	0.62619	2.31827
3.95	0.29123	0.13662	10.24965	2.13163	0.62902	2.32724
3.96	0.26011	0.13662	10.34197	2.13294	0.62984	2.33621
3.97	0.28900	0.13541	10.43504	2.13424	0.63065	2.34517
3.98	0.28790	0.13482	10.52886	2.13553	0.63146	2.35412
3.99	0.28681	0.13422	10.62343	2.13681	0.63227	2.36306
4.00	0.28571	0.13363	10.71875	2.13809	0.63306	2.37199
4.10	0.27510	0.12793	11.71465	2.15046	0.64080	2.46084
4.20	0.26502	0.12257	12.79164	2.16215	0.64810	2.54879
4.30	0.25543	0.11753	13.95490	2.17321	0.65499	2.63583
4.40	0.24631	0.11279	15.20987	2.18368	0.66149	2.72194
4.50	0.23762	0.10833	16.56219	2.19360	0.66763	2.80712
4.60	0.22936	0.10411	18.01779	2.20300	0.67345	2.89136
4.70	0.22148	0.10013	19.58283	2.21192	0.67895	2.97465
4.80	0.21398	0.09637	21.26371	2.22038	0.68417	3.05700
4.90	0.20683	0.09281	23.06712	2.22842	0.68911	3.13841
5.00	0.20000	0.08944	25.00000	2.23607	0.69330	3.21888
5.10	0.19349	0.08625	27.06957	2.24334	0.69826	3.29841
5.20	0.18727	0.08322	29.28333	2.25026	0.70249	3.37702
5.30	0.18132	0.08034	31.64905	2.25685	0.70652	3.45471
5.40	0.17564	0.07761	34.17481	2.26313	0.71035	3.53149
5.50	0.17021	0.07501	36.86896	2.26913	0.71400	3.60737
5.60	0.16502	0.07254	39.74018	2.27484	0.71748	3.68236
5.70	0.16004	0.07018	42.79743	2.28030	0.72080	3.75648
5.80	0.15528	3.06794	46.05000	2.28552	0.72397	3.82973
5.90	0.15072	0.06580	49.50747	2.29051	0.72699	3.90212

(*continued*)

Table A–22

Fanno Flow Parameters (Continued)

M	T/T^*	p/p^*	p_t/p_t^*	V/V^*	fL_{max}/D	S_{max}/R
6.00	0.14634	0.06376	53.17978	2.29528	0.72988	3.97368
6.10	0.14215	0.06181	57.07718	2.29984	0.73264	4.04440
6.20	0.13812	0.05994	61.21023	2.30421	0.73528	4.11431
6.30	0.13426	0.05816	65.58987	2.30840	0.73780	4.18342
6.40	0.13055	0.05646	70.22736	2.31241	0.74022	4.25174
6.50	0.12698	0.05482	75.13431	2.31626	0.74254	4.31928
6.60	0.12356	0.05326	80.32271	2.31996	0.74477	4.38605
6.70	0.12026	0.05176	85.80487	2.32351	0.74690	4.45208
6.80	0.11710	0.05032	91.59351	2.32691	0.74895	4.51736
6.90	0.11405	0.04894	97.70169	2.33019	0.75091	4.58192
7.00	0.11111	0.04762	104.14286	2.33333	0.75280	4.64576
7.50	0.09796	0.04173	141.84148	2.34738	0.76121	4.95471
8.00	0.08696	0.03686	190.10937	2.35907	0.76819	5.24760
8.50	0.07767	0.03279	251.08617	2.36889	0.77404	5.52580
9.00	0.06977	0.02935	327.18930	2.37722	0.77899	5.79054
9.50	0.06299	0.02642	421.13137	2.38433	0.78320	6.04294
10.00	0.05714	0.02390	535.93750	2.39046	0.78683	6.28402
∞	0.0	0.0	∞	2.4495	0.82153	∞

Table A–23

Rayleigh Flow Parameters

M	T_t/T_t^*	T/T^*	p/p^*	p_t/p_t^*	V/V^*	S_{max}/R
0.0	0.0	0.0	2.40000	1.26790	0.0	∞
0.01	0.00048	0.00058	2.39966	1.26779	0.00024	26.98422
0.02	0.00192	0.00230	2.39866	1.26752	0.00096	22.13471
0.03	0.00431	0.00517	2.39698	1.26708	0.00216	19.30065
0.04	0.00765	0.00917	2.39464	1.26646	0.00383	17.29274
0.05	0.01192	0.01430	2.39163	1.26567	0.00598	15.73828
0.06	0.01712	0.02053	2.38796	1.26470	0.00860	14.47123
0.07	0.02322	0.02784	2.38365	1.26356	0.01168	13.40303
0.08	0.03022	0.03621	2.37869	1.26226	0.01522	12.48081
0.09	0.03807	0.04562	2.37309	1.26078	0.01922	11.67046
0.10	0.04678	0.05602	2.36686	1.25915	0.02367	10.94870
0.11	0.05630	0.06739	2.36002	1.25735	0.02856	10.29890
0.12	0.06661	0.07970	2.35257	1.25539	0.03388	9.70879
0.13	0.07768	0.09290	2.34453	1.25629	0.03962	9.16904
0.14	0.08947	0.10695	2.33590	1.25103	0.04578	8.67240
0.15	0.10196	0.12181	2.32671	1.24863	0.05235	8.21311
0.16	0.11511	0.13743	2.31696	1.24608	0.05931	7.78653
0.17	0.12888	0.15377	2.30667	1.24340	0.06666	7.38886
0.18	0.14324	0.17078	2.29586	1.24059	0.07439	7.01694
0.19	0.15814	0.18841	2.28454	1.23765	0.08247	6.66813
0.20	0.17355	0.20661	2.27273	1.23460	0.09091	6.34018
0.21	0.18943	0.22533	2.26044	1.23142	0.09969	6.03118
0.22	0.20574	0.24452	2.24770	1.22814	0.10879	5.73946
0.23	0.22244	0.26413	2.23451	1.22475	0.11821	5.46359
0.24	0.23948	0.28411	2.22091	1.22126	0.12792	5.20232
0.25	0.25684	0.30440	2.20690	1.21767	0.13793	4.95454
0.26	0.27446	0.32496	2.19250	1.21400	0.14821	4.71926
0.27	0.29231	0.34573	2.17774	1.21025	0.15876	4.49561
0.28	0.31035	0.36667	2.16263	1.20642	0.16955	4.28281
0.29	0.32855	0.38774	2.14719	1.20251	0.18058	4.08016
0.30	0.34686	0.40887	2.13144	1.19855	0.19183	3.88703
0.31	0.36525	0.43004	2.11539	1.19452	0.20329	3.70283
0.32	0.38669	0.45119	2.09908	1.19045	0.21495	3.52706
0.33	0.40214	0.47228	2.08250	1.18632	0.22678	3.35922
0.34	0.42056	0.49327	2.06569	1.18215	0.23879	3.19888

(*continued*)

Table A–23

Rayleigh Flow Parameters (Continued)

M	T_t/T_t^*	T/T^*	p/p^*	p_t/p_t^*	V/V^*	S_{max}/R
0.35	0.43894	0.51413	2.04866	1.17795	0.25096	3.04565
0.36	0.45723	0.53482	2.03142	1.17371	0.26327	2.89915
0.37	0.47541	0.55529	2.01400	1.16945	0.27572	2.75904
0.38	0.49346	0.57553	1.99641	1.16517	0.28828	2.62500
0.39	0.51134	0.59549	1.97866	1.16088	0.30095	2.49673
0.40	0.52903	0.61515	1.96078	1.15656	0.31378	2.37397
0.41	0.54651	0.63448	1.94278	1.15227	0.32658	2.25645
0.42	0.56376	0.65346	1.92468	1.14796	0.33951	2.14394
0.43	0.58076	0.67205	1.90649	1.14366	0.35251	2.03622
0.44	0.59748	0.69025	1.88822	1.13936	0.36556	1.93306
0.45	0.61393	0.70804	1.86989	1.13508	0.37865	1.83429
0.46	0.63007	0.72538	1.85151	1.13082	0.39178	1.73970
0.47	0.64589	0.74228	1.83310	1.12659	0.40493	1.64912
0.48	0.66139	0.75871	1.81466	1.12238	0.41810	1.56239
0.49	0.67655	0.77466	1.79622	1.11820	0.43127	1.47935
0.50	0.69136	0.79012	1.77778	1.11405	0.44444	1.39985
0.51	0.70581	0.80509	1.75935	1.10995	0.45761	1.32374
0.52	0.71990	0.81955	1.74095	1.10588	0.47075	1.25091
0.53	0.73361	0.83351	1.72258	1.10186	0.48387	1.18121
0.54	0.74695	0.84695	1.70425	1.09789	0.49696	1.11453
0.55	0.75991	0.85987	1.68599	1.09397	0.51001	1.05076
0.56	0.77249	0.87227	1.66778	1.09011	0.52302	0.98977
0.57	0.78468	0.88416	1.64964	1.08630	0.53597	0.93148
0.58	0.79648	0.89552	1.63159	1.08256	0.54887	0.87577
0.59	0.80789	0.90637	1.61362	1.07887	0.56170	0.82255
0.60	0.81892	0.91670	1.59574	1.07525	0.57447	0.77174
0.61	0.82957	0.92653	1.57797	1.07170	0.58716	0.72323
0.62	0.83983	0.93584	1.56031	1.06822	0.59978	0.67696
0.63	0.84970	0.94466	1.54275	1.06481	0.61232	0.63284
0.64	0.85920	0.95298	1.52532	1.06147	0.62477	0.59078
0.65	0.86833	0.96081	1.50801	1.05821	0.63713	0.55073
0.66	0.87708	0.96816	1.49083	1.05503	0.64941	0.51260
0.67	0.88547	0.97503	1.47379	1.05193	0.66158	0.47634
0.68	0.89850	0.98144	1.45688	1.04890	0.67366	0.44187
0.69	0.90118	0.98739	1.44011	1.04596	0.68564	0.40913

Table A–23

Rayleigh Flow Parameters (Continued)

M	T_t/T_t^*	T/T^*	p/p^*	p_t/p_t^*	V/V^*	s_{max}/R
0.70	0.90850	0.99290	1.42349	1.04310	0.69751	0.37807
0.71	0.91548	0.99796	1.40701	1.04033	0.70928	0.34861
0.72	0.92212	1.00260	1.39069	1.03764	0.72093	0.32072
0.73	0.92843	1.00682	1.37452	1.03504	0.73248	0.29433
0.74	0.93442	1.01062	1.35851	1.03253	0.74392	0.26940
0.75	0.94009	1.01403	1.34266	1.03010	0.75524	0.24587
0.76	0.94546	1.01706	1.32696	1.02777	0.76645	0.22370
0.77	0.95052	1.01970	1.31143	1.02552	0.77755	0.20283
0.78	0.95528	1.02198	1.29606	1.02337	0.78853	0.18324
0.79	0.95975	1.02390	1.28086	1.02131	0.79939	0.16486
0.80	0.96395	1.02548	1.26582	1.01934	0.81013	0.14767
0.81	0.96787	1.02672	1.25095	1.01747	0.82075	0.13162
0.82	0.97152	1.02763	1.23625	1.01569	0.83125	0.11668
0.83	0.97492	1.02823	1.22171	1.01400	0.84164	0.10280
0.84	0.97807	1.02853	1.20734	1.01241	0.85190	0.08995
0.85	0.98097	1.02854	1.19314	1.01091	0.86204	0.07810
0.86	0.98363	1.02826	1.17911	1.00951	0.87207	0.06722
0.87	0.98607	1.02771	1.16524	1.00820	0.88197	0.05727
0.88	0.98828	1.02689	1.15154	1.00699	0.89175	0.04822
0.89	0.99028	1.02583	1.13801	1.00587	0.90142	0.04004
0.90	0.99207	1.02452	1.12465	1.00486	0.91097	0.03270
0.91	0.99366	1.02297	1.11145	1.00393	0.92039	0.02618
0.92	0.99506	1.02120	1.09842	1.00311	0.92970	0.02044
0.93	0.99627	1.01922	1.08555	1.00238	0.93889	0.01547
0.94	0.99729	1.01702	1.07285	1.00175	0.94797	0.01124
0.95	0.99814	1.01463	1.06030	1.00122	0.95693	0.00771
0.96	0.99883	1.01205	1.04793	1.00078	0.96577	0.00488
0.97	0.99935	1.00929	1.03571	1.00044	0.97450	0.00271
0.98	0.99971	1.00636	1.02365	1.00019	0.98311	0.00119
0.99	0.99993	1.00326	1.01174	1.00005	0.99161	0.00029
1.00	1.00000	1.00000	1.00000	1.00000	1.00000	0.00000
1.01	0.99993	0.99659	0.98841	1.00005	1.00828	0.00029
1.02	0.99973	0.99304	0.97698	1.00019	1.01645	0.00114
1.03	0.99940	0.98936	0.96569	1.00044	1.02450	0.00254
1.04	0.99895	0.98554	0.95456	1.00078	1.03246	0.00447

(*continued*)

Table A–23

Rayleigh Flow Parameters (Continued)

M	T_t/T_t^*	T/T^*	p/p^*	p_t/p_t^*	V/V^*	s_{max}/R
1.05	0.99838	0.98161	0.94358	1.00122	1.04030	0.00690
1.06	0.99769	0.97755	0.93275	1.00175	1.04804	0.00983
1.07	0.99690	0.97339	0.92206	1.00238	1.05567	0.01324
1.08	0.99601	0.96913	0.91152	1.00311	1.06320	0.01711
1.09	0.99501	0.96477	0.90112	1.00394	1.07063	0.02143
1.10	0.99392	0.96031	0.89087	1.00486	1.07795	0.02618
1.11	0.99275	0.95577	0.88075	1.00588	1.08518	0.03135
1.12	0.99148	0.95115	0.87078	1.00699	1.09230	0.03692
1.13	0.99013	0.94645	0.86094	1.00821	1.09933	0.04288
1.14	0.98871	0.94169	0.35123	1.00952	1.10626	0.04922
1.15	0.98721	0.93685	0.84166	1.01093	1.11310	0.05593
1.16	0.98564	0.93196	0.83222	1.01243	1.11984	0.06298
1.17	0.98400	0.92701	0.82292	1.01403	1.12649	0.07038
1.18	0.98230	0.92200	0.81374	1.01573	1.13305	0.07812
1.19	0.98054	0.91695	0.80468	1.01752	1.13951	0.08617
1.20	0.97872	0.91185	0.79576	1.01942	1.14589	0.09453
1.21	0.97684	0.90671	0.78695	1.02140	1.15218	0.10318
1.22	0.97492	0.90153	0.77827	1.02349	1.15838	0.11213
1.23	0.97294	0.89632	0.76971	1.02567	1.16449	0.12135
1.24	0.97092	0.89108	0.76127	1.02795	1.17052	0.13085
1.25	0.96886	0.88581	0.75294	1.03033	1.17647	0.14060
1.26	0.96675	0.88052	0.74473	1.03280	1.18233	0.15061
1.27	0.96461	0.87521	0.73663	1.03537	1.18812	0.16086
1.28	0.96243	0.86988	0.72865	1.03803	1.19382	0.17135
1.29	0.96022	0.86453	0.72078	1.04080	1.19945	0.18206
1.30	0.95798	0.85917	0.71301	1.04366	1.20499	0.19299
1.31	0.95571	0.85330	0.70536	1.04662	1.21046	0.20413
1.32	0.95341	0.84843	0.69780	1.04968	1.21585	0.21548
1.33	0.95108	0.84305	0.69036	1.05283	1.22117	0.22702
1.34	0.94873	0.83766	0.68301	1.05608	1.22642	0.23876
1.35	0.94637	0.83227	0.67577	1.05943	1.23159	0.25068
1.36	0.94398	0.82689	0.66863	1.06288	1.23669	0.26277
1.37	0.94157	0.82151	0.66158	1.06642	1.24173	0.27504
1.38	0.93914	0.81613	0.65464	1.07007	1.24669	0.28747
1.39	0.93671	0.81076	0.64778	1.07381	1.25158	0.30006

Table A–23

Rayleigh Flow Parameters (Continued)

M	T_t/T_t^*	T/T^*	P/P^*	P_t/P_t^*	V/V^*	S_{max}/R
1.40	0.93425	0.80539	0.64103	1.07765	1.25641	0.31281
1.41	0.93179	0.80004	0.63436	1.08159	1.26117	0.32570
1.42	0.92931	0.79469	0.62779	1.08563	1.26587	0.33874
1.43	0.92683	0.78936	0.62130	1.08977	1.27050	0.35191
1.44	0.92434	0.78405	0.61491	1.09401	1.27507	0.36522
1.45	0.92184	0.77874	0.60860	1.09835	1.27957	0.37865
1.46	0.91933	0.77346	0.60237	1.10278	1.28402	0.39221
1.47	0.91682	0.76819	0.59623	1.10732	1.28840	0.40589
1.48	0.91431	0.76294	0.59010	1.11196	1.29273	0.41968
1.49	0.91179	0.75771	0.58421	1.11670	1.29700	0.43358
1.50	0.90928	0.75250	0.57831	1.12155	1.30120	0.44758
1.51	0.90676	0.74782	0.57250	1.12649	1.30536	0.46169
1.52	0.90424	0.74215	0.56676	1.13153	1.30945	0.47589
1.53	0.90172	0.73701	0.56111	1.13668	1.31350	0.49019
1.54	0.89920	0.73139	0.55552	1.14193	1.31748	0.50458
1.55	0.89669	0.72680	0.55002	1.14729	1.32142	0.51905
1.56	0.89418	0.72173	0.54458	1.15274	1.32530	0.53361
1.57	0.89168	0.71669	0.53922	1.15830	1.32913	0.54824
1.58	0.88917	0.71168	0.53393	1.16397	1.33291	0.56295
1.59	0.83668	0.70669	0.52871	1.16974	1.33663	0.57774
1.60	0.88419	0.70174	0.52356	1.17561	1.34031	0.59259
1.61	0.88170	0.69680	0.51848	1.18159	1.34394	0.60752
1.62	0.87922	0.69190	0.51346	1.18768	1.34753	0.62250
1.63	0.87675	0.68703	0.50851	1.19387	1.35106	0.63755
1.64	0.87429	0.68219	0.50363	1.20017	1.35455	0.65265
1.65	0.87184	0.67738	0.49880	1.20657	1.35800	0.66781
1.66	0.86939	0.67259	0.49405	1.21309	1.36140	0.68303
1.67	0.86696	0.66784	0.48935	1.21971	1.36475	0.69829
1.68	0.86453	0.66312	0.48472	1.22644	1.36806	0.71360
1.69	0.86212	0.65843	0.48014	1.23328	1.37133	0.72896
1.70	0.85971	0.65377	0.47562	1.24024	1.37455	0.74436
1.71	0.85731	0.64914	0.47117	1.24730	1.37774	0.75981
1.72	0.85493	0.64455	0.46677	1.25447	1.38088	0.77529
1.73	0.85256	0.63999	0.46242	1.26175	1.38398	0.79081
1.74	0.85019	0.63545	0.45813	1.26915	1.38705	0.80636

(continued)

Rayleigh Flow Parameters (Continued)

M	T_t/T_t^*	T/T^*	p/p^*	p_t/p_t^*	V/V^*	S_{max}/R
1.75	0.84784	0.63095	0.45390	1.27666	1.39007	0.82195
1.76	0.84551	0.62649	0.44972	1.28428	1.39306	0.83757
1.77	0.84318	0.62205	0.44559	1.29202	1.39600	0.85322
1.78	0.84087	0.61765	0.44152	1.29987	1.39891	0.86889
1.79	0.83857	0.61328	0.43750	1.30784	1.40179	0.88459
1.80	0.83628	0.60894	0.43353	1.31592	1.40462	0.90031
1.81	0.83400	0.60464	0.42960	1.32413	1.40743	0.91606
1.82	0.83174	0.60036	0.42573	1.33244	1.41019	0.93183
1.83	0.82949	0.59612	0.42191	1.34088	1.41292	0.94761
1.84	0.82726	0.59191	0.41813	1.34943	1.41562	0.96342
1.85	0.82504	0.58774	0.41440	1.35811	1.41829	0.97924
1.86	0.82283	0.58359	0.41072	1.36690	1.42092	0.99507
1.87	0.82064	0.57948	0.40708	1.37582	1.42351	1.01092
1.88	0.81845	0.57540	0.40349	1.38486	1.42608	1.02678
1.89	0.81629	0.57136	0.39994	1.39402	1.42862	1.04265
1.90	0.81414	0.56734	0.39643	1.40330	1.43112	1.05853
1.91	0.81200	0.56336	0.39297	1.41271	1.43359	1.07441
1.92	0.80987	0.55941	0.38955	1.42224	1.43604	1.09031
1.93	0.80776	0.55549	0.38617	1.43190	1.43845	1.10621
1.94	0.80567	0.55160	0.38283	1.44168	1.44083	1.12211
1.95	0.80358	0.54774	0.37954	1.45159	1.44319	1.13802
1.96	0.80152	0.54392	0.37628	1.46164	1.44551	1.15393
1.97	0.79946	0.54012	0.37306	1.47180	1.44781	1.16984
1.98	0.79742	0.53636	0.36988	1.48210	1.45008	1.18575
1.99	0.79540	0.53263	0.36674	1.49253	1.45233	1.20167
2.00	0.79339	0.52893	0.36364	1.50310	1.45455	1.21758
2.01	0.79139	0.52525	0.36057	1.51379	1.45674	1.23348
2.02	0.78941	0.52161	0.35754	1.52462	1.45890	1.24939
2.03	0.78744	0.51800	0.35454	1.53558	1.46104	1.26529
2.04	0.78549	0.51442	0.35158	1.54668	1.46315	1.28118
2.05	0.78355	0.51087	0.34866	1.55791	1.46524	1.29707
2.06	0.78162	0.50735	0.34577	1.56928	1.46731	1.31296
2.07	0.77971	0.50386	0.34291	1.58079	1.46935	1.32883
2.08	0.77782	0.50040	0.34009	1.59244	1.47136	1.34470
2.09	0.77593	0.49696	0.33730	1.60423	1.47336	1.36056

Table A–23

Rayleigh Flow Parameters (Continued)

M	T_t/T_t^*	T/T^*	p/p^*	P_t/P_t^*	V/V^*	S_{max}/R
2.10	0.77406	0.49356	0.33454	1.61616	1.47533	1.37641
2.11	0.77221	0.49018	0.33182	1.62823	1.47727	1.39225
2.12	0.77037	0.48684	0.32912	1.64045	1.47920	1.40807
2.13	0.76854	0.48352	0.32646	1.65281	1.48110	1.42389
2.14	0.76673	0.48023	0.32382	1.66531	1.48298	1.43970
2.15	0.76493	0.47696	0.32122	1.67796	1.48484	1.45549
2.16	0.76314	0.47373	0.31865	1.69076	1.48668	1.47127
2.17	0.76137	0.47052	0.31610	1.70371	1.48850	1.48703
2.18	0.75961	0.46734	0.31359	1.71680	1.49029	1.50278
2.19	0.75787	0.46418	0.31110	1.73005	1.49207	1.51852
2.20	0.75613	0.46106	0.30864	1.74345	1.49383	1.53424
2.21	0.75442	0.45796	0.30621	1.75700	1.49550	1.54994
2.22	0.75271	0.45488	0.30381	1.77070	1.49728	1.56563
2.23	0.75102	0.45184	0.30143	1.78456	1.49898	1.58130
2.24	0.74934	0.44882	0.29908	1.79858	1.50066	1.59696
2.25	0.74768	0.44582	0.29675	1.81275	1.50232	1.61259
2.26	0.74602	0.44285	0.29446	1.82708	1.50396	1.62821
2.27	0.74438	0.43990	0.29218	1.84157	1.50558	1.64381
2.28	0.74276	0.43698	0.28993	1.85623	1.50719	1.65939
2.29	0.74114	0.43409	0.28771	1.87104	1.50878	1.67496
2.30	0.73954	0.43122	0.28551	1.88602	1.51035	1.69050
2.31	0.73795	0.42638	0.28333	1.90116	1.51190	1.70602
2.32	0.73638	0.42555	0.28118	1.91647	1.51344	1.72152
2.33	0.73482	0.42276	0.27905	1.93195	1.51496	1.73700
2.34	0.73326	0.41998	0.27695	1.94759	1.51646	1.75246
2.35	0.73173	0.41723	0.27487	1.96340	1.51795	1.76790
2.36	0.73020	0.41451	0.27281	1.97939	1.51942	1.78332
2.37	0.72868	0.41181	0.27077	1.99554	1.52088	1.79872
2.38	0.72718	0.40913	0.26875	2.01187	1.52232	1.81409
2.39	0.72569	0.40647	0.26676	2.02837	1.52374	1.82944
2.40	0.72421	0.40384	0.26478	2.04505	1.52515	1.84477
2.41	0.72275	0.40122	0.26283	2.06191	1.52655	1.86008
2.42	0.72129	0.39864	0.26090	2.07895	1.52793	1.87536
2.43	0.71985	0.39607	0.25899	2.09616	1.52929	1.89062
2.44	0.71842	0.39352	0.25710	2.11356	1.53065	1.90585

(continued)

Rayleigh Flow Parameters (Continued)

M	T_t/T_t^*	T/T^*	p/p^*	p_t/p_t^*	V/V^*	S_{max}/R
2.45	0.71699	0.39100	0.25522	2.13114	1.53198	1.92106
2.46	0.71558	0.38850	0.25337	2.14891	1.53331	1.93625
2.47	0.71419	0.38602	0.25154	2.16685	1.53461	1.95141
2.48	0.71280	0.38356	0.24973	2.18499	1.53591	1.96655
2.49	0.71142	0.38112	0.24793	2.20332	1.53719	1.98167
2.50	0.71006	0.37870	0.24615	2.22183	1.53846	1.99676
2.51	0.70871	0.37630	0.24440	2.24054	1.53972	2.01182
2.52	0.70736	0.37392	0.24266	2.25944	1.54096	2.02686
2.53	0.70603	0.37157	0.24093	2.27853	1.54219	2.04187
2.54	0.70471	0.36923	0.23923	2.29782	1.54341	2.05686
2.55	0.70340	0.36691	0.23754	2.31730	1.54461	2.07183
2.56	0.70210	0.36461	0.23587	2.33699	1.54581	2.08676
2.57	0.70081	0.36233	0.23422	2.35687	1.54699	2.10167
2.58	0.69952	0.36007	0.23258	2.37696	1.54816	2.11656
2.59	0.69826	0.35783	0.23096	2.39725	1.54931	2.13142
2.60	0.69700	0.35561	0.22936	2.41774	1.55046	2.14625
2.61	0.69575	0.35341	0.22777	2.43844	1.55159	2.16106
2.62	0.69451	0.35122	0.22620	2.45935	1.55272	2.17584
2.63	0.69328	0.34906	0.22464	2.48047	1.55383	2.19059
2.64	0.69206	0.34691	0.22310	2.50179	1.55493	2.20532
2.65	0.69084	0.34478	0.22158	2.52334	1.55602	2.22002
2.66	0.68964	0.34266	0.22007	2.54509	1.55710	2.23470
2.67	0.68845	0.34057	0.21857	2.56706	1.55816	2.24934
2.68	0.68727	0.33849	0.21709	2.58925	1.55922	2.26396
2.69	0.68610	0.33643	0.21562	2.61166	1.56027	2.27856
2.70	0.68494	0.33439	0.21417	2.63429	1.56131	2.29312
2.71	0.68378	0.33236	0.21273	2.65714	1.56233	2.30766
2.72	0.68264	0.33035	0.21131	2.68021	1.56335	2.32217
2.73	0.68150	0.32836	0.20990	2.70351	1.56436	2.33666
2.74	0.68037	0.32638	0.20850	2.72704	1.56536	2.35111
2.75	0.67926	0.32442	0.20712	2.75080	1.56634	2.36554
2.76	0.67815	0.32248	0.20575	2.77478	1.56732	2.37995
2.77	0.67705	0.32055	0.20439	2.79900	1.56829	2.39432
2.78	0.67595	0.31864	0.20305	2.82346	1.56925	2.40867
2.79	0.67487	0.31674	0.20172	2.84815	1.57020	2.42299

Table A–23

Rayleigh Flow Parameters (Continued)

M	T_t/T_t^*	T/T^*	p/p^*	p_t/p_t^*	V/V^*	s_{max}/R
2.80	0.67380	0.31486	0.20040	2.87308	1.57114	2.43728
2.81	0.67273	0.31299	0.19910	2.89825	1.57207	2.45154
2.82	0.67167	0.31114	0.19780	2.92366	1.57300	2.46578
2.83	0.67062	0.30931	0.19652	2.94931	1.57391	2.47999
2.84	0.66958	0.30749	0.19525	2.97521	1.57482	2.49417
2.85	0.66855	0.30568	0.19399	3.00136	1.57572	2.50833
2.86	0.66752	0.30389	0.19275	3.02775	1.57661	2.52245
2.87	0.66651	0.30211	0.19151	3.05440	1.57749	2.53655
2.88	0.66550	0.30035	0.19029	3.08129	1.57836	2.55062
2.89	0.66450	0.29860	0.18908	3.10844	1.57923	2.56467
2.90	0.66350	0.29687	0.18788	3.13585	1.58008	2.57868
2.91	0.66252	0.29515	0.18669	3.16352	1.58093	2.59267
2.92	0.66154	0.29344	0.18551	3.19145	1.58178	2.60663
2.93	0.66057	0.29175	0.18435	3.21963	1.58261	2.62057
2.94	0.65960	0.29007	0.18019	3.24809	1.58343	2.63447
2.95	0.65865	0.28841	0.18205	3.27680	1.58425	2.64835
2.96	0.65770	0.28675	0.18091	3.30579	1.58506	2.66220
2.97	0.65676	0.28512	0.17979	3.33505	1.58587	2.67602
2.98	0.65583	0.28349	0.17867	3.36457	1.58666	2.68981
2.99	0.65490	0.28188	0.17757	3.39437	1.58745	2.70358
3.00	0.65398	0.28028	0.17647	3.42445	1.58824	2.71732
3.01	0.65307	0.27869	0.17539	3.45481	1.58901	2.73103
3.02	0.65216	0.27711	0.17431	3.48544	1.58978	2.74472
3.03	0.65126	0.27555	0.17324	3.51636	1.59054	2.75837
3.04	0.65037	0.27400	0.17219	3.54756	1.59129	2.77200
3.05	0.64949	0.27246	0.17114	3.57905	1.59204	2.78560
3.06	0.64861	0.27094	0.17010	3.61082	1.59278	2.79918
3.07	0.64774	0.26942	0.16908	3.64289	1.59352	2.81272
3.08	0.64687	0.26792	0.16806	3.67524	1.59425	2.82624
3.09	0.64601	0.26643	0.16705	3.70790	1.59497	2.83974
3.10	0.64516	0.26495	0.16604	3.74084	1.59568	2.85320
3.11	0.64432	0.26349	0.16505	3.77409	1.59639	2.86664
3.12	0.64348	0.26203	0.16407	3.80764	1.59709	2.88005
3.13	0.64265	0.26059	0.16309	3.84149	1.59779	2.89343
3.14	0.64182	0.25915	0.16212	3.87565	1.59848	2.90679

(*continued*)

Rayleigh Flow Parameters (Continued)

M	T_t/T_t^*	T/T^*	p/p^*	P_t/P_t^*	V/V^*	S_{max}/R
3.15	0.64100	0.25773	0.16117	3.91011	1.59917	2.92011
3.16	0.64018	0.25632	0.16022	3.94488	1.59985	2.93342
3.17	0.63938	0.25492	0.15927	3.97997	1.60052	2.94669
3.18	0.63857	0.25353	0.15834	4.01537	1.60119	2.95994
3.19	0.63778	0.25215	0.15741	4.05108	1.60185	2.97316
3.20	0.63699	0.25078	0.15649	4.08712	1.60250	2.98635
3.21	0.63621	0.24943	0.15558	4.12347	1.60315	2.99952
3.22	0.63543	0.24808	0.15468	4.16015	1.60380	3.01266
3.23	0.63465	0.24674	0.15379	4.19715	1.60444	3.02577
3.24	0.63389	0.24541	0.15290	4.23449	1.60507	3.03885
3.25	0.63313	0.24410	0.15202	4.27215	1.60570	3.05191
3.26	0.63237	0.24279	0.15115	4.31014	1.60632	3.06495
3.27	0.63162	0.24149	0.15028	4.34847	1.60694	3.07795
3.28	0.63088	0.24021	0.14942	4.38714	1.60755	3.09093
3.29	0.63014	0.23893	0.14857	4.42614	1.60816	3.10388
3.30	0.62940	0.23766	0.14773	4.46549	1.60877	3.11681
3.31	0.62868	0.23640	0.14689	4.50518	1.60936	3.12971
3.32	0.62795	0.23515	0.14606	4.54527	1.60996	3.14258
3.33	0.62724	0.23391	0.14524	4.58561	1.61054	3.15543
3.34	0.62652	0.23268	0.14442	4.62635	1.61113	3.16825
3.35	0.62582	0.23146	0.14361	4.66744	1.61170	3.18105
3.36	0.62512	0.23025	0.14281	4.70889	1.61228	3.19382
3.37	0.62442	0.22905	0.14201	4.75070	1.61285	3.20656
3.38	0.62373	0.22785	0.14122	4.79287	1.61341	3.21928
3.39	0.62304	0.22667	0.14044	4.83540	1.61397	3.23197
3.40	0.62236	0.22549	0.13966	4.87830	1.61453	3.24463
3.41	0.62168	0.22432	0.13889	4.92157	1.61508	3.25727
3.42	0.62101	0.22317	0.13813	4.96521	1.61562	3.26988
3.43	0.62034	0.22201	0.13737	5.00923	1.61616	3.28247
3.44	0.61968	0.22087	0.13662	5.05362	1.61670	3.29503
3.45	0.61902	0.21974	0.13587	5.09839	1.61723	3.30757
3.46	0.61837	0.21861	0.13513	5.14355	1.61776	3.32008
3.47	0.61772	0.21750	0.13440	5.18909	1.61829	3.33257
3.48	0.61708	0.21639	0.13367	5.23501	1.61881	3.34503
3.49	0.61644	0.21529	0.13295	5.28133	1.61932	3.35746

Table A–23

Rayleigh Flow Parameters (Continued)

M	T_t/T_t^*	T/T^*	p/p^*	p_t/p_t^*	V/V^*	s_{max}/R
3.50	0.61580	0.21419	0.13223	5.32804	1.61983	3.36987
3.51	0.61517	0.21311	0.13152	5.37514	1.62034	3.38225
3.52	0.61455	0.21203	0.13081	5.42264	1.62085	3.39461
3.53	0.61393	0.21096	0.13011	5.47054	1.62135	3.40695
3.54	0.61331	0.20990	0.12942	5.51885	1.62184	3.41926
3.55	0.61270	0.20885	0.12873	5.56756	1.62233	3.43154
3.56	0.61209	0.20780	0.12805	5.61668	1.62282	3.44380
3.57	0.61149	0.20676	0.12737	5.66621	1.62331	3.45603
3.58	0.61089	0.20573	0.12670	5.71615	1.62379	3.46824
3.59	0.61029	0.20470	0.12603	5.76652	1.62427	3.48043
3.60	0.60970	0.20369	0.12537	5.81730	1.62474	3.49259
3.61	0.60911	0.20268	0.12471	5.86850	1.62521	3.50472
3.62	0.60853	0.20167	0.12406	5.92013	1.62567	3.51683
3.63	0.60795	0.20068	0.12341	5.97219	1.62614	3.52892
3.64	0.60738	0.19969	0.12277	6.02468	1.62660	3.54098
3.65	0.60681	0.19871	0.12213	6.07761	1.62705	3.55302
3.66	0.60624	0.19773	0.12150	6.13097	1.62750	3.56503
3.67	0.60568	0.19677	0.12087	6.18477	1.62795	3.57702
3.68	0.60512	0.19581	0.12024	6.23902	1.62840	3.58899
3.69	0.60456	0.19485	0.11963	6.29371	1.62884	3.60093
3.70	0.60401	0.19390	0.11901	6.34884	1.62928	3.61285
3.71	0.60346	0.19296	0.11840	6.40443	1.62971	3.62474
3.72	0.60292	0.19203	0.11780	6.46048	1.63014	3.63661
3.73	0.60288	0.19110	0.11720	6.51698	1.63057	3.64845
3.74	0.60184	0.19018	0.11660	6.57394	1.63100	3.66028
3.75	0.60131	0.18926	0.11601	6.63137	1.63142	3.67207
3.76	0.60078	0.18836	0.11543	6.68926	1.63184	3.68385
3.77	0.60025	0.18745	0.11484	6.74763	1.63225	3.69560
3.78	0.59973	0.18656	0.11427	6.80646	1.63267	3.70733
3.79	0.59921	0.18567	0.11369	6.86578	1.63308	3.71903
3.80	0.59870	0.18478	0.11312	6.92557	1.63348	3.73071
3.81	0.59819	0.18391	0.11256	6.98584	1.63389	3.74237
3.82	0.59768	0.18303	0.11200	7.04660	1.63429	3.75401
3.83	0.59717	0.18217	0.11144	7.10784	1.63469	3.76562
3.84	0.59667	0.18131	0.11089	7.16958	1.63508	3.77721

(continued)

Table A–23

Rayleigh Flow Parameters (Continued)

M	T_t/T_t^*	T/T^*	p/p^*	p_t/p_t^*	V/V^*	S_{max}/R
3.85	0.59617	0.18045	0.11034	7.23181	1.63547	3.78877
3.86	0.59568	0.17961	0.10979	7.29454	1.63586	3.80031
3.87	0.59519	0.17876	0.10925	7.35777	1.63625	3.81183
3.88	0.59470	0.17793	0.10871	7.42151	1.63663	3.82333
3.89	0.59421	0.17709	0.10818	7.48575	1.63701	3.83481
3.90	0.59373	0.17627	0.10765	7.55050	1.63739	3.84626
3.91	0.59325	0.17545	0.10716	7.61577	1.63777	3.85769
3.92	0.59278	0.17463	0.10661	7.68156	1.63814	3.36909
3.93	0.59231	0.17383	0.10609	7.74786	1.63851	3.88048
3.94	0.59184	0.17302	0.10557	7.81469	1.63888	3.89184
3.95	0.59137	0.17222	0.10506	7.88205	1.63924	3.90318
3.96	0.59091	0.17143	0.10456	7.94993	1.63960	3.91450
3.97	0.59045	0.17064	0.10405	8.01835	1.63996	3.92579
3.98	0.58999	0.16986	0.10355	8.08731	1.64032	3.93706
3.99	0.58954	0.16908	0.10306	8.15681	1.64067	3.94831
4.00	0.58909	0.16831	0.10256	8.22685	1.64103	3.95954
4.10	0.58473	0.16086	0.09782	8.95794	1.64441	4.07064
4.20	0.58065	0.15388	0.09340	9.74729	1.64757	4.17961
4.30	0.57682	0.14734	0.08927	10.59854	1.65052	4.28652
4.40	0.57322	0.14119	0.08540	11.51554	1.65329	4.39143
4.50	0.56982	0.13540	0.08177	12.50226	1.65588	4.49440
4.60	0.56663	0.12996	0.07837	13.56288	1.65831	4.59550
4.70	0.56362	0.12483	0.07517	14.70174	1.66059	4.69477
4.80	0.56078	0.12000	0.07217	15.92337	1.66274	4.79229
4.90	0.55809	0.11543	0.06934	17.23245	1.66476	4.88809
5.00	0.55556	0.11111	0.06667	18.63390	1.66667	4.98224
5.10	0.55315	0.10703	0.06415	20.13279	1.66847	5.07477
5.20	0.55088	0.10316	0.06177	21.73439	1.67017	5.16575
5.30	0.54872	0.09950	0.05951	23.44420	1.67178	5.25522
5.40	0.54667	0.09602	0.05738	25.26788	1.67330	5.34322
5.50	0.54473	0.09272	0.05536	27.21132	1.67474	5.42979
5.60	0.54288	0.08958	0.05345	29.28063	1.67611	5.51498
5.70	0.54112	0.08660	0.05163	31.48210	1.67741	5.59883
5.80	0.53944	0.08376	0.04990	33.82228	1.67864	5.68138
5.90	0.53785	0.08106	0.04826	36.30790	1.67982	5.76265

Table A–23

Rayleigh Flow Parameters (Continued)

M	T_t/T_t^*	T/T^*	p/p^*	p_t/p_t^*	V/V^*	S_{max}/R
6.00	0.53633	0.07849	0.04669	38.94594	1.68093	5.84270
6.10	0.53488	0.07603	0.04520	41.74362	1.68200	5.92155
6.20	0.53349	0.07369	0.04378	44.70837	1.68301	5.99924
6.30	0.53217	0.07145	0.04243	47.84787	1.68398	6.07579
6.40	0.53091	0.06931	0.04114	51.17004	1.68490	6.15124
6.50	0.52970	0.06726	0.03990	54.68303	1.68579	6.22562
6.60	0.52854	0.06531	0.03872	58.39527	1.68663	6.29896
6.70	0.52743	0.06343	0.03759	62.31541	1.68744	6.37128
6.80	0.52637	0.06164	0.03651	66.45238	1.68821	6.44261
6.90	0.52535	0.05991	0.03547	70.81536	1.68895	6.51298
7.00	0.52438	0.05826	0.03448	75.41379	1.68966	6.58240
7.50	0.52004	0.05094	0.03009	102.28748	1.69279	6.91625
8.00	0.51647	0.04491	0.02649	136.62352	1.69536	7.22982
8.50	0.51349	0.03988	0.02349	179.92363	1.69750	7.52538
9.00	0.51098	0.03565	0.02098	233.88395	1.69930	7.80482
9.50	0.50885	0.03205	0.01885	300.40722	1.70082	8.06978
10.00	0.50702	0.02897	0.01702	381.61488	1.70213	8.32165
∞	0.48980	0.0	0.0	∞	1.7143	∞

Table A–24

Oblique Shock Functions

	Flow turning angle θ, deg													
M_1	0.0	2.0	4.0	6.0	8.10	10.0	12.0	14.0	16.0	18.0	20.0	22.0	β	θ_{max}
1.00														
1.01	81.93												85.36	0.05
1.02	78.64												83.49	0.14
1.03	76.14												82.08	0.26
1.04	74.06												80.93	0.40
1.05	72.25												79.94	0.56
1.06	70.63												79.06	0.73
1.07	69.16												78.27	0.91
1.08	67.81												77.56	1.10
1.09	66.55												76.90	1.30
1.10	65.38												76.30	1.52
1.11	64.28												75.73	1.73
1.12	63.23												75.21	1.96

(continued)

Table A–24

Oblique Shock Functions (Continued)

M_1	0.0	2.0	4.0	6.0	8.10	10.0	12.0	14.0	16.0	18.0	20.0	22.0	β	θ_{max}
					Flow turning angle θ, deg									
1.13	62.25	70.93											74.72	2.19
1.14	61.31	68.71											74.26	2.43
1.15	60.41	67.00											73.82	2.67
1.16	59.55	65.56											73.41	2.92
1.17	58.73	64.28											73.02	3.17
1.18	57.94	63.11											72.66	3.42
1.19	57.18	62.05											72.31	3.68
1.20	56.44	61.05											71.98	3.94
1.21	55.74	60.12	68.09										71.66	4.21
1.22	55.05	59.24	66.03										71.36	4.47
1.23	54.39	58.40	64.47										71.07	4.74
1.24	53.75	57.60	63.15										70.80	5.01
1.25	53.13	56.84	61.99										70.54	5.29
1.26	52.53	56.12	60.93										70.29	5.56
1.27	51.94	55.42	59.97										70.05	5.83
1.28	51.38	54.75	59.06	67.38									69.82	6.11
1.29	50.82	54.10	58.22	65.05									69.60	6.39
1.30	50.28	53.47	57.42	63.46									69.40	6.66
1.31	49.76	52.87	56.67	62.16									69.19	6.94
1.32	49.25	52.28	55.95	61.03									69.00	7.22
1.33	48.75	51.72	55.26	60.02									68.82	7.49
1.34	48.27	51.17	54.60	59.09									68.64	7.77
1.35	47.79	50.63	53.97	58.23	66.91								68.47	8.05
1.36	47.33	50.11	53.36	57.43	64.29								68.31	8.33
1.37	46.88	49.61	52.77	56.68	62.70								68.15	8.60
1.38	46.44	49.12	52.20	55.96	61.43								68.00	8.88
1.39	46.01	48.64	51.65	55.28	60.34								67.85	9.15
1.40	45.58	48.17	51.12	54.63	59.37								67.72	9.43
1.41	45.17	47.72	50.60	54.01	58.48								67.58	9.70
1.42	44.77	47.27	50.10	53.42	57.67								67.45	9.97
1.43	44.37	46.84	49.61	52.84	56.91	63.95							67.33	10.25
1.44	43.98	46.42	49.14	52.29	56.19	62.31							67.21	10.52
1.45	43.60	46.00	48.68	51.76	55.52	61.05							67.10	10.79
1.46	43.23	45.60	48.23	51.24	54.87	59.98							66.99	11.05
1.47	42.86	45.20	47.79	50.74	54.26	59.04							66.88	11.32
1.48	42.51	44.82	47.37	50.25	53.68	58.19							66.78	11.59
1.49	42.16	44.44	46.95	49.78	53.11	57.41							66.68	11.85
1.50	41.81	44.06	46.54	49.33	52.57	56.68	64.36						66.59	12.11

Table A–24

Oblique Shock Functions (Continued)

| | Flow turning angle θ, deg | | | | | | | | | | | | | |
M_1	0.0	2.0	4.0	6.0	8.10	10.0	12.0	14.0	16.0	18.0	20.0	22.0	β	θ_{max}
1.50	41.81	44.06	46.54	49.33	52.57	56.68	64.36						66.59	12.11
1.51	41.47	43.70	46.15	48.88	52.05	56.00	62.42						66.50	12.37
1.52	41.14	43.34	45.76	48.45	51.55	55.35	61.10						66.41	12.63
1.53	40.81	42.99	45.38	48.03	51.06	54.74	60.02						66.33	12.89
1.54	40.49	42.65	45.01	47.62	50.59	54.16	59.08						66.25	13.15
1.55	40.18	42.32	44.64	47.21	50.13	53.60	58.24						66.17	13.40
1.56	39.87	41.99	44.29	46.82	49.69	53.06	57.47						66.10	13.66
1.57	39.56	41.66	43.94	46.44	49.26	52.55	56.77						66.03	13.91
1.58	39.27	41.34	43.59	46.07	48.84	52.06	56.10	63.38					65.96	14.16
1.59	38.97	41.03	43.26	45.70	48.43	51.58	55.48	61.73					65.89	14.41
1.60	38.68	40.72	42.93	45.34	48.03	51.12	54.89	60.54					65.83	14.65
1.61	38.40	40.42	42.61	44.99	47.64	50.67	54.33	59.55					65.77	14.90
1.62	38.12	40.13	42.29	44.65	47.26	50.23	53.79	58.69					65.71	15.14
1.63	37.84	39.84	41.98	44.32	46.89	49.81	53.28	57.91					65.65	15.38
1.64	37.57	39.55	41.68	43.99	46.53	49.41	52.79	57.20					65.60	15.62
1.65	37.31	39.27	41.38	43.67	46.18	49.01	52.31	56.54					65.55	15.86
1.66	37.04	38.99	41.08	43.35	45.84	48.62	51.85	55.93	63.58				65.50	16.09
1.67	36.78	38.72	40.79	43.04	45.50	48.24	51.41	55.34	61.80				65.45	16.32
1.68	36.53	38.45	40.51	42.74	45.17	47.88	50.98	54.79	60.60				65.40	16.55
1.69	36.28	38.19	40.23	42.44	44.85	47.52	50.57	54.27	59.63				65.36	16.78
1.70	36.03	37.93	39.96	42.14	44.53	47.17	50.17	53.77	58.79				65.32	17.01
1.71	35.79	37.67	39.69	41.86	44.22	46.82	49.78	53.29	58.05				65.28	17.24
1.72	35.55	37.42	39.42	41.57	43.91	46.49	49.40	52.83	57.36				65.24	17.46
1.73	35.31	37.17	39.16	41.30	43.61	46.16	49.03	52.39	56.73				65.20	17.68
1.74	35.08	36.93	38.90	41.02	43.32	45.84	48.67	51.96	56.14				65.17	17.90
1.75	34.85	36.69	38.65	40.76	43.03	45.53	48.32	51.55	55.59	62.94			65.13	18.12
1.76	34.62	36.45	38.40	40.49	42.75	45.22	47.98	51.15	55.06	61.42			65.10	18.34
1.77	34.40	36.22	38.16	40.23	42.48	44.92	47.64	50.76	54.57	60.34			65.07	18.55
1.78	34.18	35.99	37.91	39.98	42.20	44.63	47.32	50.38	54.09	59.46			65.04	18.76
1.79	33.96	35.76	37.68	39.73	41.94	44.34	47.00	50.02	53.63	58.69			65.01	18.97
1.80	33.75	35.54	37.44	39.48	41.67	44.06	46.69	49.66	53.20	57.99			64.99	19.18
1.81	33.54	35.32	37.21	39.24	41.42	43.78	46.38	49.31	52.78	57.36			64.96	19.39
1.82	33.33	35.10	36.99	39.00	41.16	43.51	46.08	48.98	52.37	56.78			64.94	19.59
1.83	33.12	34.89	36.76	38.76	40.91	43.24	45.79	48.65	51.98	56.23			64.91	19.80
1.84	32.92	34.68	36.54	38.53	40.67	42.98	45.50	48.33	51.60	55.71			64.89	20.00
1.85	32.72	34.47	36.32	38.30	40.42	42.72	45.22	48.01	51.23	55.23	62.10		64.87	20.20
1.86	32.52	34.26	36.11	38.08	40.19	42.46	44.95	47.71	50.88	54.76	60.91		64.85	20.40

(continued)

Table A–24

Oblique Shock Functions (Continued)

M_1	\multicolumn{12}{c}{Flow turning angle θ, deg}	β	θ_{max}
	0.0	2.0	4.0	6.0	8.10	10.0	12.0	14.0	16.0	18.0	20.0	22.0		
1.87	32.33	34.06	35.90	37.86	39.95	42.21	44.68	47.41	50.53	54.32	59.99		64.83	20.59
1.88	32.13	33.86	35.69	37.64	39.72	41.97	44.41	47.12	50.19	53.90	59.21		64.82	20.79
1.89	31.94	33.66	35.48	37.42	39.50	41.73	44.15	46.83	49.86	53.49	58.52		64.80	20.98
1.90	31.76	33.47	35.28	37.21	39.27	41.49	43.90	46.55	49.54	53.10	57.90		64.78	21.17
1.91	31.57	33.27	35.08	37.00	39.05	41.26	43.65	46.28	49.23	52.72	57.33		64.77	21.36
1.92	31.39	33.08	34.88	36.79	38.84	41.03	43.40	46.01	48.93	52.35	56.80		64.75	21.54
1.93	31.21	32.90	34.69	36.59	38.62	40.80	43.16	45.74	48.63	52.00	56.30		64.74	21.73
1.94	31.03	32.71	34.49	36.39	38.41	40.58	42.92	45.48	48.34	51.65	55.83		64.73	21.91
1.95	30.85	32.53	34.30	36.19	38.20	40.36	42.69	45.23	48.06	51.32	55.38	62.86	64.72	22.09
1.96	30.68	32.35	34.12	36.00	38.00	40.14	42.46	44.98	47.78	51.00	54.96	61.49	64.71	22.27
1.97	30.51	32.17	33.93	35.80	37.80	39.93	42.23	44.74	47.51	50.68	54.55	60.53	64.70	22.45
1.98	30.33	31.99	33.75	35.61	37.60	39.72	42.01	44.50	47.25	50.38	54.16	59.74	64.69	22.63
1.99	30.17	31.82	33.57	35.43	37.40	39.52	41.79	44.26	46.99	50.08	53.78	59.06	64.68	22.80
2.00	30.00	31.65	33.39	35.24	37.21	39.31	41.58	44.03	46.73	49.79	53.42	58.46	64.67	22.97

Table A–24

Oblique Shock Functions (Continued)

	Flow turning angle θ, deg																				
M_1	0.0	2.0	4.0	6.0	8.0	10.0	12.0	14.0	16.0	18.0	20.0	22.0	24.0	26.0	28.0	30.0	32.0	34.0	36.0	β	θ_{max}
2.00	30.00	31.65	33.39	35.24	37.21	39.31	41.58	44.03	46.73	49.79	53.42	58.46								64.67	22.97
2.02	29.67	31.31	33.04	34.88	36.83	38.92	41.15	43.58	46.24	49.22	52.74	57.39								64.65	23.31
2.04	29.35	30.98	32.70	34.52	36.46	38.53	40.75	43.14	45.76	48.69	52.09	56.46								64.64	23.65
2.06	29.04	30.66	32.37	34.18	36.10	38.15	40.35	42.72	45.30	48.17	51.49	55.63								64.63	23.98
2.08	28.74	30.34	32.04	33.84	35.75	37.79	39.97	42.31	44.86	47.68	50.91	54.87	61.28							64.63	24.30
2.10	28.44	30.03	31.72	33.51	35.41	37.43	39.59	41.91	44.43	47.21	50.36	54.17	59.77							64.62	24.61
2.12	28.14	29.73	31.41	33.19	35.08	37.09	39.23	41.53	44.02	46.75	49.84	53.52	58.62							64.62	24.92
2.14	27.86	29.44	31.11	32.88	34.76	36.75	38.87	41.15	43.62	46.32	49.35	52.91	57.65							64.62	25.23
2.16	27.58	29.15	30.81	32.57	34.44	36.42	38.53	40.79	43.23	45.89	48.87	52.34	56.81							64.62	25.52
2.18	27.30	28.87	30.52	32.27	34.13	36.10	38.20	40.44	42.85	45.49	48.41	51.79	56.05							64.62	25.82
2.20	27.04	28.59	30.24	31.98	33.83	35.79	37.87	40.09	42.49	45.09	47.98	51.28	55.36	62.70						64.62	26.10
2.22	26.77	28.32	29.96	31.69	33.53	35.48	37.55	39.76	42.14	44.71	47.55	50.79	54.72	60.85						64.62	26.38
2.24	26.51	28.06	29.69	31.42	33.24	35.18	37.24	39.44	41.79	44.34	47.15	50.32	54.12	59.63						64.63	26.66
2.26	26.26	27.80	29.42	31.14	32.96	34.89	36.94	39.12	41.46	43.98	46.75	49.87	53.56	58.65						64.64	26.93
2.28	26.01	27.54	29.16	30.87	32.69	34.60	36.64	38.81	41.13	43.64	46.37	49.44	53.03	57.82						64.64	27.19
2.30	25.77	27.29	28.91	30.61	32.42	34.33	36.35	38.51	40.82	43.30	46.01	49.03	52.54	57.08						64.65	27.45
2.32	25.53	27.05	28.66	30.35	32.15	34.05	36.07	38.22	40.51	42.97	45.65	48.63	52.06	56.41						64.66	27.71
2.34	25.30	26.81	28.41	30.10	31.89	33.79	35.80	37.93	40.21	42.65	45.31	48.25	51.61	55.79						64.67	27.96
2.36	25.07	26.58	28.17	29.86	31.64	33.53	35.53	37.65	39.91	42.34	44.97	47.88	51.18	55.22	61.97					64.68	28.20
2.38	24.85	26.35	27.93	29.61	31.39	33.27	35.26	37.38	39.63	42.04	44.65	47.52	50.77	54.69	60.69					64.70	28.45
2.40	24.62	26.12	27.70	29.38	31.15	33.02	35.01	37.11	39.35	41.75	44.34	47.17	50.37	54.18	59.66					64.71	28.68
2.42	24.41	25.90	27.48	29.14	30.91	32.78	34.76	36.85	39.08	41.46	44.03	46.84	49.99	53.71	58.83					64.72	28.91
2.44	24.19	25.68	27.25	28.92	30.68	32.54	34.51	36.60	38.82	41.18	43.73	46.52	49.62	53.26	58.11					64.74	29.14
2.46	23.99	25.47	27.03	28.69	30.45	32.31	34.27	36.35	38.56	40.91	43.45	46.20	49.27	52.83	57.47					64.75	29.36
2.48	23.78	25.26	26.82	28.47	30.23	32.08	34.03	36.10	38.30	40.65	43.16	45.90	48.93	52.43	56.88					64.77	29.58
2.50	23.58	25.05	26.61	28.26	30.01	31.85	33.80	35.87	38.06	40.39	42.89	45.60	48.60	52.04	56.33					64.78	29.80
2.52	23.38	24.85	26.40	28.05	29.79	31.63	33.58	35.63	37.82	40.14	42.62	45.31	48.28	51.66	55.83	64.27				64.80	30.01
2.54	23.18	24.65	26.20	27.84	29.58	31.41	33.35	35.41	37.58	39.89	42.36	45.04	47.97	51.30	55.36	62.05				64.81	30.22
2.56	22.99	24.45	26.00	27.64	29.37	31.20	33.14	35.18	37.35	39.65	42.11	44.76	47.67	50.96	54.91	60.94				64.83	30.42
2.58	22.81	24.26	25.80	27.44	29.17	30.99	32.92	34.96	37.12	39.41	41.86	44.50	47.38	50.63	54.49	60.08				64.85	30.62
2.60	22.62	24.07	25.61	27.24	28.97	30.79	32.71	34.75	36.90	39.19	41.62	44.24	47.10	50.31	54.09	59.35				64.87	30.81
2.62	22.44	23.89	25.42	27.05	28.77	30.59	32.51	34.54	36.68	38.96	41.39	43.99	46.83	50.00	53.71	58.72				64.88	31.01
2.64	22.26	23.70	25.24	26.86	28.58	30.39	32.31	34.33	36.47	38.74	41.16	43.75	46.56	49.70	53.34	58.14				64.90	31.19

Index

Note: Page numbers followed by *f* denote figures.

A

Absorptivity, 345–346, 345*f*, 350–351
Adiabatic isolation, 40, 44–45, 48. *See also* Isentropic process
  adiabatic flow:
    assumption, 171–172
    reversible, 233
    total relative properties of, 164, 166–167
    in two-stage machine, 158–159
  adiabatic mixture, 246
  in Brayton cycle, 90
  expansion in, 39–41
  Mach number and, 140
  in refrigeration, 79
  stator and rotor in, 168*f*
  work absorption in, 48
Adverse pressure gradient, 125–127
  over cylinder, 127*f*
  in exhaust diffuser, 172
  over solid surface, 126*f*
  in turbine, 157–158, 158*f*
Air standard, 84, 90, 102
Aircraft:
  separation in, 113–114, 113*f*
  similitude in, 337
Airfoil:
  angle, 154
  blade cascade, 150, 151*f*, 152
  boundary layer for, 128
  separation from, 113–114, 113*f*
  in turbomachine, 150, 151*f*, 152
  vorticity around, 135
  wake of, 114*f*
Alternating current, 344
Analogy:
  Chilton-Colburn, 309
  electrical, 286, 286*f*, 287, 318
  Reynolds, 308
Angle:
  of attack, 113–114
  deflection, 224, 231
  incidence and deviation, 169*f*–170*f*, 170
  Mach, 233–234
  metal, 154
  shock, 224–225, 231
  sign convention for, 154, 155*f*, 166
  solid, 352–354, 355*f*, 356

Annulus, 333
  constant-area, 265–268, 272, 278
  effective, 266–267, 267*f*
  in heat exchanger, 324–325
  height, 163*f*, 164, 184*f*, 266
APU. *See* Auxiliary power unit
Archimedes' principle, 310
Arithmetic mean temperature difference, 328
Ash, 332
Auxiliary power unit (APU), 179–180, 182*f*
Axial-flow turbine, 150, 152, 153*f*, 173
  multistage, 158–159, 159*f*, 173–174, 174*f*
  rotor, 267
  stages of, 151*f*, 152*f*, 165*f*, 175*f*
  stator, 162, 163*f*, 164
  variable geometry in, 208
Axial force, 156
Axial momentum, 156, 339
Axial velocity, 156, 162, 163*f*

B

BDC. *See* Bottom dead center
Bernoulli's equation, 131, 139–140
Bessel function, 295
Biot number, 294–295, 318, 320
Blackbody:
  absorptivity of, 351
  as diffuse, 357
  isothermal enclosure as, 351*f*
  properties of, 345*f*
  radiation, 344–350, 345*f*, 349*t*
    wavelengths, 344, 344*f*
  temperature of, 346–349, 349*t*
Blade cascade, 150, 151*f*, 152
Blasius, Paul Richard Heinrich, 120
Bluff body, 113–114
Body force, 242–243, 309
Body of revolution, 227*f*, 228*f*
Boiler, 335–336
Bottleneck, 331
Bottom dead center (BDC), 84
Boundary, 4
  condition:
    convection, 298–299
    flow direction and, 227–228
    heat flux, 246–247, 247*f*

    isothermal, 293–295
    Laplace equation and, 128–129
    no-slip, 115–116, 125, 125*f*, 339
  heat penetrating, 272, 272*f*
  work, 7, 29, 29*f*
    examples of, 4*f*
    in polytropic process, 31, 31*f*
    reversibility and, 46
Boundary layer, 115–116, 118–120
  for airfoil, 128
  buildup, 117*f*, 123, 131
  edge of, 118*f*, 261*f*
  equation, 338–341
  flat-plate, 339
  flow, 112–113, 125, 338
  heat conduction across, 307
  separation:
    from adverse pressure gradient, 125–127, 126*f*
    in aircraft, 113–114, 113*f*
    in cascade, 157–158, 158*f*
  thickness, 304*f*
    comparison, 306–307
    growth of, 121
    hydrodynamic, 112, 115–116, 339
    hydrodynamic, in natural convection, 311, 312*f*
    thermal, 112, 118–119
  velocity components in, 340
Brayton cycle:
  ideal, 89–90, 89*f*
  real, 90, 91*f*, 92*f*, 150*f*
  with regeneration, 101–102, 101*f*
  thermal efficiency of, 90, 101–102
Buckingham theorem, 338
Buoyancy force, 309–311

C

Camber line, mean, 199*f*
  definition of, 169*f*–170*f*, 170
Capillary tube, 77–78
Carburetor, 87
Carnot cycle:
  ideal-gas, 42*f*
  isentropic process in, 52, 62, 69, 75
  for pure substance, 40–41, 41*f*
  reversed, 75, 76*f*–77*f*, 77

Carnot cycle (*Cont.*):
  reversible process in, 40–41, 41*f*, 45–46
  thermal efficiency of, 44, 62
Cascade:
  airfoil blade, 150, 151*f*, 152
  compressor, 154, 155*f*, 157–158
  rotating, 172–173
  stator, 159, 160*f*, 161*f*
    supersonic, 180–181, 184*f*
  turbine, 157–158, 157*f*, 158*f*
Cauchy-Riemann equations, 129
Centrifugal compressor, 191
Centripetal acceleration, 190
Chemical element, 12
Chilton-Colburn analogy, 309
Choking:
  at exit station, 266
  friction, 260–261, 261*f*
  rotor, 193, 195–196, 196*f*
  on *T-s* diagram, 261*f*, 262
Churchill, 313
CI engine. *See* Compression-ignition engine
Circulation, 135
Clausius inequality, 41
  reversible process and, 49–50
Closed system, 38
  open and, 1, 4, 4*f*, 5*f*
Coefficient:
  dynamic viscosity, 197–198, 197*f*
  heat transfer, 305–308, 324
    overall, 328–331, 336
    Reynolds number and, 306, 316–317
  loss, 172–173, 173*f*
    Reynolds number and, 192–193
  of performance (COP), 78, 79
    for reversed cycle, 74–75
  pressure, 127
  of volumetric expansion, 313
Combustion, 90, 92*f*
  product of, 84
Combustor, 278, 279*f*
Complex velocity potential, 129
Compressed liquid, 62
Compressibility:
  of flow field, 136, 140, 144–145
  ideal gas and, 32–34
  Mach number and, 242
  pressure and, 33*f*
  simple, 8, 16–17, 22–23
  sonic speed and, 135–136
Compression-ignition (CI) engine, 87
Compressor:
  cascade, 154, 155*f*, 157–158
  centrifugal, 191
  efficiency of, 173–174, 174*f*
  stages, 165*f*

Condensation:
  constant wall temperature through, 249
  in heat exchanger, 325, 325*f*, 327
Condenser, 335–336
Conduction, 284–285
  electrical, 344, 352
  heat, 27–28, 284–285
    across boundary layer, 307
    laws of thermodynamics and, 284, 335
    transient, 293–295, 294*f*, 299
  as interaction mode, 27–28
  resistance, 287, 319, 329–330
Conductivity, fluid thermal, 119
Configuration factor:
  for concentric cylinders, 352, 353*f*
  as directional property, 352–356
  reciprocity rule for, 348
Conservation. *See also* Law of thermodynamics
  of energy, 1, 38, 69
    convection and, 248, 249*f*, 250
    enthalpy and, 247, 249
    Fanno flow and, 254, 256, 257
    in internal flow, 245–250
    Mach number and, 138
    mechanical, 244–245
    in refrigeration, 79
    reversed cycle and, 74, 74*f*
  of mass, 242
    continuity equation and, 108, 339
    control volume and, 242, 242*f*
    laminar flow and, 115–116, 118–121
    mean temperature and, 246
    mechanical energy and, 244
    meridional flow and, 162
  of momentum, 172, 243, 244, 246
Constant-temperature line:
  inflection point in, 32, 33*f*, 34
  on *p-v* diagram, 17, 18*f*
Continuity equation, 128–129, 142–143
  Fanno flow and, 254–256, 257–258
  mass conservation and, 108, 339
  Rayleigh flow and, 274–275, 277
  reference frame and, 177
  in shock analysis, 216
  stage flow and, 162, 164
  turning flow and, 230
Control volume:
  control-volume (CV) analysis, 14–15, 14*f*, 69, 105
  Fanno flow and, 254*f*
  heat flux and, 245–247, 246*f*, 249, 284–285
  mass conservation and, 242, 242*f*

pressure drop and, 242–243, 243*f*
  in shock analysis, 216, 217*f*
Convection, 28, 304–305, 331
  boundary condition, 298–299
  energy conservation and, 248, 249*f*, 250
  forced, 304, 309, 311, 321–322
  natural, 304, 309–311, 310*f*, 311*f*
  resistance, 286–287, 329–330
COP. *See* Coefficient, of performance
Coriolis acceleration, 190–191, 202, 269
Corrosion, 332
Critical isotherm:
  inflection point in, 32, 33*f*, 34
  on *p-v* diagram, 17, 18*f*
Critical Mach number, 140–143, 141*f*, 177
Critical point, 16, 32–34, 33*f*
  in refrigeration, 80, 81*f*
  in shock analysis, 219*f*, 220
Critical temperature, 141
Cutoff ratio, 87, 87*f*
CV analysis, 14–15, 14*f*, 69, 105
Cycle. *See* Engine

D

Darcy friction factor, 244–245
Darcy-Weisbach equation, 245
Deflection angle, 224, 231
DeLaval nozzle, 144, 146, 180
  as isentropic, 142–143
  shock in, 216, 219, 219*f*
Density, 119, 138, 162, 163*f*
  buoyancy and, 309–310
  Fanno flow and, 254–256
Deviation angle, 169*f*–170*f*, 170
Diagram. *See also T-s* diagram
  equilibrium, 15–16
  h-s (Mollier), 254–255, 276
  h-v, 254–255
  Moody, 259*f*, 260, 262
  p-T (phase), 16, 16*f*
  p-v, 15*f*, 18*f*, 26, 26*f*
    critical isotherm on, 17, 18*f*
    discontinuity on, 216
    for refrigeration, 80, 81*f*
Diesel cycle, 87, 87*f*
Diffuse emitting area, 354, 356–357
Diffuser. *See also* DeLaval nozzle
  diffuser-like passage, 165
  exhaust, 172
  frictionless, 137–138, 137*f*
  shock in, 220
  subsonic, 108, 108*f*, 145–146, 146*f*
Diffusion, 193
Diffusivity, thermal, 293
Dimensionless quantity:
  equations with, 338–341
  for infinite cylinder, 297–298

for irregular body, 319–320
nondimensionalization, 294
plot of, 118*f*, 119–120
shape factor as, 352–353
solid angle as, 352–353
Directional property, 350, 352–357, 354*f*, 357
Discontinuity, tangential, 216, 216*f*
Displacement thickness, 261*f*, 262, 265–266, 265*f*
Distribution:
 temperature, 246, 248, 248*f*
  transient, 293–295, 294*f*, 299
 velocity, 242, 242*f*, 246
Distributor, 131, 133*f*
Downstream effect, 136
Drag, 112, 115, 118–119, 339
Dryness factor:
 p-v-T surface and, 17
 in refrigeration, 77–79
 reheat and, 71–72
 in saturation, 1
Dynamic similarity, 337–338

E

Effectiveness, 102, 334
Efficiency, 47–48
 entropy-based, 173–174, 174*f*
 of heat engine, 43–44, 47–48, 62
 isentropic, 98–99, 99*f*, 100*f*
 thermal, 43–44
  of Brayton cycle, 90, 101–102
  of Carnot cycle, 44, 62
  of Rankine cycle, 70–72, 71*f*
 total-to-static, 171*f*, 172
 total-to-total, 170–172, 171*f*
Electrical analogy, 286, 286*f*, 287, 318
Electrical conduction, 344, 352
Electromagnetic wave, 344, 344*f*
Element, 12
Elevation:
 equilibrium and, 5–6
 in first law of thermodynamics, 38
 as negligible, 248
 potential energy and, 13, 247
 simple compressible system and, 8
Emissivity, 28, 351
 directional, 352, 354*f*
Endwall, 163*f*, 164, 180
 movable, 211*f*
 swirl and, 267
Energy:
 conservation of, 1, 38, 69
  convection and, 248, 249*f*, 250
  enthalpy and, 247, 249
  Fanno flow and, 254, 256, 257
  in internal flow, 245–250
  Mach number and, 138
  mechanical, 244–245

in refrigeration, 79
 reversed cycle and, 74, 74*f*
equation, 109, 217
forms of, 12–13
internal, 13–14, 14*f*, 26–27
kinetic, 13, 172–173, 173*f*, 245
 in first law of thermodynamics, 38
maximum, 320
mechanical, 248
monochromatic, 346–350
potential, 13, 247
Engine. *See also* Brayton cycle; Carnot cycle; Heat; Rankine cycle
compression-ignition, 87
Diesel cycle, 87, 87*f*
gas power cycle:
 assumptions for, 84, 90, 102
 ideal, 89–90, 89*f*, 91*f*
 turbine, 174, 208
heat, 43*f*, 50*f*
 Carnot, 41*f*, 45–46
 efficiency of, 43–44, 47–48, 62
 friction in, 46–49
 reversible, 43–46, 48
 work from, 47–48
jet, 89–90, 92*f*, 94*f*, 140
 Brayton cycle in, 150*f*
Otto cycle, 84
 ideal, 84*f*, 85, 85*f*
 real, 85*f*
reciprocating, 84
reversed:
 Carnot, 75, 76*f*–77*f*, 77
 coefficient of performance for, 74–75
 energy conservation and, 74, 74*f*
turbofan, 96, 96*f*, 97*f*
turboprop, 94, 95*f*, 172
vapor-compression cycle, 78–79, 78*f*
Enthalpy:
 in control-volume analysis, 14–15, 14*f*
 energy conservation and, 247, 249
 Fanno flow and, 254–256
 friction and, 248
 of ideal gas, 26–27
 maximum, 275–276, 275*f*
 mean, 249–250
 in Prandtl-Meyer flow, 237
 of rotor, 164
 specific heat and, 22–23
 T-ds equations and, 51
 total, 164–165
Entropy, 1. *See also* Isentropic process
 in Brayton cycle, 90
 Clausius inequality and, 49–50
 compressed liquid and, 62
 efficiency and, 173–174, 174*f*
 in Fanno flow, 254–255, 258

heat and, 273, 274*f*
heat transfer rate and, 75, 77
of ideal gas, 26–27, 51–52
maximum, 276, 278
in multistage turbomachine, 158–159, 159*f*, 173–174, 174*f*
pure substance and, 53
specific, 52, 53*f*, 173–174
T-ds equations and, 50–51
Equation of state, 27
 Bernoulli's equation and, 139–140
 compressibility and, 32–34
 entropy and, 52
 frictionless diffuser and, 138
 Mach number and, 144–145
 Rayleigh flow and, 277
 shock and, 218, 219
 sonic speed and, 136
 turning flow and, 230
 van der Waals, 34–35
Equilibrium, 5–6, 5*f*. *See also* *p-T* diagram; *p-v* diagram; *p-v-T* diagram
 in Carnot cycle, 75
 diagram, 15–16
 isentropic flow and, 144
 thermal, 10, 284, 294
Error function, 293
Euler's equation:
 DeLaval nozzle and, 144
 shaft work and, 186
 swirl and, 172
 for turbomachine, 166
Evaporation, 325, 325*f*, 327
Exact similarity solution, 121
Exhaust diffuser, 172
Expansion:
 adiabatic, 39–41
 coefficient of volumetric, 313
 overexpansion and underexpansion, 220
 Prandtl-Meyer, 234, 234*f*, 238
 wave, 237
External flow, 112, 125–127

F

F109 engine, 131, 134*f*
Failure, mechanical, 156, 156*f*, 157
Fanno flow. *See specific topics*
Fanno line, 254–256, 255*f*, 260–261, 266–267
Fanno table, 258–260
Favorable pressure gradient, 126–127
 over cylinder, 127*f*
 in turbine, 157–158, 158*f*
Finned wall, 330, 332, 334
First critical, 219*f*, 220
First law of thermodynamics, 38, 51, 109

First law of thermodynamics (*Cont.*):
  heat conduction and, 284, 335
  for irregular body, 319
Flat plate:
  boundary layer, 339
  horizontal, 314
  vertical, 313–314
Flow. *See also specific topics*
  adiabatic:
    assumption, 171–172
    reversible, 233
    total relative properties of, 164,
      166–167
    in two-stage machine, 158–159
  axial, 150, 152, 153*f*, 173
    multistage turbine, 158–159, 159*f*,
      173–174, 174*f*
    stages of turbine, 151*f*, 152*f*, 165*f*,
      175*f*
    turbine rotor, 267
    turbine stator, 162, 163*f*, 164
    in variable-geometry
      turbomachines, 208
  boundary layer, 112–113, 125, 338
  choked, 260–261, 261*f*
  direction, 227–228, 227*f*, 327
  doubly connected domain, 131, 132*f*
  external, 112, 125–127
  field:
    compressibility of, 136, 140,
      144–145
    subsonic, 145
    supersonic, 145–146
  free-stream, 113, 115–116,
    118–119
  internal:
    assumptions for, 242
    energy conservation in, 245–250
    gravity and, 242–244, 243*f*
    inviscid, 127*f*, 128, 131, 144
    isentropic, 144–146, 220, 235
      equilibrium and, 144
    isothermal and nonisothermal, 245,
      248
  laminar, 313–315, 331
    mass conservation and, 115–116,
      118–121
  mass, 166, 254, 260–261, 322
  meridional, 185–186, 187*f*
    mass conservation and, 162
  mixed, 123*f*, 155*f*
  parallel, 324–325, 327, 329*f*
  potential, 128–129, 131, 133*f*
  Prandtl-Meyer, 234–237
  Rayleigh, 272–273, 274*f*
    continuity equation and, 274–275,
      277
    equation of state and, 277
    Mach number in, 275*f*, 277

reverse, 114*f*, 126, 172
rotational and irrotational, 128–129,
  131, 132*f*, 133*f*
  around airfoil, 135
separation:
  from adverse pressure gradient,
    125–127, 126*f*
  in aircraft, 113–114, 113*f*
  in cascade, 157–158, 158*f*
stage, 163*f*
  continuity equation in, 162, 164
steady, 242, 334
turbulent, 314, 331
turning, 131, 132*f*, 133*f*
  isentropic, 230–234, 232*f*, 233*f*,
    234*f*
Fluid:
  interface, 288*f*, 291
  thermal conductivity, 119
Force:
  axial, 156
  body, 242–243, 304
  buoyancy, 309–311
  shear, 46–47, 115–116, 118
    no-slip condition and, 125, 125*f*,
      339
    rotation from, 131, 133*f*
    wall, 242–244, 250
  viscous, 242–243, 304
Forced convection, 304, 309, 311,
  321–322
Fouling factor, 331–332
Fourier number, 294, 320
Fourier's law, 284–285, 284*f*, 305
Frame of reference, 13
  continuity equation and, 177
Free convection, 304, 309–311, 310*f*,
  311*f*
Free-stream flow, 113, 115–116,
  118–119
Friction, 118, 119
  choking, 260–261, 261*f*
  Darcy factor, 244–245
  enthalpy and, 248
  in Fanno flow, 254–257, 260*f*
  frictionless diffuser, 137–138, 137*f*
  in heat engine, 46–49
  reversibility and, 75, 166–167, 173
  swirl and, 172
  wall, 242, 242*f*
Fuel injector, 87

G

Gas power cycle. *See also specific topics*
  assumptions for, 84, 90, 102
  ideal, 89–90, 89*f*, 91*f*
  turbine, 174, 208
Gauss error function, 293
Geometric similarity, 337

Grashof number, 313
Gravity:
  buoyancy and, 309–310
  equilibrium and, 5–6
  in first law of thermodynamics, 38
  gravitational work, 32
  internal flow and, 242–244, 243*f*
  potential energy and, 13, 247
  simple compressible system and, 8

H

h-s (Mollier) diagram, 254–255, 276
h-v diagram, 254–255
Heat. *See also* Adiabatic isolation;
      Carnot cycle
  capacity rate, 335
  conduction, 27–28, 284–285
    across boundary layer, 307
    laws of thermodynamics and,
      284, 335
    transient, 293–295, 294*f*, 299
  convection, 28, 304–305, 331
    boundary condition, 298–299
    energy conservation and, 248, 249*f*,
      250
    forced, 304, 309, 311, 321–322
    natural, 304, 309–311,
      310*f*, 311*f*
    resistance, 286–287, 329–330
  conversion to work, 68
  engine, 43*f*, 50*f*
    Carnot, 41*f*, 45–46
    efficiency of, 43–44, 47–48, 62
    friction in, 46–49
    reversible, 43–46, 48
    work from, 47–48
  entropy and, 273, 274*f*
  exchanger, 330, 330*f*, 334–336
    annulus in, 324–325
    condensation and evaporation in,
      325, 325*f*, 327
    examples of, 324*f*, 325*f*, 326*f*,
      329*f*
    fouling of, 331–332
  flux, 115, 118–119
    boundary condition, 246–247,
      247*f*
    control volume and, 245–247, 246*f*,
      249, 284–285
    uniform, 335, 336*f*
  isothermal process, 62, 75, 345*f*
    as blackbody, 351*f*
    boundary condition, 293–295
    flow, 245, 248
  Joule's heating, 319
  penetrating boundary, 272, 272*f*
  pump, 75, 77*f*, 78–79
  radiation, 28
  reheat, 71–72, 71*f*

reservoir, 40, 47–48, 74
 temperature of, 43–44
sources of, 246–247
specific, 14–15, 14*f*
 enthalpy and, 22–23
 LMTD and, 327
 in shock analysis, 218
superheating, 77–78, 79
thermal equilibrium, 10, 284, 294
transfer coefficient, 305–308, 324
 overall, 328–331, 336
 Reynolds number and, 306,
 316–317
transfer rate, 115, 119, 317, 327
 approximate, 249
 entropy and, 75, 77
 radiative, 353
Homogeneous problem, 299
Hydrodynamic boundary layer
 thickness, 112, 115–116, 339
 in natural convection, 311, 312*f*

I

Ideal gas, 1, 26. *See also* Equation
 of state
 in Carnot cycle, 42*f*
 compressibility and, 32–34
 enthalpy of, 26–27
 entropy of, 26–27, 51–52
 Grashof number and, 313
 shock in, 216, 218, 237
 sonic speed in, 136, 224
 turning flow in, 230
Incidence angle, 169*f*–170*f*, 170
Independent property, 8, 16–17
Index of refraction, 347
Indirect tip leakage, 156, 156*f*
Infinity:
 infinite cylinder, 297–298
 infinitely long slab, 299
 infinitesimal emitting area, 354–355
 infinitesimal shock, 231–234, 232*f*,
 233*f*, 234*f*
 semi-infinite solid, 293
Inflection point, 32, 33*f*, 34
Intensity, of radiation, 356–357
Interface, 289, 289*t*, 290*f*
 fluid, 288*f*, 291
Internal and external reversibility,
 45–46, 49
Internal energy, 13–14, 14*f*, 26–27
 in first law of thermodynamics, 38
Internal flow:
 assumptions for, 242
 energy conservation in, 245–250
 gravity and, 242–244, 243*f*
Inverse process, 45
Inviscid flow, 127*f*, 128, 131, 144
Irregular body, 319–320

Irreversible process. *See also* Reversible
 process
 boundary work and, 46
 definition of, 44
 examples of, 44–45
 in Fanno flow, 255, 255*f*
 friction and, 75, 166–167, 173
 heat engine using, 48
 internally and externally, 45–46, 49
 viscosity in, 266–267
Irrotational flow, 128–129, 131, 132*f*,
 133*f*
 around airfoil, 135
Isentropic process, 52, 98–99, 135
 in Brayton cycle, 90
 in Carnot cycle, 52, 62, 69, 75
 DeLaval nozzle as, 142–143
 efficiency of, 98–99, 99*f*, 100*f*
 total-to-total, 170–172, 171*f*
 flow, 144–146, 220, 235
 equilibrium and, 144
 frictionless diffuser as, 138
 in Rankine cycle, 69
 in refrigeration, 78
 reversible:
 in Carnot cycle, 52, 62, 69, 75
 in Rankine cycle, 62
 turn, 233
 total relative properties in, 167
 turn, 231–234, 232*f*, 233*f*, 234*f*
Isotherm, critical:
 inflection point in, 32, 33*f*, 34
 on *p-v* diagram, 17, 18*f*
Isothermal process, 62, 75, 345*f*
 as blackbody, 351*f*
 boundary condition, 293–295
 flow, 245, 248
Isotropic radiation, 345

J

Jet engine, 89–90, 92*f*, 94*f*, 140
 Brayton cycle in, 150*f*
Joule, James Prescott, 319
Joule's heating, 319

K

von Kármán, Theodore, 120
Kelvin-Planck statement, 41, 43–44
Kinetic energy, 13, 172–173,
 173*f*, 245
 in first law of thermodynamics, 38
Kirchhoff's law, 345*f*

L

Lambert's cosine law, 356
Laminar flow, 313–315, 331
 mass conservation and, 115–116,
 118–121

Lampblack, 346
Laplace equation, 128–129
Lattice, 12
Law of thermodynamics, 1. *See also*
 Conservation
 first, 38, 51, 109
 heat conduction and, 284, 335
 for irregular body, 319
 perpetual motion machine and,
 42–45, 42*f*, 47–49
 second, 39–45, 47–49, 158–159
 Clausius statement of, 41, 49–50
 efficiency and, 173–174, 174*f*
 heat conduction and, 284, 335
 Kelvin-Planck statement of, 41,
 43–44
 pressure and, 166–167
 reversible process and, 43–44, 48
 zeroth, 10
Leakage, tip, 156, 156*f*
Limiting point, 256–257, 256*f*, 274*f*,
 275*f*
Line:
 constant-temperature:
 inflection point in, 32, 33*f*, 34
 on *p-v* diagram, 17, 18*f*
 Fanno, 254–256, 255*f*, 260–261,
 266–267
 mean camber, 199*f*
 definition of, 169*f*–170*f*, 170
 Rayleigh, 273, 273*f*, 276
 streamline, 129, 130–131, 157–158,
 158*f*
 master, 173
 relative, 269, 269*f*
Linear problem, 299
Liquid:
 compressed, 62
 in Rankine cycle, 71–72
Logarithmic mean temperature
 difference (LMTD), 324–325,
 326*f*, 327–328, 334
Loss coefficient, 172–173, 173*f*
 Reynolds number and, 192–193

M

Mach angle, 233–234
Mach number, 136–138, 181
 absolute, 185, 193
 adiabatic isolation and, 140
 compressibility and, 242
 critical, 140–143, 141*f*, 177
 energy conservation and, 138
 equation of state and, 144–145
 in Fanno flow, 256*f*, 258–261
 inlet and exit, 219, 263
 in Rayleigh flow, 275*f*, 277
 in shock analysis, 218, 220, 224–225,
 232–238

Mach wave, 225, 232–235
Mass:
 conservation of, 242
  continuity equation and, 108, 339
  control volume and, 242, 242*f*
  laminar flow and, 115–116,
   118–121
  mean temperature and, 246
  mechanical energy and, 244
  meridional flow and, 162
 flow rate, 166, 254, 260–261, 322
 flux, 131, 254
Mean camber line, 199*f*
 definition of, 169*f*–170*f*, 170
Mean effective pressure (MEP), 85, 85*f*
Mechanical energy, 248
 conservation of, 244–245
Mechanical failure, 156, 156*f*, 157
MEP. *See* Mean effective pressure
Meridional flow, 185–186, 187*f*
 mass conservation and, 162
Metal angle, 154
Model, 337
Molecule, 12, 13
Mollier diagram, 254–255, 276
Momentum:
 axial, 156, 339
 conservation of, 172, 243, 244, 246
 equation, 217–218, 257, 259, 277
 radial, 157
 in turning flow, 230
Monochromatic energy, 346–350
Moody diagram, 259*f*, 260, 262
Multilayer wall, 287–289

N

Natural convection, 304, 309–311, 310*f*,
  311*f*
Newton's law of cooling, 28
No-slip boundary condition, 115–116
 shear force and, 125, 125*f*, 339
Nonregainable work, 7, 7*f*, 30*f*
Nondimensionalization, 294
Nonisothermal flow, 245, 248
Normal shock, 216, 217*f*, 218–220,
  228*f*
Nozzle, 126–127, 140–141
 DeLaval, 144, 146, 180
  as isentropic, 142–143
  shock in, 216, 219, 219*f*
 Fanno flow and, 261*f*, 262
 nozzle-like stator, 163*f*, 164
 shape of, 157–158, 157*f*
 shock in, 220
 subsonic and supersonic, 145–146,
  146*f*, 260
Nusselt number, 120, 306–309, 312,
  314–317
 in forced convection, 321

O

Oblique shock, 224–225, 224*f*,
  227–228, 227*f*
Opaque surface, 350
Open system, 14–15, 14*f*
 closed and, 1, 4, 4*f*, 5*f*
Otto cycle, 84
 ideal, 84*f*, 85, 85*f*
 real, 85*f*
Overexpansion, 220

P

p-T (phase) diagram, 16, 16*f*
p-v diagram, 15*f*, 18*f*, 26, 26*f*
 critical isotherm on, 17, 18*f*
 discontinuity on, 216
 for refrigerant, 80, 81*f*
p-v-T diagram, 16–17, 17*f*
Performance, coefficient of, 78, 79
 for reversed cycle, 74–75
Perpetual motion machine (PMM),
  42–45, 42*f*, 47–49
Phase, 1, 4, 15–17
 diagram, 16, 16*f*
 of pure substance, 12
Pi theorem, 338
Pitot tube, 228
Planck, Max, 346
Planck's law, 346, 348–349
Plate:
 boundary layer, 339
 horizontal, 314
 vertical, 313–314
PMM. *See* Perpetual motion machine
Polytropic process, 31, 31*f*
Potential energy, 13, 247
Potential flow, 128–129, 131, 133*f*
Potential velocity, 128, 130*f*, 131–132,
  135
 complex, 129
Power plant, 89–90, 92*f*
Prandtl, Ludwig, 112–113, 339
Prandtl number, 120, 308
 in forced convection, 321
Prandtl-Meyer:
 compression, 233*f*, 234, 238
 expansion, 234, 234*f*, 238
 flow, 234–237
 function, 237
 table, 238
Pressure, 15–17, 16*f*, 17*f*, 18*f*
 in Brayton cycle, 90
 in choked flow, 261
 coefficient, 127
 compressibility and, 33*f*
 constant, 14–15, 15*f*, 19*f*, 21*f*
 drop, 242–244, 243*f*, 250
 entropy and, 52, 53*f*
 in Fanno flow, 255–256, 255*f*, 258

gradient, 125–127
 over cylinder, 127*f*
 in exhaust diffuser, 172
 over solid surface, 126*f*
 in turbine, 157–158, 158*f*
 mean effective, 85, 85*f*
 in Otto cycle, 84, 84*f*
 second law of thermodynamics and,
  166–167
 shock and, 219–220, 230
 static, 145, 158, 172, 225
 sufficiently low, 26
 total, 158–159, 173, 173*f*, 258
 along tube, 243–244
Principle of corresponding states,
  32–33
Profile penalty, 114*f*
Property, 4. *See also* Total property
 directional, 350, 352–357, 354*f*, 357
 independent, 8, 16–17
 radiative, 350, 353
 spectral, 350
 thermodynamic, 13–14
Pure substance, 68
 Carnot cycle for, 40–41, 41*f*
 entropy change for, 53
 phase change of, 12

Q

Quasi-static process, 6, 6*f*

R

Radial momentum, 157
Radial turbomachine, 185–186, 186*f*,
  187*f*, 188
Radiant emittance, 345, 346
Radiant flux, 354
Radiation:
 analysis, 348
 blackbody, 344–350, 344*f*, 345*f*, 349*t*
 heat, 28
 intensity of, 356–357
 isotropic, 345
 shape factor:
  for concentric cylinders,
   352, 353*f*
  as directional property, 352–356
  reciprocity rule for, 348
 from sun, 348, 351–352
 thermal, 344–346, 344*f*, 345*f*
 total, 350, 351*f*
Radiative property, 350, 353
Rankine cycle:
 ideal, 23–24, 23*f*, 24*f*, 69
 isentropic process in, 62, 69
 real, 68*f*, 69
 regeneration in, 70–71, 71*f*
 reheat for, 71–72, 71*f*
 reversibility in, 23*f*, 62

T-s diagram and, 68–69, 68*f*
thermal efficiency of, 70–72, 71*f*
γ ray, 344
Rayleigh flow, 272–273, 274*f*
  continuity equation and, 274–275,
    277
  equation of state and, 277
  Mach number in, 275*f*, 277
Rayleigh line, 273, 273*f*, 276
Rayleigh number, 313–317
Reciprocating engine, 84
Reciprocity rule, 348, 359
Recirculation, 114–115, 126*f*
Reduced specific volume, 34
Reference frame, 13
  continuity equation and, 177
Reference state, 258–260
Reflectivity, 345*f*, 350
Refraction, 347
Refrigeration, 74–75, 76*f*–77*f*
  adiabatic, 79
  critical point in, 80, 81*f*
  isentropic, 78
  steam quality in, 77–79
  vapor-compression, 78–79, 78*f*
Regeneration:
  in Brayton cycle, 101–102, 101*f*
  effectiveness of, 102
  in Rankine cycle, 70–71, 71*f*
Reheat, 71–72, 71*f*
Reservoir, heat, 40, 47–48, 74
  temperature of, 43–44
Resistance:
  conduction, 287, 319, 329–330
  convection, 286–287, 329–330
  thermal, 286–289, 287*f*, 330
    contact, 288*f*, 289, 289*t*, 290*f*
Reversed cycle:
  Carnot, 75, 76*f*–77*f*, 77
  coefficient of performance for, 74–75
  energy conservation and, 74, 74*f*
Reversible process, 39–40. *See also*
    Carnot cycle; Irreversible
    process
  adiabatic flow, 233
  boundary work and, 46
  in Carnot cycle, 40–41, 41*f*, 45–46
  Clausius inequality and, 49–50
  entropy and, 50–51
  friction and, 75, 166–167, 173
  in gas power cycle, 84
  heat engine using, 43–46, 48
  internally and externally, 45–46, 49
  isentropic:
    in Carnot cycle, 52, 62, 69, 75
    in Rankine cycle, 62
    turn, 233
  as limiting case, 46–47
  in Rankine cycle, 23*f*

second law of thermodynamics and,
  43–44, 48
T-ds equations and, 50–51
totally, 45–46, 49
Reynolds analogy, 308
Reynolds number, 119, 121, 128, 331
  in forced convection, 321
  heat transfer coefficient and, 306,
    316–317
  loss coefficient and, 192–193
  similarity and, 338
Rotating cascade, 172–173
Rotating vane, 160*f*, 209*f*
Rotational flow, 128–129, 131, 132*f*,
  133*f*
Rotational velocity, 152, 154*f*, 162, 166
Rotor, 114*f*, 115*f*, 123*f*
  adiabatic, 168*f*
  choking, 193, 195–196, 196*f*
  direction of, 154, 155*f*
  efficiency of, 173–174
  enthalpy of, 164
  inlet, 185–186, 188
  speed, 162, 166
  stagnation and, 137*f*
  turbine, 154, 155*f*, 267, 268
  in turbomachine, 150, 151*f*, 152

S

Scroll, 131, 132*f*
Second critical, 219*f*, 220
Second law of thermodynamics, 39–45,
  47–49, 158–159
  Clausius statement of, 41, 49–50
  efficiency and, 173–174, 174*f*
  heat conduction and, 284, 335
  Kelvin-Planck statement of, 41,
    43–44
  pressure and, 166–167
  reversible process and, 43–44, 48
Semi-infinite solid, 293
Separation:
  from adverse pressure gradient,
    125–127, 126*f*
  from airfoil, 113–114, 113*f*
  in cascade, 157–158, 158*f*
Separation-of-variables method, 299
Shaft:
  motion, 157
  torque, 166
  work, 32, 171–172, 237, 272
    Euler's equation and, 186
Shape factor:
  for concentric cylinders, 352, 353*f*
  as directional property, 352–356
  reciprocity rule for, 348
Shear force, 46–47, 115–116, 118
  no-slip condition and, 125,
    125*f*, 339

rotation from, 131, 133*f*
wall, 242–244, 250
Shock:
  angle, 224–225, 231
  in DeLaval nozzle, 216, 219, 219*f*
  infinitesimal, 230–238, 232*f*, 233*f*,
    234*f*
  normal, 216, 217*f*, 218–220, 228*f*
  oblique, 224–225, 224*f*, 227–228,
    227*f*
  as tangential discontinuity, 216,
    216*f*
  weak and strong, 227–228, 227*f*, 231
Sign convention:
  angle, 154, 155*f*, 166
  radial turbomachine, 185–186, 186*f*,
    187*f*, 188
  work, 29
Similarity:
  dynamic, 337–338
  geometric, 337
  Reynolds number and, 338
  solution, 120
    exact, 121
    variable, 121
Similitude principle, 119, 337
Simple compressible system, 8, 16–17,
  22–23
Solid angle, 352–354, 355*f*, 356
Sonic speed, 258–259. *See also* Mach
    number
  compressibility and, 135–136
  in ideal gas, 136, 224
  temperature and, 140–141
Sonic state, 146
Space Shuttle, 157
Spark plug, 84, 87
Specific entropy, 52, 53*f*, 173–174
Specific heat, 14–15, 14*f*
  enthalpy and, 22–23
  LMTD and, 327
  in shock analysis, 218
Specific volume, 34
Spectral property, 350
Speed. *See* Velocity
Speed of sound. *See* Sonic speed
Stage:
  of axial-flow turbine, 151*f*, 152*f*, 165*f*,
    175*f*
  flow, 163*f*
    continuity equation in, 162, 164
  multistage turbomachine, 158–159,
    159*f*, 173–174, 174*f*
  reaction, 164, 165*f*
  symmetric, 176, 178
  work, 164
Stagnation, 137–138, 137*f*
  point, 127, 127*f*
Stanton number, 308

State, 4. *See also* Equation of state
  postulate, 8, 22–23
  principle of corresponding, 32–33
  reference, 258–260
  sonic, 146
Stator, 114*f*, 115*f*, 123*f*, 131
  adiabatic, 168*f*
  in axial-flow turbine, 162, 163*f*, 164
  cascade, 159, 160*f*, 161*f*
    supersonic, 180–181, 184*f*
  efficiency of, 173–174
  nozzle-like, 163*f*, 164
  performance assessor for, 172–173
  stagnation and, 137*f*
  in turbomachine, 150, 151*f*, 152
  variable-geometry, 159, 160*f*, 161*f*,
    211*f*
Steam quality:
  p-v-T surface and, 17
  in refrigeration, 77–79
  reheat and, 71–72
  in saturation, 1
Stefan-Boltzmann constant, 346
Stefan-Boltzmann law, 28, 346,
  351–352
Steradian, 356
Stream function, 128–129, 130*f*
Streamline, 129–131, 157–158, 158*f*
  master, 173
  relative, 269, 269*f*
Subsonic diffuser, 108, 108*f*, 145–146,
  146*f*
Suction. *See* Pressure
Sun, 348, 351–352
Superheating, 77–79
Surface emissivity, 28, 351
  directional, 352, 354*f*
Swirl:
  endwall and, 267
  friction and, 172
  in radial turbomachine, 188
System:
  closed, 38
  closed and open, 1, 4, 4*f*, 5*f*
  interaction with surroundings, 7, 7*f*,
    27–28
  open, 14–15, 14*f*
  simple compressible, 8, 16–17,
    22–23

T

T-ds equations, 50–51, 69
T-s diagram:
  for Brayton cycle, 101*f*
  choking on, 261*f*, 262
  for heat pump, 77*f*
  isentropic efficiency and, 98–99
  for Rankine cycle, 68–69, 68*f*
  for refrigeration, 78, 78*f*

specific entropy on, 52, 53*f*
  total pressure on, 173
Table:
  Fanno, 258–260
  Prandtl-Meyer, 238
Tangential discontinuity, 216, 216*f*
Teapot, 331–332
Temperature, 15–17, 16*f*, 17*f*
  ambient, 49
  of blackbody, 346–349, 349*t*
  boundary layer thickness, 112,
    118–119
  constant, line of:
    inflection point in, 32, 33*f*, 34
    on *p-v* diagram, 17, 18*f*
  critical, 141
  definition of, 10
  distribution, 246, 248, 248*f*
    transient, 293–295, 294*f*, 299
  entropy and, 52, 53*f*
  free-stream, 118–119
  isothermal process, 62, 75, 345*f*
    as blackbody, 351*f*
    boundary condition, 293–295
    flow, 245, 248
  mean, 245–249, 317
    arithmetic, 328
    logarithmic, 324–325, 326*f*,
      327–328, 334
    mass conservation and, 246
  in Otto cycle, 84, 84*f*
  relative, 166, 168*f*
  of reservoir, 43–44
  sonic speed and, 140–141
  static, 218, 219, 225, 273
  thermal equilibrium, 10, 284, 294
  total, 158–159, 159*f*
  total-to-static ratio, 138, 142
  wall, 248–250
Thermal diffusivity, 293
Thermal efficiency, 43–44
  of Brayton cycle, 90, 101–102
  of Carnot cycle, 44, 62
  of Rankine cycle, 70–72, 71*f*
Thermal equilibrium, 10, 284, 294
Thermal grease, 289
Thermal radiation, 344–346, 344*f*, 345*f*
Thermal reservoir, 40, 47–48, 74
  temperature of, 43–44
Thermal resistance, 286–289, 287*f*, 330
  contact, 288*f*, 289, 289*t*, 290*f*
Thermodynamic property, 13–14
Thermodynamics, law of. *See* Law of
  thermodynamics
Third critical, 219*f*, 220
Throat, 140–141, 142–143, 144
  condition for, 145–146, 146*f*
  shock and, 220
Thrust bearing, 156

Tip:
  leakage, 156, 156*f*
  reaction, 165
  slippage, 189*f*, 191
Ton, 75
Torque, 166
Total property, 350
  enthalpy, 164–165
  irradiation, 350, 351*f*
  pressure, 158–159, 173, 173*f*, 258
  relative, 164, 166–167
  reversibility, 45–46, 49
  temperature, 158–159, 159*f*
Total-to-static efficiency, 171*f*, 172
Total-to-static temperature ratio, 138,
  142
Total-to-total efficiency, 170–172,
  171*f*
Transmissivity, 345*f*, 350
Tube:
  capillary, 77–78
  Pitot, 228
  pressure along, 243–244
Turbine:
  axial-flow, 150, 152, 153*f*, 173
    multistage, 158–159, 159*f*,
      173–174, 174*f*
    rotor, 267
    stages of, 151*f*, 152*f*, 165*f*,
      175*f*
    stator, 162, 163*f*, 164
    variable-geometry in, 208
  blade, 157–158, 157*f*, 158*f*
  cascade, 157–158, 157*f*, 158*f*
  gas, 174, 208
  rotor, 154, 155*f*, 268
Turbofan engine, 96, 96*f*, 97*f*
Turbojet engine, 89–90, 92*f*, 94*f*, 140
  Brayton cycle in, 150*f*
Turbomachine:
  Euler's equation for, 166
  multistage, 158–159, 159*f*, 173–174,
    174*f*
  parts of, 150, 151*f*, 152
  radial, 185–186, 186*f*, 187*f*, 188
  variable-geometry, 208
Turboprop engine, 94, 95*f*, 172
Turbopump, 157
Turbulent flow, 314, 331
Turning flow, 131, 132*f*, 133*f*
  isentropic, 230–234, 232*f*, 233*f*,
    234*f*

U

Underexpansion, 220
Unfavorable pressure gradient,
  125–127
  over cylinder, 127*f*
  in exhaust diffuser, 172

over solid surface, 126*f*
in turbine, 157–158, 158*f*
Universal gas constant, 27

V

van der Waals equation of state,
34–35
Vane, 160*f*, 209*f*
Vapor:
compression of, 78–79, 78*f*
evaporation, 325, 325*f*, 327
Vapor-compression cycle, 78–79, 78*f*
Variable-geometry stator, 159, 160*f*,
161*f*, 211*f*
Variable-geometry turbomachine, 208
Velocity:
axial, 156, 162, 163*f*
components of, 224–225, 224*f*, 225*f*
in boundary layer, 340
in Prandtl-Meyer flow, 235
distribution, 242, 242*f*, 246
field, 128
free-stream, 115–116
gradient, 118*f*, 119–120, 125, 135
incremental change in, 235–236,
235*f*
potential, 128, 130*f*, 131–132, 135
complex, 129
profile, 115–116, 121, 261*f*
penalty, 114*f*
rotational, 152, 154*f*, 162, 166
Venturi meter, 113–114, 117*f*, 181
View factor:

for concentric cylinders, 352, 353*f*
as directional property, 352–356
reciprocity rule for, 348
Viscosity, 112, 120, 125, 131
dynamic viscosity coefficient,
197–198, 197*f*
irreversibility from, 266–267
isentropic flow and, 144
viscous force, 242–243, 304
Volume, 15–17, 15*f*, 17*f*, 18*f*
constant, 14–15, 19*f*, 20*f*, 22–23
control, 242–243, 243*f*
analysis, 14–15, 14*f*, 69, 105
Fanno flow and, 254*f*
heat flux and, 245–247, 246*f*, 249,
284–285
in shock analysis, 216, 217*f*
entropy and, 52, 53*f*
reduced specific, 34
Volumetric expansion, coefficient of,
313
Vortex, 158*f*
Vorticity, 128, 129, 131, 135

W

Wake, 113–114
of airfoil, 114*f*
Wall:
endwall, 154, 163*f*, 180
movable, 211*f*
swirl and, 267
finned, 330, 332, 334
friction, 242, 242*f*

multilayer, 287–289
plane, 294–295, 299
shear force, 242–244, 250
temperature, 248–250
Wave:
electromagnetic, 344, 344*f*
expansion, 237
Mach, 225, 232–235
wavelength, 347–348, 352
blackbody, 344, 344*f*
Wet-mixture dome, 79
Whirl, 157
Wien, Willy, 348
Wien's displacement law, 345*f*,
346–348
Wind tunnel, 337
Work. *See also* Boundary
adiabatic absorption of, 48
conversion of heat to, 68
definition of, 12, 29
gravitational, 32
heat engine producing, 47–48
nonregainable, 7, 7*f*, 30*f*
shaft, 32, 171–172, 237, 272
Euler's equation and, 186
sign convention for, 29

X

x ray, 344

Z

Zero incidence, 113–114
Zeroth law of thermodynamics, 10

CPSIA information can be obtained
at www.ICGtesting.com
Printed in the USA
LVHW061553030222
710167LV00006B/396